BIOMEDICAL OPTICS

THE WILEY BICENTENNIAL–KNOWLEDGE FOR GENERATIONS

*E*ach generation has its unique needs and aspirations. When Charles Wiley first opened his small printing shop in lower Manhattan in 1807, it was a generation of boundless potential searching for an identity. And we were there, helping to define a new American literary tradition. Over half a century later, in the midst of the Second Industrial Revolution, it was a generation focused on building the future. Once again, we were there, supplying the critical scientific, technical, and engineering knowledge that helped frame the world. Throughout the 20th Century, and into the new millennium, nations began to reach out beyond their own borders and a new international community was born. Wiley was there, expanding its operations around the world to enable a global exchange of ideas, opinions, and know-how.

For 200 years, Wiley has been an integral part of each generation's journey, enabling the flow of information and understanding necessary to meet their needs and fulfill their aspirations. Today, bold new technologies are changing the way we live and learn. Wiley will be there, providing you the must-have knowledge you need to imagine new worlds, new possibilities, and new opportunities.

Generations come and go, but you can always count on Wiley to provide you the knowledge you need, when and where you need it!

WILLIAM J. PESCE
PRESIDENT AND CHIEF EXECUTIVE OFFICER

PETER BOOTH WILEY
CHAIRMAN OF THE BOARD

BIOMEDICAL OPTICS

PRINCIPLES AND IMAGING

Lihong V. Wang

Hsin-i Wu

WILEY-INTERSCIENCE
A John Wiley & Sons, Inc., Publication

Copyright © 2007 by John Wiley & Sons, Inc. All rights reserved.

Published by John Wiley & Sons, Inc., Hoboken, New Jersey.
Published simultaneously in Canada.

No part of this publication may be reproduced, stored in a retrieval system, or transmitted in any form or by any means, electronic, mechanical, photocopying, recording, scanning, or otherwise, except as permitted under Section 107 or 108 of the 1976 United States Copyright Act, without either the prior written permission of the Publisher, or authorization through payment of the appropriate per-copy fee to the Copyright Clearance Center, Inc., 222 Rosewood Drive, Danvers, MA 01923, (978) 750-8400, fax (978) 750-4470, or on the web at www.copyright.com. Requests to the Publisher for permission should be addressed to the Permissions Department, John Wiley & Sons, Inc., 111 River Street, Hoboken, NJ 07030, (201) 748-6011, fax (201) 748-6008, or online at http://www.wiley.com/go/permission.

Limit of Liability/Disclaimer of Warranty: While the publisher and author have used their best efforts in preparing this book, they make no representations or warranties with respect to the accuracy or completeness of the contents of this book and specifically disclaim any implied warranties of merchantability or fitness for a particular purpose. No warranty may be created or extended by sales representatives or written sales materials. The advice and strategies contained herein may not be suitable for your situation. You should consult with a professional where appropriate. Neither the publisher nor author shall be liable for any loss of profit or any other commercial damages, including but not limited to special, incidental, consequential, or other damages.

For general information on our other products and services or for technical support, please contact our Customer Care Department within the United States at (800) 762-2974, outside the United States at (317) 572-3993 or fax (317) 572-4002.

Wiley also publishes its books in a variety of electronic formats. Some content that appears in print may not be available in electronic formats. For more information about Wiley products, visit our web site at www.wiley.com.

Wiley Bicentennial Logo: Richard J. Pacifico

Library of Congress Cataloging-in-Publication Data:

Wang, Lihong V.
    Biomedical optics : principles and imaging / Lihong V. Wang, Hsin-i Wu.
        p. ; cm.
    Includes bibliographical references and index.
    ISBN: 978-0-471-74304-0 (cloth)
    1. Imaging systems in medicine. 2. Lasers in medicine. 3. Optical detectors. I. Wu, Hsin-i. II. Title.
    [DNLM: 1. Optics. 2. Diagnostic Imaging—methods. 3. Light. 4. Models, Theoretical. 5. Tomography, Optical—methods. WB 117 W246b 2007]
    R857.O6W36 2007
    616.07'54–dc22

2006030754

Printed in the United States of America.

10 9 8 7 6

To our families, mentors, students, and friends

CONTENTS

Preface		**xiii**
1.	**Introduction**	**1**
	1.1. Motivation for Optical Imaging	1
	1.2. General Behavior of Light in Biological Tissue	2
	1.3. Basic Physics of Light–Matter Interaction	3
	1.4. Absorption and its Biological Origins	5
	1.5. Scattering and its Biological Origins	7
	1.6. Polarization and its Biological Origins	9
	1.7. Fluorescence and its Biological Origins	9
	1.8. Image Characterization	10
	Problems	14
	Reading	15
	Further Reading	15
2.	**Rayleigh Theory and Mie Theory for a Single Scatterer**	**17**
	2.1. Introduction	17
	2.2. Summary of Rayleigh Theory	17
	2.3. Numerical Example of Rayleigh Theory	19
	2.4. Summary of Mie Theory	20
	2.5. Numerical Example of Mie Theory	21
	Appendix 2A. Derivation of Rayleigh Theory	23
	Appendix 2B. Derivation of Mie Theory	26
	Problems	34
	Reading	35
	Further Reading	35

3. Monte Carlo Modeling of Photon Transport in Biological Tissue — 37

- 3.1. Introduction — 37
- 3.2. Monte Carlo Method — 37
- 3.3. Definition of Problem — 38
- 3.4. Propagation of Photons — 39
- 3.5. Physical Quantities — 50
- 3.6. Computational Examples — 55
- Appendix 3A. Summary of MCML — 58
- Appendix 3B. Probability Density Function — 60
- Problems — 60
- Reading — 62
- Further Reading — 62

4. Convolution for Broadbeam Responses — 67

- 4.1. Introduction — 67
- 4.2. General Formulation of Convolution — 67
- 4.3. Convolution over a Gaussian Beam — 69
- 4.4. Convolution over a Top-Hat Beam — 71
- 4.5. Numerical Solution to Convolution — 72
- 4.6. Computational Examples — 77
- Appendix 4A. Summary of CONV — 77
- Problems — 80
- Reading — 81
- Further Reading — 81

5. Radiative Transfer Equation and Diffusion Theory — 83

- 5.1. Introduction — 83
- 5.2. Definitions of Physical Quantities — 83
- 5.3. Derivation of Radiative Transport Equation — 85
- 5.4. Diffusion Theory — 88
- 5.5. Boundary Conditions — 101
- 5.6. Diffuse Reflectance — 106

	5.7.	Photon Propagation Regimes	114
		Problems	116
		Reading	117
		Further Reading	118
6.	**Hybrid Model of Monte Carlo Method and Diffusion Theory**		**119**
	6.1.	Introduction	119
	6.2.	Definition of Problem	119
	6.3.	Diffusion Theory	119
	6.4.	Hybrid Model	122
	6.5.	Numerical Computation	124
	6.6.	Computational Examples	125
		Problems	132
		Reading	133
		Further Reading	133
7.	**Sensing of Optical Properties and Spectroscopy**		**135**
	7.1.	Introduction	135
	7.2.	Collimated Transmission Method	135
	7.3.	Spectrophotometry	139
	7.4.	Oblique-Incidence Reflectometry	140
	7.5.	White-Light Spectroscopy	144
	7.6.	Time-Resolved Measurement	145
	7.7.	Fluorescence Spectroscopy	146
	7.8.	Fluorescence Modeling	147
		Problems	148
		Reading	149
		Further Reading	149
8.	**Ballistic Imaging and Microscopy**		**153**
	8.1.	Introduction	153
	8.2.	Characteristics of Ballistic Light	153

8.3.	Time-Gated Imaging	154
8.4.	Spatiofrequency-Filtered Imaging	156
8.5.	Polarization-Difference Imaging	157
8.6.	Coherence-Gated Holographic Imaging	158
8.7.	Optical Heterodyne Imaging	160
8.8.	Radon Transformation and Computed Tomography	163
8.9.	Confocal Microscopy	164
8.10.	Two-Photon Microscopy	169
Appendix 8A. Holography		171
Problems		175
Reading		177
Further Reading		177

9. Optical Coherence Tomography — 181

9.1.	Introduction	181
9.2.	Michelson Interferometry	181
9.3.	Coherence Length and Coherence Time	184
9.4.	Time-Domain OCT	185
9.5.	Fourier-Domain Rapid-Scanning Optical Delay Line	195
9.6.	Fourier-Domain OCT	198
9.7.	Doppler OCT	206
9.8.	Group Velocity Dispersion	207
9.9.	Monte Carlo Modeling of OCT	210
Problems		213
Reading		215
Further Reading		215

10. Mueller Optical Coherence Tomography — 219

10.1.	Introduction	219
10.2.	Mueller Calculus versus Jones Calculus	219
10.3.	Polarization State	219
10.4.	Stokes Vector	222

10.5.	Mueller Matrix	224
10.6.	Mueller Matrices for a Rotator, a Polarizer, and a Retarder	225
10.7.	Measurement of Mueller Matrix	227
10.8.	Jones Vector	229
10.9.	Jones Matrix	230
10.10.	Jones Matrices for a Rotator, a Polarizer, and a Retarder	230
10.11.	Eigenvectors and Eigenvalues of Jones Matrix	231
10.12.	Conversion from Jones Calculus to Mueller Calculus	235
10.13.	Degree of Polarization in OCT	236
10.14.	Serial Mueller OCT	237
10.15.	Parallel Mueller OCT	237
	Problems	243
	Reading	244
	Further Reading	245

11. Diffuse Optical Tomography — 249

11.1.	Introduction	249
11.2.	Modes of Diffuse Optical Tomography	249
11.3.	Time-Domain System	251
11.4.	Direct-Current System	252
11.5.	Frequency-Domain System	253
11.6.	Frequency-Domain Theory: Basics	256
11.7.	Frequency-Domain Theory: Linear Image Reconstruction	261
11.8.	Frequency-Domain Theory: General Image Reconstruction	267
	Appendix 11A. ART and SIRT	275
	Problems	276
	Reading	279
	Further Reading	279

12. Photoacoustic Tomography — 283

12.1.	Introduction	283
12.2.	Motivation for Photoacoustic Tomography	283

12.3.	Initial Photoacoustic Pressure	284
12.4.	General Photoacoustic Equation	287
12.5.	General Forward Solution	288
12.6.	Delta-Pulse Excitation of a Slab	293
12.7.	Delta-Pulse Excitation of a Sphere	297
12.8.	Finite-Duration Pulse Excitation of a Thin Slab	302
12.9.	Finite-Duration Pulse Excitation of a Small Sphere	303
12.10.	Dark-Field Confocal Photoacoustic Microscopy	303
12.11.	Synthetic Aperture Image Reconstruction	307
12.12.	General Image Reconstruction	309
	Appendix 12A. Derivation of Acoustic Wave Equation	313
	Appendix 12B. Green Function Approach	316
	Problems	317
	Reading	319
	Further Reading	319

13. Ultrasound-Modulated Optical Tomography — **323**

13.1.	Introduction	323
13.2.	Mechanisms of Ultrasonic Modulation of Coherent Light	323
13.3.	Time-Resolved Frequency-Swept UOT	326
13.4.	Frequency-Swept UOT with Parallel-Speckle Detection	329
13.5.	Ultrasonically Modulated Virtual Optical Source	331
13.6.	Reconstruction-Based UOT	332
13.7.	UOT with Fabry–Perot Interferometry	335
	Problems	338
	Reading	339
	Further Reading	339

Appendix A. Definitions of Optical Properties — **343**

Appendix B. List of Acronyms — **345**

Index — **347**

PREFACE

Biomedical optics is a rapidly growing area of research. Although many universities have begun to offer courses on the topic, a textbook containing examples and homework problems has not been available. The need to fill this void prompted us to write this book.

This book is based on our lecture notes for a one-semester (45 lecture hours) entry-level course, which we have taught since 1998. The contents are divided into two major parts: (1) fundamentals of photon transport in biological tissue and (2) optical imaging. In the first part (Chapters 1–7), we start with a brief introduction to biomedical optics and then cover single-scatterer theories, Monte Carlo modeling of photon transport, convolution for broadbeam responses, radiative transfer equation and diffusion theory, hybrid Monte Carlo method and diffusion theory, and sensing of optical properties and spectroscopy. In the second part (Chapters 8–13), we cover ballistic imaging, optical coherence tomography, diffuse optical tomography, photoacoustic tomography, and ultrasound-modulated optical tomography.

When the book is used as the textbook in a course, the instructor may request a solution manual containing homework solutions from the publisher. To benefit from this text, students are expected to have a background in calculus and differential equations. Experience in MATLAB® or C/C++ is also helpful. Source codes and other information can be found at ftp://ftp.wiley.com/public/sci_tech_med/biomedical_optics.

Although our multilayered Monte Carlo model is in the public domain, we have found that students are able to better grasp the concept of photon transport in biological tissue when they implement simple semiinfinite versions of the model. For this reason, we encourage the use of simulations whenever appropriate.

Because a great deal of material beyond our original lecture notes has been added, two semesters are recommended to cover the complete textbook. Alternatively, selected chapters can be covered in a one-semester course. In addition to serving as a textbook, this book can also be used as a reference for professionals and a supplement for trainees engaged in short courses in the field of biomedical optics.

We are grateful to Mary Ann Dickson for editing the text and to Elizabeth Smith for redrawing the figures. We appreciate Sancy Wu's close reading of the manuscript. We are also thankful to the many students who contributed to the homework solutions. Finally, we wish to thank our students Li Li, Manojit Pramanik, and Sava Sakadzic for proofreading the book.

LIHONG V. WANG, PH.D.
HSIN-I WU, PH.D.

CHAPTER 1

Introduction

1.1. MOTIVATION FOR OPTICAL IMAGING

The most common medical imaging modalities include X-ray radiography, ultrasound imaging (ultrasonography), X-ray computed tomography (CT), and magnetic resonance imaging (MRI). The discovery of X rays in 1895, for which Roentgen received the first Nobel Prize in Physics in 1901, marked the advent of medical imaging. Ultrasonography, which is based on sonar, was introduced into medicine in the 1940s after World War II. The invention of CT in the 1970s, for which Cormack and Hounsfield received the Nobel Prize in Medicine in 1979, initiated digital cross-sectional imaging (tomography). The invention of MRI, also in the 1970s, for which Lauterbur and Mansfield received the Nobel Prize in Medicine in 2003, enabled functional imaging with high spatial resolution. Optical imaging, which is compared with the other modalities in Table 1.1, is currently emerging as a promising new addition to medical imaging.

Reasons for optical imaging of biological tissue include

1. Optical photons provide nonionizing and safe radiation for medical applications.
2. Optical spectra—based on absorption, fluorescence, or Raman scattering—provide biochemical information because they are related to molecular conformation.
3. Optical absorption, in particular, reveals angiogenesis and hypermetabolism, both of which are hallmarks of cancer; the former is related to the concentration of hemoglobin and the latter, to the oxygen saturation of hemoglobin. Therefore, optical absorption provides contrast for functional imaging.
4. Optical scattering spectra provide information about the size distribution of optical scatterers, such as cell nuclei.
5. Optical polarization provides information about structurally anisotropic tissue components, such as collagen and muscle fiber.

Biomedical Optics: Principles and Imaging, by Lihong V. Wang and Hsin-i Wu
Copyright © 2007 John Wiley & Sons, Inc.

TABLE 1.1. Comparison of Various Medical Imaging Modalities

Characteristics	X-ray Imaging	Ultrasonography	MRI	Optical Imaging
Soft-tissue contrast	Poor	Good	Excellent	Excellent
Spatial resolution	Excellent	Good	Good	Mixed[a]
Maximum imaging depth	Excellent	Good	Excellent	Good
Function	None	Good	Excellent	Excellent
Nonionizing radiation	No	Yes	Yes	Yes
Data acquisition	Fast	Fast	Slow	Fast
Cost	Low	Low	High	Low

[a] High in ballistic imaging (see Chapters 8–10) and photoacoustic tomography (see Chapter 12); low in diffuse optical tomography (see Chapter 11).

6. Optical frequency shifts due to the optical Doppler effect provide information about blood flow.
7. Optical properties of targeted contrast agents provide contrast for the molecular imaging of biomarkers.
8. Optical properties or bioluminescence of products from gene expression provide contrast for the molecular imaging of gene activities.
9. Optical spectroscopy permits simultaneous detection of multiple contrast agents.
10. Optical transparency in the eye provides a unique opportunity for high-resolution imaging of the retina.

1.2. GENERAL BEHAVIOR OF LIGHT IN BIOLOGICAL TISSUE

Most biological tissues are characterized by strong optical scattering and hence are referred to as either *scattering media* or *turbid media*. By contrast, optical absorption is weak in the 400–1350-nm spectral region. The mean free path between photon scattering events is on the order of 0.1 mm, whereas the mean absorption length (mean path length before photon absorption) can extend to 10–100 mm.

Photon propagation in biological tissue is illustrated in Figure 1.1. The light source is spatially a pencil beam (an infinitely narrow collimated beam) and temporally a Dirac delta pulse. The optical properties (see Appendix A) of the tissue include the following: refractive index $n = 1.37$, absorption coefficient $\mu_a = 1.4$ cm^{-1}, scattering coefficient $\mu_s = 350$ cm^{-1}, and scattering anisotropy $g = 0.8$. The mean free path equals 28 μm, corresponding to a propagation time of 0.13 ps. The transport mean free path equals 140 μm, corresponding to a propagation time of 0.64 ps. Note how widely the photons spread versus time in relation to the two time constants mentioned above. This diffusion-like behavior of light in biological tissue presents a key challenge for optical imaging. Various techniques have been designed to meet this challenge.

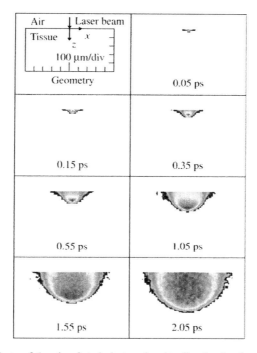

Figure 1.1. Snapshots of the simulated photon density distribution in a piece of biological tissue projected along the y axis, which points out of the paper.

1.3. BASIC PHYSICS OF LIGHT–MATTER INTERACTION

Absorption of a photon can elevate an electron of a molecule from the ground state to an excited state, which is termed *excitation*. Excitation can also be caused by other mechanisms, which are either mechanical (frictional) or chemical in nature. When an electron is raised to an excited state, there are several possible outcomes. The excited electron may relax to the ground state and give off luminescence (another photon) or heat. If another photon is produced, the emission process is referred to as *fluorescence* or *phosphorescence*, depending on the lifetime of the excited electron; otherwise, it is referred to as *nonradiative relaxation*. *Lifetime* is defined as the average time that an excited molecule spends in the excited state before returning to the ground state. The ratio of the number of photons emitted to the number of photons absorbed is referred to as the *quantum yield of fluorescence*. If the excited molecule is near another molecule with a similar electronic configuration, the energy may be transferred by excitation energy transfer—the excited electron in one molecule drops to the ground state while the energy is transferred to the neighboring molecule, raising an electron in that molecule to an excited state with a longer lifetime. Another possible outcome is photochemistry, in which an excited electron is actually transferred to another

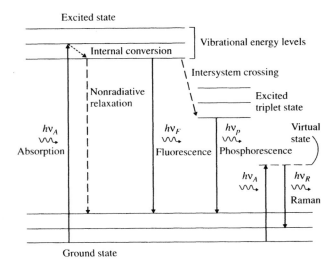

Figure 1.2. Jablonski energy diagram showing excitation and various possible relaxation mechanisms. Each $h\nu$ denotes the photon energy, where subscripts A, F, P, and R denote absorption, fluorescence, phosphorescence, and Raman scattering, respectively.

molecule. This type of electron transfer alters the chemical properties of both the electron donor and the electron acceptor, as in photosynthesis.

A Jablonski energy diagram describing electronic transitions between ground states and excited states is shown in Figure 1.2. Molecules can absorb photons that match the energy difference between two of their discrete energy levels, provided the transitions are allowed. These energy levels define the absorption and the emission bands.

Fluorescence involves three events with vastly different timescales. Excitation by a photon takes place in femtoseconds (1 fs = 10^{-15} s, about one optical period). Vibrational relaxation (also referred to as *internal conversion*) of an excited-state electron to the lowest vibrational energy level in the excited state lasts for picoseconds (1 ps = 10^{-12} s) and does not result in emission of a photon (nonradiative). Fluorescence emission lingers over nanoseconds (1 ns = 10^{-9} s). Consequently, fluorescence lifetime is on the order of a nanosecond.

Phosphorescence is similar to fluorescence, but the excited molecule further transitions to a metastable state by intersystem crossing, which alters the electron spin. Because relaxation from the metastable state to the ground state is spin-forbidden, emission occurs only when thermal energy raises the electron to a state where relaxation is allowed. Consequently, phosphorescence depends on temperature and has a long lifetime (milliseconds or longer).

Two types of photon scattering by a molecule exist: elastic and inelastic (or Raman) scattering. The former involves no energy exchange between the photon and the molecule, whereas the latter does. Although both Raman scattering and fluorescence alter the optical wavelength, they have different mechanisms. In

Raman scattering, the molecule is excited to a virtual state; in fluorescence, the molecule is excited to a real stationary state. In both cases, the excited molecule relaxes to an energy level of the ground state and emits a photon. The molecule may either gain energy from, or lose energy to, the photon. If the molecule gains energy, the transition is known as a *Stokes transition*. Otherwise, the transition is known as an *anti-Stokes transition*. The scattered photon shifts its frequency accordingly since the total energy is conserved. Raman scattering can reveal the specific chemical composition and molecular structure of biological tissue, whereas elastic scattering can reveal the size distribution of the scatterers.

1.4. ABSORPTION AND ITS BIOLOGICAL ORIGINS

The *absorption coefficient* μ_a is defined as the probability of photon absorption in a medium per unit path length (strictly speaking, per unit infinitesimal path length). It has a representative value of 0.1 cm^{-1} in biological tissue. The reciprocal of μ_a is referred to as the *mean absorption length*.

For a single absorber, the absorption cross section σ_a, which indicates the absorbing capability, is related to its geometric cross-sectional area σ_g through the absorption efficiency $Q_a : \sigma_a = Q_a \sigma_g$. In a medium containing many absorbers with number density N_a, the absorption coefficient can be considered as the total cross-sectional area for absorption per unit volume:

$$\mu_a = N_a \sigma_a. \tag{1.1}$$

Here, absorption by different absorbers is considered to be independent.

According to the definition of the absorption coefficient, light attenuates as it propagates in an absorbing-only medium according to the following equation:

$$\frac{dI}{I} = -\mu_a \, dx, \tag{1.2}$$

where I denotes the light intensity and x denotes the distance along the light propagation direction. Equation (1.2) means that the percentage of light being absorbed in interval $(x, x + dx)$ is proportional to the product of μ_a and dx; the negative sign is due to the decrease of I as x increases. Integrating Eq. (1.2) leads to the well-known Beer law

$$I(x) = I_0 \exp(-\mu_a x), \tag{1.3}$$

where I_0 is the light intensity at $x = 0$. Beer's law actually holds even for a tortuous path. The transmittance is defined by

$$T(x) = \frac{I(x)}{I_0}, \tag{1.4}$$

which represents the probability of survival after propagation over x.

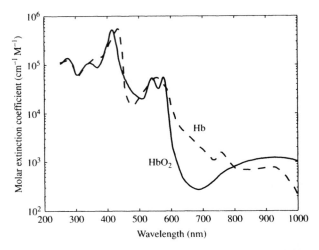

Figure 1.3. Molar extinction coefficients of oxy- and deoxyhemoglobin versus wavelength.

Optical absorption in biological tissue originates primarily from hemoglobin, melanin, and water. Hemoglobin has two forms: oxygenated and deoxygenated. Figure 1.3 shows the molar extinction coefficients—the extinction coefficient divided by ln(10) (see Section 7.3) per unit molar concentration—of oxy- and deoxyhemoglobin as a function of wavelength, where the extinction coefficient is defined as the probability of photon interaction with a medium per unit path length. Although extinction includes both absorption and scattering, absorption dominates scattering in hemoglobin. The molar extinction spectra of oxy- and deoxyhemoglobin are distinct but share a few intersections, termed *isosbestic points*. At these points, the absorption coefficient of an oxy- and deoxyhemoglobin mixture depends only on the total concentration, regardless of the oxygen saturation.

The absorption coefficients of some primary absorbing biological tissue components are plotted as a function of wavelength in Figure 1.4. Melanin absorbs ultraviolet (UV) light strongly but longer-wavelength light less strongly. Even water can be highly absorbing in some spectral regions. At the 2.95-μm water absorption peak, the penetration depth is less than 1 μm since $\mu_a = 12{,}694$ cm^{-1}.

The absorption coefficients of biological tissue at two wavelengths can be used to estimate the concentrations of the two forms of hemoglobin based on the following equations:

$$\mu_a(\lambda_1) = \ln(10)\varepsilon_{ox}(\lambda_1)C_{ox} + \ln(10)\varepsilon_{de}(\lambda_1)C_{de}, \quad (1.5)$$

$$\mu_a(\lambda_2) = \ln(10)\varepsilon_{ox}(\lambda_2)C_{ox} + \ln(10)\varepsilon_{de}(\lambda_2)C_{de}. \quad (1.6)$$

Here, λ_1 and λ_2 are the two wavelengths; ε_{ox} and ε_{de} are the known molar extinction coefficients of oxy- and deoxyhemoglobin, respectively; C_{ox} and C_{de}

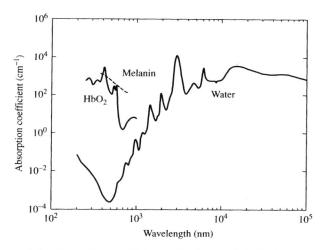

Figure 1.4. Absorption coefficients of primary biological absorbers.

are the molar concentrations of oxy- and deoxyhemoglobin, respectively, in the tissue. Once C_{ox} and C_{de} are obtained, the oxygen saturation (SO_2) and the total concentration (C_{Hb}) of hemoglobin can be computed as follows:

$$SO_2 = \frac{C_{ox}}{C_{ox} + C_{de}}, \tag{1.7}$$

$$C_{Hb} = C_{ox} + C_{de}. \tag{1.8}$$

This principle provides the basis for pulse oximetry and functional imaging. Angiogenesis can increase C_{Hb}, whereas tumor hypermetabolism can decrease SO_2.

1.5. SCATTERING AND ITS BIOLOGICAL ORIGINS

Scattering of light by a spherical particle of any size can be modeled exactly by the Mie theory, which reduces to the simpler Rayleigh theory if the spherical particle is much smaller than the wavelength. In a scattering medium containing many scatterers that are distributed randomly in space, photons usually encounter multiple scattering events. If scatterers are sparsely distributed (where the mean distance between particles is much greater than both the scatterer size and the wavelength), the medium is considered to be loosely packed. In this case, scattering events are considered to be independent; hence, single-scattering theory applies to each scattering event. Otherwise, the medium is considered to be densely packed. In this case, scattering events are coupled; thus, single-scattering theory does not apply. In this book, we consider only loosely packed scattering media. Keep in mind that one must differentiate a single coupled-scattering event (which involves multiple particles) from successive independent-scattering events (each of which involves a single particle).

The *scattering coefficient* μ_s is defined as the probability of photon scattering in a medium per unit path length. It has a representative value of 100 cm^{-1} in biological tissue. The reciprocal of μ_s is referred to as the *scattering mean free path*.

For a single scatterer, the scattering cross section σ_s, which indicates the scattering capability, is related to its geometric cross-sectional area σ_g through the scattering efficiency Q_s : $\sigma_s = Q_s \sigma_g$. For a medium containing many scatterers with number density N_s, the scattering coefficient can be considered as the total cross-sectional area for scattering per unit volume:

$$\mu_s = N_s \sigma_s. \tag{1.9}$$

The probability of no scattering (or ballistic transmittance T) after a photon propagates over path length x can be computed by Beer's law:

$$T(x) = \exp(-\mu_s x). \tag{1.10}$$

Optical scattering originates from light interaction with biological structures, which range from cell membranes to whole cells (Figure 1.5). Photons are scattered most strongly by a structure whose size matches the optical wavelength and whose refractive index mismatches that of the surrounding medium. The indices of refraction of common tissue components are 1.35–1.36 for extracellular fluid, 1.36–1.375 for cytoplasm, 1.38–1.41 for nuclei, 1.38–1.41 for mitochondria and organelles, and 1.6–1.7 for melanin. Cell nuclei and mitochondria are primary scatterers. The volume-averaged refractive index of most biological tissue falls within 1.34–1.62, which is greater than the refractive index of water (1.33).

The *extinction coefficient* μ_t, also referred to as the *total interaction coefficient*, is given by

$$\mu_t = \mu_a + \mu_s. \tag{1.11}$$

The reciprocal of μ_t is the mean free path between interaction events.

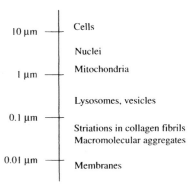

Figure 1.5. Biological structures of various sizes for photon scattering.

1.6. POLARIZATION AND ITS BIOLOGICAL ORIGINS

Linear birefringence (or simply birefringence), which is also known as *double refraction*, is the most important polarization property. A linearly birefringent material has dual principal indices of refraction associated with two linear polarization states of light (orientations of the electric field). The index of refraction for light polarization that is parallel with the optical axis of the material (e.g., the orientation of collagen fibers) is commonly denoted by n_e, while the light is referred to as the *extraordinary ray*. By contrast, the index of refraction for light polarization that is perpendicular to the optical axis is commonly denoted by n_o, while the light is referred to as the *ordinary ray*. If $n_e > n_o$, the birefringence is said to be positive. Conversely, if $n_e < n_o$, the birefringence is said to be negative.

Similarly, a circularly birefringent material has dual principal indices of refraction associated with the left and the right circular polarization states of light; as a result, it can rotate a linear polarization. The amount of rotation depends on the properties and the concentration of the active material, the optical wavelength, and the path length. If the other parameters are known, the amount of rotation can reveal the concentration.

Collagen, muscle fibers, myelin, retina, and keratin have linear birefringence. Collagen I is intensely positively birefringent, whereas collagen III is weakly negatively birefringent. Amino acids and glucose have circular birefringence; amino acids are levorotatory (exhibit left rotation) to linearly polarized light, whereas glucose is dextrorotatory (exhibits right rotation).

1.7. FLUORESCENCE AND ITS BIOLOGICAL ORIGINS

Fluorescence has the following characteristics:

1. Fluorescence light is red-shifted (wavelength is increased or frequency is reduced) relative to the excitation light; this phenomenon is known as the *Stokes shift*. The primary origins include the initial vibrational relaxations and the subsequent inclined fluorescence transitions to higher vibrational energy levels of the ground state. Other origins include excited-state reactions, complex formations, and resonance energy transfers.
2. Emission wavelengths are not only longer than but also independent of the excitation wavelength. Although the initial excited state is related to the excitation wavelength, a vibrational relaxation to the same intermediate state terminates the memory of such a relationship.
3. Fluorescence light is incoherent even if the excitation light is coherent because the uncertain delays in the vibrational relaxations spread over more than one light period.
4. Fluorescence spectrum, when plotted against the frequency, is generally a mirror image of the absorption spectrum for the following reasons: (a) before excitation, almost all the molecules are at the lowest vibrational energy level of the ground state; (b) before emission, almost all the

molecules are at the lowest vibrational energy level of the first excited state; (c) the least photon energy for excitation equals the greatest emission photon energy; (d) the vibrational energy levels in the ground and first excited states have similar spacing structures; and (e) the probability of a ground-state electron excited to a particular vibrational energy level in the first excited state is similar to that of an excited electron returning to a corresponding vibrational energy level in the ground state.

The properties of some endogenous fluorophores are listed in Table 1.2 (where λ_a denotes maximum absorption wavelength; ε denotes molar extinction coefficient; λ_x denotes maximum excitation wavelength; λ_m denotes maximum emission wavelength; Y denotes quantum yield of fluorescence). Fluorescence can provide information about the structure, dynamics, and interaction of a bioassembly. For example, mitochondrial fluorophore NADH (nicotinamide adenine dinucleotide, reduced form) is a key discriminator in cancer detection; it tends to be more abundant in cancer cells owing to their higher metabolic rate. NAD(P)H (nicotinamide adenine dinucleotide phosphate, reduced form) has a lifetime of 0.4 ns when free but a longer lifetime of 1–3 ns when bound.

1.8. IMAGE CHARACTERIZATION

Several parameters are important in the characterization of medical images. In this section, the discussion is limited primarily to two-dimensional (2D) images, but the principles involved can be extended to one-dimensional (1D) or three-dimensional (3D) images.

When a high-contrast point target is imaged, the point appears as a blurred blob in the image because any practical imaging system is imperfect. The spatial distribution of this blob in the image is referred to as the *point spread function* (PSF). The PSF is sometimes called the *impulse response* (or Green's function) because

TABLE 1.2. Properties of Endogenous Fluorophores at Physiologic pH

Fluorophore	λ_a(nm)	ε(cm^{-1}M^{-1})	λ_x(nm)	λ_m(nm)	Y
Ceroid	—	—	340–395	430–460	—
				540–640	
Collagen, elastin	—	—	325	400	—
FAD	—	—	450	515	—
Lipofuscin	—	—	340–395	430–460	—
				540–540	
NAD$^+$	260	18×10^3	—	—	—
NADH	260	14.4×10^3	290	440	—
	340	6.2×10^3	340	450	—
Phenylalanine	260	0.2×10^3	—	280	0.04
Tryptophan	280	5.6×10^3	280	350	0.2
Tyrosine	275	1.4×10^3	—	300	0.1

a geometric point can be represented by a spatial Dirac delta function (an impulse function). When two point targets are too close to each other, the combined blob in the image can no longer be clearly resolved into two entities. The full width at half maximum (FWHM) of the PSF is often defined as the spatial resolution. Even though an ideal geometric point target cannot be constructed or detected in reality, a point target needs only to be much smaller than the spatial resolution.

Sometimes, a line spread function (LSF), which is the system response to a high-contrast geometric line, is measured instead of a PSF. For a linear system, an LSF can be related to a PSF on the (x, y) plane by

$$\text{LSF}(x) = \int \text{PSF}(x, y)\,dy. \tag{1.12}$$

Likewise, an edge spread function (ESF), which is the system response to a high-contrast semiinfinite straight edge, can be measured as well. For a linear system, an ESF can be related to an LSF as follows (Figure 1.6):

$$\text{ESF}(x) = \int_{-\infty}^{x} \text{LSF}(x')\,dx', \tag{1.13}$$

$$\text{LSF}(x) = \frac{d}{dx}\text{ESF}(x). \tag{1.14}$$

In a linear, stationary, and spatially translation-invariant system, image function $i(\vec{r})$ equals the convolution of object function $o(\vec{r})$ with point spread function $\text{PSF}(\vec{r})$:

$$i(\vec{r}) = o(\vec{r}) ** \text{PSF}(\vec{r}), \tag{1.15}$$

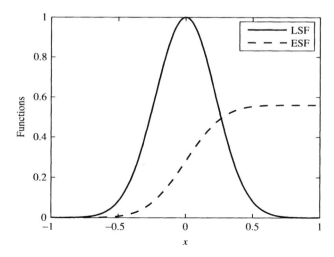

Figure 1.6. Illustration of an LSF and an ESF.

where $\vec{r} = (x, y)$ and $**$ represents 2D spatial convolution. Equation (1.15) can be expressed in several forms:

$$\begin{aligned} i(\vec{r}) &= \iint o(\vec{r}')\mathrm{PSF}(\vec{r} - \vec{r}')\, d\vec{r}' \\ &= \iint o(x', y')\mathrm{PSF}(x - x', y - y')\, dx'\, dy' \\ &= \iint o(\vec{r} - \vec{r}'')\mathrm{PSF}(\vec{r}'')\, d\vec{r}''. \end{aligned} \qquad (1.16)$$

Taking the 2D Fourier transformation of Eq. (1.15) yields

$$I(\rho, \xi) = O(\rho, \xi) H(\rho, \xi). \qquad (1.17)$$

Here, ρ and ξ represent the spatial frequencies; I represents the image spectrum; O represents the object spectrum; and H represents the PSF spectrum, which is the system transfer function (STF). The amplitude of the STF is referred to as the *modulation transfer function* (MTF):

$$\mathrm{MTF}(\rho, \xi) = |H(\rho, \xi)|. \qquad (1.18)$$

Similarly, for an LSF, the MTF is based on the 1D Fourier transformation:

$$\mathrm{MTF}(\rho) = \left| \int_{-\infty}^{+\infty} \exp(-j2\pi\rho x)[\mathrm{LSF}(x)]\, dx \right|. \qquad (1.19)$$

Most imaging systems act as lowpass filters, resulting in blurring of the fine structures.

The visibility of a structure in an image depends on, among other factors, the contrast C:

$$C = \frac{\Delta I}{\langle I \rangle}. \qquad (1.20)$$

While $\langle I \rangle$ is the average background image intensity, ΔI is the intensity variation in the region of interest (Figure 1.7).

Contrast does not represent a fundamental limitation on visualization since it can be artificially enhanced by, for example, subtracting part of the background (thresholding) or raising the intensity to some power. Statistical noise does, however, represent a fundamental limitation. The signal-to-noise ratio (SNR) is defined as

$$\mathrm{SNR} = \frac{\langle I \rangle}{\sigma_I}, \qquad (1.21)$$

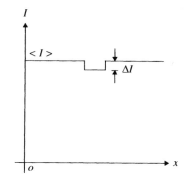

Figure 1.7. Illustration of image contrast.

where σ_I denotes the standard deviation of the background intensity, that is, the noise representing the root-mean-squared (rms) value of the intensity fluctuations.

Ultimately, the ability to visualize a structure depends on the contrast-to-noise ratio (CNR), which is defined as

$$\text{CNR} = \frac{\Delta I}{\sigma_I}, \quad (1.22)$$

which can be rewritten as

$$\text{CNR} = C \cdot \text{SNR}. \quad (1.23)$$

The *field of view* (FOV) in an image refers to the extent of the image field that can be seen all at once. A tradeoff often exists between FOV and spatial resolution. For example, "zooming in" with a camera compromises the FOV for resolution.

The maximum imaging depth in tomography is the depth limit at which the SNR or the CNR is acceptable. A tradeoff often exists between maximum imaging depth and depth resolution. The ratio of maximum imaging depth to depth resolution, referred to as the *depth-to-resolution ratio* (DRR), represents the number of effective pixels in the depth dimension. A DRR of 100 or greater is considered to indicate high resolution in terms of pixel count.

The *frame rate* is defined as the number of frames of an animation that are displayed per second, measured in frames per second (fps); it measures how rapidly an imaging system produces consecutive 2D images. At or above the video rate (30 fps), the human eye cannot resolve the transition of images; hence, the animation appears smooth.

In this book, the object to be imaged is typically a scattering medium, which can be a biological tissue phantom, a sample (specimen) of biological tissue, or an insitu or invivo biological entity. Sometimes, "sample" refers broadly to the object to be imaged.

Example 1.1. Derive Eq. (1.13).

On the basis of 1D convolution followed by a change of variable, we derive

$$\text{ESF}(x) = \int_0^{+\infty} \text{LSF}(x - x')\, dx' = \int_x^{-\infty} \text{LSF}(x'')\, d(-x'')$$
$$= \int_{-\infty}^x \text{LSF}(x')\, dx'. \tag{1.24}$$

PROBLEMS

1.1 Derive the following relationship between electromagnetic wavelength λ in the unit of μm and photon energy $h\nu$ in electron volts (eV): $\lambda h\nu = 1.24$, where h denotes the Planck constant and ν denotes the electromagnetic frequency.

1.2 In a purely absorbing (nonscattering) medium with absorption coefficient μ_a, what percentage of light is left after a lightbeam propagates a length of L? Plot this percentage as a function of L in MATLAB.

1.3 In a purely absorbing (nonscattering) medium with absorption coefficient μ_a, derive the average length of survival of a photon.

1.4 In a purely scattering (nonabsorbing) medium with scattering coefficient μ_s, what percentage of light has not been scattered after the original lightbeam propagates a length of L?

1.5 In a purely scattering (nonabsorbing) medium with scattering coefficient μ_s, derive the average length of survival of a photon.

1.6 In a scattering medium with absorption coefficient μ_a and scattering coefficient μ_s, what percentage of light has survived scattering and absorption after the original lightbeam propagates a length of L? Of the percentage that has been absorbed and scattered, what is the percentage that has been absorbed?

1.7 In MATLAB, draw a 2D diagram to simulate a random walk by following the subsequent steps: (1) start the point at (0,0); (2) sample a random number x_1 that is evenly distributed in interval (0,1]; (3) determine a step size by $s = 100 \ln(x_1)$; (2) sample a random number x_2 that is evenly distributed in interval (0,1]; (4) determine an angle by $\alpha = 2\pi x_2$; (5) move the point by step size s along angle α; (6) repeat steps 2–5 20 times to obtain a trajectory; (7) repeat steps 1–6 3 times to trace multiple trajectories.

1.8 Derive the oxygen saturation SO_2 and the total concentration of hemoglobin C_{Hb} based on Eqs. (1.5) and (1.6).

1.9 Download the data for the molar extinction coefficients of oxy- and deoxyhemoglobin as a function of wavelength from the Web (URL: http://omlc.ogi.edu/spectra/) and plot the two curves in MATLAB.

1.10 Download the data for the molar extinction coefficients of oxy- and deoxyhemoglobin as a function of wavelength from the Web (URL: http://omlc.ogi.edu/spectra/). Download the data for the absorption coefficient of pure water as a function of wavelength as well. Using physiologically representative values for both oxygen saturation SO_2 and total concentration of hemoglobin C_{Hb}, compute the corresponding absorption coefficients. Plot the three absorption spectra on the same plot in MATLAB. Identify the low-absorption near-IR window that provides deep penetration.

READING

Drezek R, Dunn A, and Richards-Kortum R (1999): Light scattering from cells: Finite-difference time-domain simulations and goniometric measurements, *Appl. Opt.* **38**: 3651–3661. [See Section 1.5, above (in this book).]

Jacques SL (2005): From http://omlc.ogi.edu/spectra/ and http://omlc.ogi.edu/classroom/. (See Sections 1.5 and 1.6, above.)

Richards-Kortum R and Sevick-Muraca E (1996): Quantitative optical spectroscopy for tissue diagnosis, *Ann. Rev. Phys. Chem.* **47**: 555–606. (See Section 1.7, above.)

Wang LHV and Jacques SL (1994): Animated simulation of light transport in tissues. Laser-tissue interaction V, *SPIE* **2134**: 247–254. (See Section 1.2, above.)

Wang LHV, Jacques SL, and Zheng LQ (1995): MCML—Monte Carlo modeling of photon transport in multi-layered tissues, *Comput. Meth. Prog. Biomed.* **47**(2): 131–146. (See Section 1.2, above.)

Wang LHV (2003): Ultrasound-mediated biophotonic imaging: A review of acousto-optical tomography and photo-acoustic tomography, *Disease Markers* **19**(2–3): 123–138. (See Section 1.1, above.)

FURTHER READING

Hecht E (2002): *Optics*, Addison-Wesley, Reading, MA.

Lakowicz JR (1999): *Principles of Fluorescence Spectroscopy*, Kluwer Academic/Plenum, New York.

Macovski A (1983): *Medical Imaging Systems*, Prentice-Hall, Englewood Cliffs, NJ.

Mourant JR, Freyer JP, Hielscher AH, Eick AA, Shen D, and Johnson TM (1998): Mechanisms of light scattering from biological cells relevant to noninvasive optical-tissue diagnostics, *Appl Opt.* **37**(16): 3586–3593.

Shung KK, Smith MB, and Tsui BMW (1992): *Principles of Medical Imaging*, Academic Press, San Diego.

Tuchin VV (2000): *Tissue Optics: Light Scattering Methods and Instruments for Medical Diagnosis*, SPIE Optical Engineering Press, Bellingham, WA.

Welch AJ and van Gemert MJC (1995): *Optical-Thermal Response of Laser-Irradiated Tissue*, Plenum Press, New York.

CHAPTER 2
Rayleigh Theory and Mie Theory for a Single Scatterer

2.1. INTRODUCTION

Both the Rayleigh and the Mie theories, which are based on the Maxwell equations, model the scattering of a plane monochromatic optical wave by a single particle. Even if the particle size is much greater than the optical wavelength, the wave is diffracted by the particle with an effective cross section that is usually different from the geometric cross section. The Rayleigh theory is applicable only to particles that are much smaller than the optical wavelength, whereas the Mie theory is valid for homogeneous isotropic spheres of any size. The Mie theory reduces to the Rayleigh theory when the particle is much smaller than the wavelength.

2.2. SUMMARY OF RAYLEIGH THEORY

The Rayleigh theory (Appendix 2A) models the scattering of light by particles that are much smaller than the optical wavelength. Figure 2.1 shows the spherical polar coordinates used for light scattering. The incident light propagates along the z axis; the scatterer is located at the origin; field point P is located at (r, θ, ϕ). The distribution of the scattered light intensity for unpolarized incident light is given by

$$I(r, \theta) = \frac{(1 + \cos^2 \theta) k^4 |\alpha|^2}{2r^2} I_0. \tag{2.1}$$

Here, α denotes the polarizability of the particle; I_0 denotes the incident light intensity; k denotes the propagation constant (also referred to as the *magnitude of the wavevector* or the *angular wavenumber*) in the background medium. We have

$$k = \frac{2\pi n_b}{\lambda}, \tag{2.2}$$

Biomedical Optics: Principles and Imaging, by Lihong V. Wang and Hsin-i Wu
Copyright © 2007 John Wiley & Sons, Inc.

18 RAYLEIGH THEORY AND MIE THEORY FOR A SINGLE SCATTERER

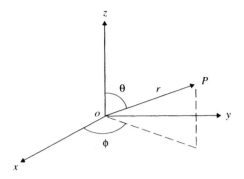

Figure 2.1. Spherical polar coordinates used for light scattering, where θ denotes the polar angle and ϕ denotes the azimuthal angle.

where n_b denotes the refractive index of the background medium and λ denotes the wavelength in vacuum. Substituting Eq. (2.2) into Eq. (2.1), we obtain $I(r, \theta) \propto 1/\lambda^4$. This strong wavelength dependence explains the blue sky in broad daylight because blue light is scattered much more strongly than red light.

The scattering cross section is given by

$$\sigma_s = \frac{8\pi k^4 |\alpha|^2}{3}. \tag{2.3}$$

The polarizability of a sphere with radius a is given by

$$\alpha = \frac{n_{rel}^2 - 1}{n_{rel}^2 + 2} a^3, \tag{2.4}$$

where n_{rel} is the relative refractive index of the particle:

$$n_{rel} = \frac{n_s}{n_b}. \tag{2.5}$$

Here, n_s is the refractive index of the sphere and n_b is the refractive index of the background. Substituting Eq. (2.4) into Eq. (2.3), we obtain

$$\sigma_s = \frac{8\pi a^2 x^4}{3} \left| \frac{n_{rel}^2 - 1}{n_{rel}^2 + 2} \right|^2, \tag{2.6}$$

where size parameter x is defined as

$$x = ka. \tag{2.7}$$

Substituting Eqs. (2.7) and (2.2) into Eq. (2.6), we obtain $\sigma_s \propto a^6/\lambda^4$. From Eq. (2.6), the scattering efficiency is given by

$$Q_s = \frac{8x^4}{3} \left| \frac{n_{rel}^2 - 1}{n_{rel}^2 + 2} \right|^2. \qquad (2.8)$$

The scattering efficiency depends only on x and n_{rel}. If n_{rel} is close to unity, Eq. (2.8) reduces to

$$Q_s = \frac{32x^4}{27} |n_{rel} - 1|^2. \qquad (2.9)$$

Note that n_{rel} can be complex, in which case the imaginary part is responsible for absorption.

2.3. NUMERICAL EXAMPLE OF RAYLEIGH THEORY

The Rayleigh theory can compute scattering cross section σ_s and scattering efficiency Q_s. As an example, the following parameters are given:

1. Diameter of sphere: $2a = 20$ nm
2. Wavelength in vacuum: $\lambda = 400$ nm
3. Refractive index of sphere: $n_s = 1.57$
4. Refractive index of background: $n_b = 1.33$
5. Specific weight of sphere: $\rho_s = 1.05$ g/cm^3
6. Specific weight of background: $\rho_b = 1.00$ g/cm^3
7. Concentration of spheres in background by weight: $C_{wt} = 1 \times 10^{-5}$

We compute in SI units as follows:

1. Propagation constant in background medium: $k = 2\pi n_b/\lambda = 2.09 \times 10^7$ m^{-1}.
2. Relative refractive index of sphere: $n_{rel} = n_s/n_b = 1.18$.
3. Size parameter: $x = ka = 0.209$.
4. From Eq. (2.6), $\sigma_s = 2.15 \times 10^{-20}$ m^2.
5. From Eq. (2.8), $Q_s = 6.83 \times 10^{-5}$.
6. Compute the number density of scatterers N_s. For a sphere, the mass density is $\rho_s = m_s/V_s$, where m_s denotes the mass and V_s denotes the volume — $V_s = (4/3)\pi a^3$. For the background, the mass density is $\rho_b = m_b/V_b$, where m_b is the total mass of the background and V_b is the total volume of the background.

The concentration by weight is $C_{wt} = (N_s V_b) m_s/m_b$. Therefore, we have $N_s = C_{wt}(\rho_b/\rho_s)/V_s = 2.27 \times 10^{18}$ m^{-3}.

7. From Section 1.5, $\mu_s = N_s \sigma_s = 0.0488$ m^{-1}.

We can implement the Rayleigh theory with the following MATLAB script:

```
% Rayleigh scattering
% Use SI units

diameter=input('Diameter of sphere (e.g., 20 nm):')*1e-9;
radius=diameter/2;
lambda=input('Wavelength (e.g., 400 nm):')*1e-9;
n_sphere=input('Refractive index of sphere (e.g., 1.57):');
n_background=input('Refractive index of background (e.g., 1.33):');
w_sphere=input('Specific weight of sphere(e.g., 1.05 g/cc):')*1e3;
w_background=input('Specific weight of background(e.g., 1 g/cc):')*1e3;
concentration=input('Concentration by weight (e.g., 1e-5):');

k=2*pi*n_background/lambda
x=k*radius
n_rel=n_sphere/n_background

Qs = 8*x^4/3*abs((n_rel^2 - 1)/(n_rel^2 + 2))^2
sigma_s=Qs*pi*radius^2

vol_sphere = 4*pi/3*radius^3
N_s=concentration*w_background/(vol_sphere*w_sphere)
mu_s=N_s*sigma_s

% Output results
{'wavelength(nm)','Qs (-)','mus (/cm)'; lambda*1e9, Qs, mu_s/1e2}
```

2.4. SUMMARY OF MIE THEORY

The Mie theory (Appendix 2B) models the scattering of light by a spherical particle of any radius a. The sphere is made of homogeneous and isotropic material and is irradiated by a plane monochromatic wave. In practice, we can treat the incident wave as a plane wave if the wavefront is much wider than both the wavelength and the particle size.

Application of the Mie theory is straightforward. The scattering efficiency Q_s and the scattering anisotropy g (defined by $g = \langle \cos\theta \rangle$) can be computed as follows:

$$Q_s = \frac{2}{x^2} \sum_{l=1}^{\infty} (2l+1)(|a_l|^2 + |b_l|^2), \tag{2.10}$$

$$g = \frac{4}{Q_s x^2} \sum_{l=1}^{\infty} \left[\frac{l(l+2)}{l+1} \text{Re}(a_l a_{l+1}^* + b_l b_{l+1}^*) + \frac{2l+1}{l(l+1)} \text{Re}(a_l b_l^*) \right]. \tag{2.11}$$

Size parameter $x = ka$. Coefficients a_l and b_l are given by

$$a_l = \frac{\psi_l'(y)\psi_l(x) - n_{\text{rel}}\psi_l(y)\psi_l'(x)}{\psi_l'(y)\zeta_l(x) - n_{\text{rel}}\psi_l(y)\zeta_l'(x)},$$

$$b_l = \frac{n_{\text{rel}}\psi_l'(y)\psi_l(x) - \psi_l(y)\psi_l'(x)}{n_{\text{rel}}\psi_l'(y)\zeta_l(x) - \psi_l(y)\zeta_l'(x)},$$
(2.12)

where superscript prime denotes first-order differentiation and size parameter y is defined by

$$y = n_{\text{rel}} x = \frac{2\pi n_s a}{\lambda}.$$
(2.13)

The Riccati–Bessel functions are defined by

$$\psi_l(z) = z j_l(z) = \left(\frac{\pi z}{2}\right)^{1/2} J_{l+1/2}(z) = S_l(z),$$
(2.14)

$$\chi_l(z) = -z y_l(z) = -\left(\frac{\pi z}{2}\right)^{1/2} Y_{l+1/2}(z) = C_l(z),$$
(2.15)

$$\zeta_l(z) = \psi_l(z) + i\chi_l(z) = z h_l^{(2)}(z) = \left(\frac{\pi z}{2}\right)^{1/2} H_{l+1/2}^{(2)}(z).$$
(2.16)

Here, l and $l + \frac{1}{2}$ are the orders; $j_l(\ )$ and $y_l(\ )$ denote the spherical Bessel functions of the first and second kind, respectively; $J_l(\ )$ and $Y_l(\ )$ denote the Bessel functions of the first and second kind, respectively; $h_l^{(2)}(\ )$ denotes the spherical Hankel function of the second kind; $H_l^{(2)}(\ )$ denotes the Hankel function of the second kind; and $S_l(\ )$ and $C_l(\ )$ are alternative symbols that are commonly used. Note that

$$h_l^{(2)}(\ ) = j_l(\ ) - i y_l(\ ),$$
(2.17)

$$H_l^{(2)}(\ ) = J_l(\ ) - i Y_l(\ ).$$
(2.18)

If n_{rel} is complex, the extinction instead of scattering efficiency that also contains a component representing absorption can be computed.

2.5. NUMERICAL EXAMPLE OF MIE THEORY

For a spherical particle of any size, the Mie theory can compute the scattering efficiency Q_s, the scattering anisotropy g, and the scattering cross section σ_s. For a scattering medium, we can further compute the scattering coefficient μ_s and the reduced scattering coefficient μ_s'. When the Mie theory is implemented in MATLAB or another high-level computer language, the following derivative identities are useful:

$$J'_l(z) = -\frac{n}{z}J_l(z) + J_{l-1}(z).\tag{2.19}$$

$$Y'_l(z) = -\frac{n}{z}Y_l(z) + Y_{l-1}(z).\tag{2.20}$$

An example MATLAB script is shown below:

```
% Mie theory
% Use SI units

diameter = input('Diameter of sphere (e.g., 579 nm):')*1e-9;
radius = diameter/2;
lambda = input('Wavelength (e.g., 400 nm):')*1e-9;
n_s = input('Refractive index of sphere (e.g., 1.57):');
n_b = input('Refractive index of background (e.g., 1.33):');
w_s = input('Specific weight of sphere(e.g., 1.05 g/cc):')*1e3;
w_b = input('Specific weight of background(e.g., 1.0 g/cc):')*1e3;
concentration = input('Concentration by weight (e.g., 0.002):');

k = 2*pi*n_b/lambda
x = k*radius
n_rel = n_s/n_b
y = n_rel*x

% Calculate the summations
err = 1e-8;
Qs = 0;
gQs = 0;
for n = 1:100000
   Snx = sqrt(pi*x/2)*besselj(n+0.5,x);
   Sny = sqrt(pi*y/2)*besselj(n+0.5,y);
   Cnx = -sqrt(pi*x/2)*bessely(n+0.5,x);
   Zetax = Snx+i*Cnx;

   % Calculate the first-order derivatives
   Snx_prime = - (n/x)*Snx+sqrt(pi*x/2)*besselj(n-0.5,x);
   Sny_prime = - (n/y)*Sny+sqrt(pi*y/2)*besselj(n-0.5,y);
   Cnx_prime = - (n/x)*Cnx-sqrt(pi*x/2)*bessely(n-0.5,x);
   Zetax_prime = Snx_prime + i*Cnx_prime;

   an_num = Sny_prime*Snx-n_rel*Sny*Snx_prime;
   an_den = Sny_prime*Zetax-n_rel*Sny*Zetax_prime;
   an = an_num/an_den;

   bn_num = n_rel*Sny_prime*Snx-Sny*Snx_prime;
   bn_den = n_rel*Sny_prime*Zetax-Sny*Zetax_prime;
   bn = bn_num/bn_den;

   Qs1 = (2*n+1)*(abs(an)^2+abs(bn)^2);
   Qs = Qs+Qs1;
```

```
    if n > 1
        gQs1 = (n-1)*(n+1)/n*real(an_1*conj(an)+bn_1*conj(bn))...
               +(2*n-1)/((n-1)*n)*real(an_1*conj(bn_1));
        gQs = gQs+gQs1;
    end

    an_1 = an;
    bn_1 = bn;

    if abs(Qs1)<(err*Qs) & abs(gQs1)<(err*gQs)
        break;
    end
end

Qs = (2/x^2)*Qs;
gQs = (4/x^2)*gQs;
g = gQs/Qs;

vol_s = 4*pi/3*radius^3
N_s = concentration*w_b/(vol_s*w_s)
sigma_s = Qs*pi*radius^2;
mu_s = N_s*sigma_s

mu_s_prime = mu_s*(1-g);

% Output results
{'wavelength(nm)','Qs (-)','g (-)','mus (/cm)','mus_prime(/cm)';...
    lambda*1e9,Qs,g,mu_s*1e-2,mu_s_prime*1e-2}
```

Below, we present a numerical example with the following inputs:

1. Diameter of sphere: $2a = 579$ nm
2. Wavelength: $\lambda = 400$ nm
3. Refractive index of sphere: $n_s = 1.57$
4. Refractive index of background: $n_b = 1.33$
5. Specific weight of sphere: $\rho_s = 1.05$ g/cm^3
6. Specific weight of background: $\rho_b = 1.0$ g/cm^3
7. Concentration of spheres in the solution by weight: $C_{wt} = 0.002$

The MATLAB script gives the following outputs: $Q_s = 2.03$, $g = 0.916$, $\mu_s = 100$ cm^{-1}, and $\mu'_s = 8.40$ cm^{-1}.

In Figure 2.2, Q_s and g, which are calculated using a modified version of the MATLAB program presented above, are plotted against ka, where $n_s = 1.40$ and $n_b = 1.33$. Note that g is less than unity even for large x values.

APPENDIX 2A. DERIVATION OF RAYLEIGH THEORY

The Rayleigh theory is derived here. The polarizability α is defined as the pro-portionality constant between the induced oscillating dipole moment $\vec{p} \exp(i\omega t)$

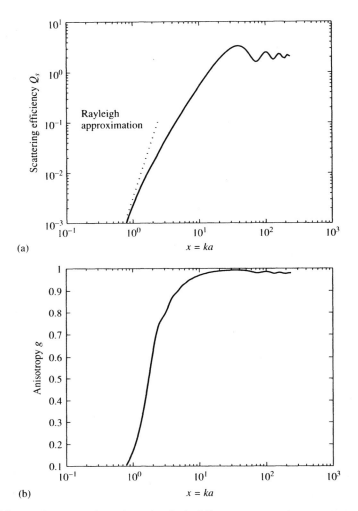

Figure 2.2. (a) Q_s versus ka, where the dashed line represents the asymptote from the Rayleigh theory; (b) anisotropy g versus ka.

and the electric field of the incident linearly polarized wave $\vec{E}_0 \exp(i\omega t)$, where ω denotes angular frequency and t denotes time:

$$\vec{p} = \alpha \vec{E}_0. \tag{2.21}$$

On the basis of the dipole radiation theory for $a \ll \lambda$, the electric field of the scattered wave in the far field ($r \gg \lambda$) is

$$E = \frac{k^2 p \sin \gamma}{r} e^{-ikr}, \tag{2.22}$$

APPENDIX 2A. DERIVATION OF RAYLEIGH THEORY

where γ is the angle between the directions of the scattered light propagation and the dipole oscillation. From Eq. (2.21) and (2.22), the scattered light intensity is

$$I = |E|^2 = \frac{k^4|p|^2 \sin^2 \gamma}{r^2} = \frac{k^4|\alpha|^2 \sin^2 \gamma}{r^2}|E_0|^2 = \frac{k^4|\alpha|^2 \sin^2 \gamma}{r^2} I_0, \quad (2.23)$$

where I_0 is the incident light intensity.

The incident light is assumed to propagate in the positive z direction. Thus, its electric field lies on the xy plane. We express both the unit vector of polarization $\hat{p}$ and the unit vector that points to the field point P from the scatterer at the origin $\hat{r}$ in terms of the unit vectors of the Cartesian coordinates, $(\hat{e}_x, \hat{e}_y, \hat{e}_z)$:

$$\hat{p} = \hat{e}_x \cos \psi + \hat{e}_y \sin \psi, \quad (2.24)$$

$$\hat{r} = \hat{e}_x \sin \theta \cos \phi + \hat{e}_y \sin \theta \sin \phi + \hat{e}_z \cos \theta. \quad (2.25)$$

Thus, we obtain

$$\cos \gamma = \hat{p} \cdot \hat{r} = \sin \theta \cos(\phi - \psi), \quad (2.26)$$

which leads to

$$\sin^2 \gamma = 1 - \cos^2 \gamma = 1 - \sin^2 \theta \cos^2(\phi - \psi). \quad (2.27)$$

If the incident light is unpolarized, $\sin^2 \gamma$ needs to be averaged over angle ψ:

$$\langle \sin^2 \gamma \rangle = 1 - \frac{1}{2} \sin^2 \theta = \frac{1}{2}(1 + \cos^2 \theta). \quad (2.28)$$

Hence

$$I(r, \theta) = \frac{k^4|\alpha|^2 (1 + \cos^2 \theta)}{2r^2} I_0, \quad (2.29)$$

which is Eq. (2.1).

The total scattering cross section σ_s is defined as

$$\sigma_s = \frac{1}{I_0} \int_{4\pi} I(r, \theta) r^2 \, d\Omega. \quad (2.30)$$

Evaluating Eq. (2.30) with differential solid angle element $d\Omega = \sin \theta d\theta d\phi$, we obtain

$$\sigma_s = 2\pi k^4 |\alpha|^2 \int_0^\pi \frac{1 + \cos^2 \theta}{2} \sin \theta d\theta = \frac{8\pi k^4 |\alpha|^2}{3}, \quad (2.31)$$

which is Eq. (2.3).

26 RAYLEIGH THEORY AND MIE THEORY FOR A SINGLE SCATTERER

APPENDIX 2B. DERIVATION OF MIE THEORY

The Mie theory is an exact solution of the Maxwell equations for a plane monochromatic electromagnetic wave scattered by a homogeneous sphere of radius a with an isotropic relative index of refraction n_{rel}. An abbreviated derivation of the Mie theory is presented here. The general idea is to (1) solve the Maxwell equations inside and outside the sphere with undetermined coefficients in the solution expansions and (2) determine these coefficients by applying boundary conditions on the spherical surface. The Mie theory for a cylindrical scatterer is beyond the scope of this book.

We assume that all conditions for the following Maxwell equations are met:

$$\nabla \cdot \vec{E} = 0, \tag{2.32}$$

$$\nabla \cdot \vec{B} = 0, \tag{2.33}$$

$$\nabla \times \vec{E} = -\frac{\partial \vec{B}}{\partial t}, \tag{2.34}$$

$$\nabla \times \vec{B} = \left(\frac{n}{c}\right)^2 \frac{\partial \vec{E}}{\partial t}. \tag{2.35}$$

Here, $\vec{E}$ and $\vec{B}$ are the electric and magnetic fields, respectively; n is the refractive index of the medium; c is the speed of light in vacuum.

2B.1. Vector and Scalar Wave Equations

The following vector wave equation for both $\vec{E}$ and $\vec{B}$ can be obtained from Eqs. (2.32)–(2.35):

$$\nabla^2 \vec{A} = \left(\frac{n}{c}\right)^2 \frac{\partial^2 \vec{A}}{\partial t^2}, \tag{2.36}$$

where $\vec{A}$ represents either $\vec{E}$ or $\vec{B}$. Each vector component satisfies the following scalar wave equation:

$$\nabla^2 \Psi = \left(\frac{n}{c}\right)^2 \frac{\partial^2 \Psi}{\partial t^2}. \tag{2.37}$$

Example 2.1. Derive Eq. (2.36) for $\vec{E}$.

Operating $\nabla \times$ from the left on both sides of Eq. (2.34), we obtain $\nabla \times (\nabla \times \vec{E}) = -\partial(\nabla \times \vec{B})/\partial t$. From $\nabla \times (\nabla \times \vec{E}) = \nabla(\nabla \cdot \vec{E}) - \nabla^2 \vec{E} = -\nabla^2 \vec{E}$, we get

$$\nabla^2 \vec{E} = \frac{\partial \nabla \times \vec{B}}{\partial t}. \tag{2.38}$$

Substituting Eq. (2.35) into Eq. (2.38), we obtain Eq. (2.36) for $\vec{E}$.

2B.2. Solution of Scalar Wave Equation

The standard procedure for solving Eq. (2.37) is by separation of the variables. Let $\Psi(\vec{x}, t) = X(\vec{x})T(t)$, where $\vec{x}$ represents the spatial coordinates. Substituting this expression into Eq. (2.37) and dividing both sides by XT, we obtain

$$\frac{\nabla^2 X}{X} = \left(\frac{n}{c}\right)^2 \frac{1}{T}\left(\frac{\partial^2 T}{\partial t^2}\right). \tag{2.39}$$

The left-hand side of this equation is a function of only $\vec{x}$, whereas the right-hand side is a function of only t; thus, the two sides must equal a constant, which is termed a *separation variable constant*; this constant is denoted by $-\beta^2$. The time-dependent part can be expressed as

$$\left(\frac{\partial^2 T}{\partial t^2}\right) + \omega^2 T = 0, \tag{2.40}$$

where $\omega = \beta c/n$. Since ω is the angular frequency of the wave, we have $\beta = k = nk_0$, where k and k_0 are the propagation constants in the medium and in vacuum, respectively. The solution of Eq. (2.40) is

$$T \sim \left\{ \begin{array}{c} \cos \omega t \\ \sin \omega t \end{array} \right\}. \tag{2.41}$$

The pair of braces represents a linear combination of the two functions inside. The spatially dependent part is a scalar Helmholtz equation

$$\nabla^2 X + k^2 X = 0, \tag{2.42}$$

which can be expressed in spherical coordinates as

$$\left[\frac{1}{r^2}\frac{\partial}{\partial r}\left(r^2 \frac{\partial}{\partial r}\right) + \frac{1}{r^2 \sin\theta}\frac{\partial}{\partial \theta}\left(\sin\theta \frac{\partial}{\partial \theta}\right) + \frac{1}{r^2 \sin^2\theta}\frac{\partial^2}{\partial \phi^2}\right] X + k^2 X = 0. \tag{2.43}$$

Letting $X = R(r)\Theta(\theta)\Phi(\phi)$, substituting it into Eq. (2.43), and then dividing the equation by $R(r)\Theta(\theta)\Phi(\phi)$, we obtain

$$\frac{1}{r^2 R}\frac{\partial}{\partial r}\left(r^2 \frac{\partial R}{\partial r}\right) + \frac{1}{r^2 \Theta \sin\theta}\frac{\partial}{\partial \theta}\left(\sin\theta \frac{\partial \Theta}{\partial \theta}\right) + \frac{1}{r^2 \Phi \sin^2\theta}\frac{\partial^2 \Phi}{\partial \phi^2} + k^2 = 0. \tag{2.44}$$

Multiplying Eq. (2.44) by $r^2 \sin^2\theta$, we see that the third term on the left-hand side depends only on ϕ, whereas the other terms are independent of ϕ. Thus, we let

$$\frac{1}{\Phi}\frac{\partial^2 \Phi}{\partial \phi^2} = \text{constant} = -m^2. \tag{2.45}$$

Thus, we have

$$\frac{\partial^2 \Phi}{\partial \phi^2} + m^2 \Phi = 0, \qquad (2.46)$$

which has the following solutions:

$$\Phi \sim \begin{Bmatrix} \cos(m\phi) \\ \sin(m\phi) \end{Bmatrix}. \qquad (2.47)$$

From condition $\Phi(0) = \Phi(2\pi)$, we have $m = 0, \pm 1, \pm 2, \ldots$. The $r\theta$-dependent part from Eq. (2.44) is

$$\frac{1}{r^2 R}\frac{\partial}{\partial r}\left(r^2 \frac{\partial R}{\partial r}\right) + \frac{1}{r^2 \Theta \sin\theta}\frac{\partial}{\partial \theta}\left(\sin\theta \frac{\partial \Theta}{\partial \theta}\right) - \frac{m^2}{r^2 \sin^2\theta} + k^2 = 0. \qquad (2.48)$$

Multiplying this equation by r^2, we see that the first and fourth terms depend only on r, whereas the other terms are independent of r. Letting

$$\frac{1}{\Theta \sin\theta}\frac{\partial}{\partial \theta}\left(\sin\theta \frac{\partial \Theta}{\partial \theta}\right) - \frac{m^2}{\sin^2\theta} = \text{constant} = -l(l+1), \qquad (2.49)$$

we have

$$\frac{1}{r^2}\frac{\partial}{\partial r}\left(r^2 \frac{\partial R}{\partial r}\right) + \left[k^2 - \frac{l(l+1)}{r^2}\right] R = 0 \qquad (2.50)$$

and

$$(1-z^2)\frac{\partial^2 \Theta}{\partial z^2} - 2z\frac{d\Theta}{dz} + \left[l(l+1) - \frac{m^2}{1-z^2}\right]\Theta = 0, \qquad (2.51)$$

where $z = \cos\theta$. The solutions of Eq. (2.50) are

$$R \sim \begin{Bmatrix} j_l(kr) \\ y_l(kr) \end{Bmatrix} = \begin{Bmatrix} j_l(nk_0 r) \\ y_l(nk_0 r) \end{Bmatrix}. \qquad (2.52)$$

Here, l is an integer; $j_l(x)$ and $y_l(x)$ are the spherical Bessel functions of the first and second kinds, respectively, given by

$$j_l(x) = \sqrt{\frac{\pi}{2x}} J_{l+(1/2)}(x), \qquad (2.53)$$

$$y_l(x) = \sqrt{\frac{\pi}{2x}} Y_{l+(1/2)}(x). \qquad (2.54)$$

APPENDIX 2B. DERIVATION OF MIE THEORY   29

The solutions of Eq. (2.51) are

$$\Theta \sim \begin{Bmatrix} P_{l,m}(\cos\theta) \\ Q_{l,m}(\cos\theta) \end{Bmatrix}, \tag{2.55}$$

where $P_{l,m}(\cos\theta)$ and $Q_{l,m}(\cos\theta)$ are the associated Legendre polynomials of the first and second kinds, respectively.

Recalling that $\Psi(\vec{x}, t) = X(\vec{x})T(t)$ and $X = R(r)\Theta(\theta)\Phi(\phi)$, we obtain the solution of the scalar wave equation [Eq. (2.37)] in spherical polar coordinates (see Figure 2.1):

$$\Psi \sim \begin{Bmatrix} \cos(\omega t) \\ \sin(\omega t) \end{Bmatrix} \begin{Bmatrix} \cos(m\phi) \\ \sin(m\phi) \end{Bmatrix} \begin{Bmatrix} j_l(nk_0 r) \\ y_l(nk_0 r) \end{Bmatrix} \begin{Bmatrix} P_{l,m}(\cos\theta) \\ Q_{l,m}(\cos\theta) \end{Bmatrix}. \tag{2.56}$$

We use $\exp(i\omega t)$ from the linear combination of the time-dependent factors for the phasor expressions of the waves. Of course, we can also use $\exp(-i\omega t)$ consistently. Function $Q_{l,m}(\cos\theta)$ is dropped off because it has singularities at $\theta \in \{0, \pi\}$. For a wave inside the sphere, function $y_l(nk_0 r)$ is dropped off because $y_l(nk_0 r) \to -\infty$ when $r \to 0$. For an outgoing spherical wave outside the sphere, the spherical Hankel function of the second kind $h_l^{(2)}(nk_0 r)$ is chosen because of its asymptotic behavior:

$$h_l^{(2)}(nk_0 r) \sim \frac{i^{l+1}}{nk_0 r} \exp(-ink_0 r). \tag{2.57}$$

Therefore, we use

$$\Psi \sim \exp(i\omega t) \begin{Bmatrix} \cos(m\phi) \\ \sin(m\phi) \end{Bmatrix} \begin{Bmatrix} j_l(nk_0 r) \\ h_l^{(2)}(nk_0 r) \end{Bmatrix} P_{l,m}(\cos\theta). \tag{2.58}$$

If $\exp(-i\omega t)$ is used, the spherical Hankel function of the first kind $h_l^{(1)}(nk_0 r)$ should be used instead for the outgoing spherical wave.

2B.3. Theorem Relating Solutions of Scalar and Vector Wave Equations

From Eq. (2.37), the time-independent scalar wave equation (the scalar Helmholtz equation) is given by

$$\nabla^2 X + n^2 k_0^2 X = 0, \tag{2.59}$$

which has the following solutions:

$$X \sim \begin{Bmatrix} \cos(m\phi) \\ \sin(m\phi) \end{Bmatrix} \begin{Bmatrix} j_l(nk_0 r) \\ h_l^{(2)}(nk_0 r) \end{Bmatrix} P_{l,m}(\cos\theta). \tag{2.60}$$

Similarly, from Eq. (2.36), the time-independent vector equation (the vector Helmholtz equation) is given by

$$\nabla^2 \vec{A} + n^2 k_0^2 \vec{A} = 0. \tag{2.61}$$

The solution to the vector Helmholtz equation can be found from the following theorem. If X satisfies the scalar wave equation [Eq. (2.59)], vectors $\vec{M}_X$ and $\vec{N}_X$, defined by

$$\vec{M}_X = \nabla \times (\vec{r} X) \quad \text{and} \quad n k_0 \vec{N}_X = \nabla \times \vec{M}_X, \tag{2.62}$$

must satisfy the vector Helmholtz equation [Eq. (2.61)], where $\vec{M}_X$ and $\vec{N}_X$ are related by

$$n k_0 \vec{M}_X = \nabla \times \vec{N}_X. \tag{2.63}$$

The full components of $\vec{M}_X$ and $\vec{N}_X$ are

$$M_r = 0, \tag{2.64}$$

$$M_\theta = \frac{1}{r \sin \theta} \frac{\partial (rX)}{\partial \phi}, \tag{2.65}$$

$$M_\phi = -\frac{1}{r} \frac{\partial (rX)}{\partial \theta}, \tag{2.66}$$

$$n k_0 N_r = \frac{1}{r^2} \frac{\partial}{\partial r} \left[r^2 \frac{\partial (rX)}{\partial r} \right] + n^2 k_0^2 r X, \tag{2.67}$$

$$n k_0 N_\theta = \frac{1}{r} \frac{\partial^2 (rX)}{\partial r \partial \theta}, \tag{2.68}$$

$$n k_0 N_\phi = \frac{1}{r \sin \theta} \frac{\partial^2 (rX)}{\partial r \partial \phi}. \tag{2.69}$$

Equations (2.67) can also be expressed as

$$n k_0 N_r = -\frac{1}{r \sin \theta} \frac{\partial}{\partial \theta} \left(\sin \theta \frac{\partial X}{\partial \theta} \right) - \frac{1}{r \sin^2 \theta} \frac{\partial^2 X}{\partial \phi^2}. \tag{2.70}$$

If u and v are two solutions to the scalar wave equation, and $\vec{M}_u$, $\vec{N}_u$, $\vec{M}_v$, and $\vec{N}_v$ are the derived vector fields, the spatial components of the solutions of the vector wave equations are

$$\vec{E} = \vec{M}_v + i \vec{N}_u \quad \text{and} \quad \vec{B} = \frac{n}{c} (-\vec{M}_u + i \vec{N}_v). \tag{2.71}$$

APPENDIX 2B. DERIVATION OF MIE THEORY   31

The components of $\vec{E}$ and $\vec{B}$ can thus be written in terms of the scalar solutions u and v and their first and second derivatives.

Example 2.2. Show that vector $\vec{M}_X$ defined by Eq. (2.62) satisfies the vector wave equation given by Eq. (2.61).

Multiplying $\vec{r}$ and then operating $\nabla \times$ from the left through Eq. (2.59), we obtain

$$\nabla \times (\vec{r}\nabla^2 X) + n^2 k_0^2 \vec{M}_X = 0. \tag{2.72}$$

From the vector identity

$$\nabla[\nabla \cdot (\nabla X \times \vec{r})] = \vec{r}(\nabla \times \nabla X) - \nabla X(\nabla \times \vec{r}) = 0, \tag{2.73}$$

we derive

$$\nabla \times \nabla \times (\nabla X \times \vec{r}) = \nabla[\nabla \cdot (\nabla X \times \vec{r})] - \nabla^2(\nabla X \times \vec{r})$$
$$= -\nabla^2(\nabla X \times \vec{r}). \tag{2.74}$$

Also

$$\nabla \times [\nabla \times (\nabla X \times \vec{r})] = \nabla \times [(\vec{r} \cdot \nabla)\nabla X - \vec{r}\nabla^2 X - (\nabla X \cdot \nabla)\vec{r} + \nabla X(\nabla \cdot \vec{r})]$$
$$= \nabla \times [(\vec{r} \cdot \nabla)\nabla X] - \nabla \times [\vec{r}\nabla^2 X] - \nabla \times [(\nabla X \cdot \nabla)\vec{r}]$$
$$+ \nabla \times [\nabla X(\nabla \cdot \vec{r})], \tag{2.75}$$

The first, third, and fourth terms on the right-hand side can be evaluated as follows:

$$\vec{r} \cdot \nabla = r\frac{\partial}{\partial r} \Rightarrow \nabla \times [(\vec{r} \cdot \nabla)\nabla X] = \nabla \times \left(r\frac{\partial}{\partial r}\nabla X\right) = 0, \tag{2.76}$$

$$(\nabla X \cdot \nabla)\vec{r} = \nabla X \Rightarrow \nabla \times [(\nabla X \cdot \nabla)\vec{r}] = \nabla \times \nabla X = 0, \tag{2.77}$$

$$\nabla \cdot \vec{r} = 3 \Rightarrow \nabla \times [\nabla X(\nabla \cdot \vec{r})] = 0. \tag{2.78}$$

Therefore, Eq. (2.75) becomes

$$\nabla \times [\nabla \times (\nabla X \times \vec{r})] = -\nabla \times (\vec{r}\nabla^2 X). \tag{2.79}$$

From Eqs. (2.74) and (2.79), we obtain

$$\nabla^2(\nabla X \times \vec{r}) = \nabla \times (\vec{r}\nabla^2 X). \tag{2.80}$$

We also obtain

$$\nabla^2 \vec{M}_X = \nabla^2[\nabla \times (\vec{r}X)] = \nabla^2[\nabla X \times \vec{r} + X(\nabla \times \vec{r})] = \nabla^2(\nabla X \times \vec{r}). \tag{2.81}$$

32 RAYLEIGH THEORY AND MIE THEORY FOR A SINGLE SCATTERER

Merging Eqs. (2.72), (2.80), and (2.81) yields

$$\nabla^2 \vec{M}_X + n^2 k_0^2 \vec{M}_X = 0, \tag{2.82}$$

which shows that $\vec{M}_X$ is a valid solution of the vector wave equation.

Example 2.3. Verify the relationships in Eqs. (2.64)–(2.66).

From the vector operation of curl in spherical coordinates

$$\nabla \times \vec{V} = \frac{\hat{r}}{r \sin\theta} \left[\frac{\partial}{\partial \theta}(\sin\theta V_\phi) - \frac{\partial V_\theta}{\partial \phi} \right] + \frac{\hat{\theta}}{r} \left[\frac{1}{\sin\theta} \frac{\partial V_r}{\partial \phi} - \frac{\partial}{\partial r}(r V_\phi) \right]$$
$$+ \frac{\hat{\phi}}{r} \left[\frac{\partial}{\partial r}(r V_\theta) - \frac{\partial V_r}{\partial \theta} \right], \tag{2.83}$$

we obtain

$$\vec{M}_X = \nabla \times (\vec{r} X) = \frac{\hat{\theta}}{\sin\theta} \frac{\partial X}{\partial \phi} - \hat{\phi} \frac{\partial X}{\partial \theta}, \tag{2.84}$$

which can be rewritten as Eqs. (2.64)–(2.66).

2B.4 Solution for Coefficients from Boundary Conditions

The origin of the coordinates is set at the center of the spherical scatterer. The positive z axis is set along the propagation direction of the incident wave. The x axis is set in the plane of electric vibration of the linearly polarized incident wave. Solutions u and v are chosen in the following forms:

1. For the incident plane wave outside the spherical particle, we have

$$u = n_b \exp(i\omega t) \cos\phi \sum_{l=1}^{\infty} (-i)^l \frac{2l+1}{l(l+1)} P_{l,1}(\cos\theta) j_l(kr), \tag{2.85}$$

$$v = n_b \exp(i\omega t) \sin\phi \sum_{l=1}^{\infty} (-i)^l \frac{2l+1}{l(l+1)} P_{l,1}(\cos\theta) j_l(kr), \tag{2.86}$$

where k represents the propagation constant of the background.

2. For the scattered wave outside the spherical particle, we have

$$u = n_b \exp(i\omega t) \cos\phi \sum_{l=1}^{\infty} -a_l(-i)^l \frac{2l+1}{l(l+1)} P_{l,1}(\cos\theta) h_l^{(2)}(kr), \tag{2.87}$$

$$v = n_b \exp(i\omega t) \sin\phi \sum_{l=1}^{\infty} -b_l(-i)^l \frac{2l+1}{l(l+1)} P_{l,1}(\cos\theta) h_l^{(2)}(kr), \tag{2.88}$$

where a_l and b_l are coefficients to be determined.

3. For the wave inside the spherical particle, we have

$$u = n_s \exp(i\omega t) \cos\phi \sum_{l=1}^{\infty} c_l(-i)^l \frac{2l+1}{l(l+1)} P_{l,1}(\cos\theta) j_l(n_{\text{rel}}kr), \quad (2.89)$$

$$v = n_s \exp(i\omega t) \sin\phi \sum_{l=1}^{\infty} d_l(-i)^l \frac{2l+1}{l(l+1)} P_{l,1}(\cos\theta) j_l(n_{\text{rel}}kr), \quad (2.90)$$

where c_l and d_l are also coefficients to be determined.

To determine these coefficients, we substitute solutions $\vec{E}$ and $\vec{B}$ from Eqs. (2.85)–(2.90) into the following boundary conditions:

$$\hat{u}_n \times (\vec{E}_o - \vec{E}_i) = \hat{u}_n \times (\vec{B}_o - \vec{B}_i) = 0. \quad (2.91)$$

Here, $\hat{u}_n$ is the unit vector perpendicular to the boundary surface; subscripts o and i represent the outside and inside, respectively. The obtained coefficients a_l and b_l are shown in Eq. (2.12).

2B.5 Scattering Efficiency and Anisotropy

Substituting the asymptotic expression of the spherical Hankel function of the second kind into Eqs. (2.87) and (2.88) leads to

$$u = -i \frac{\exp(-ikr + i\omega t)}{kr} \cos\phi \sum_{l=1}^{\infty} a_l \frac{2l+1}{l(l+1)} P_{l,1}(\cos\theta), \quad (2.92)$$

$$v = -i \frac{\exp(-ikr + i\omega t)}{kr} \sin\phi \sum_{l=1}^{\infty} b_l \frac{2l+1}{l(l+1)} P_{l,1}(\cos\theta). \quad (2.93)$$

The resulting field components are

$$E_\theta = B_\phi = -i \frac{\exp(-ikr + i\omega t)}{kr} \cos\phi \, S_2(\theta), \quad (2.94)$$

$$-E_\phi = B_\theta = -i \frac{\exp(-ikr + i\omega t)}{kr} \sin\phi \, S_1(\theta). \quad (2.95)$$

The amplitude functions are given by

$$S_1(\theta) = \sum_{l=1}^{\infty} \frac{2l+1}{l(l+1)} [a_l \pi_l(\cos\theta) + b_l \tau_l(\cos\theta)], \quad (2.96)$$

$$S_2(\theta) = \sum_{l=1}^{\infty} \frac{2l+1}{l(l+1)} [b_l \pi_l(\cos\theta) + a_l \tau_l(\cos\theta)], \quad (2.97)$$

where

$$\pi_l(\cos\theta) = \frac{P_{l,1}(\cos\theta)}{\sin\theta}, \tag{2.98}$$

$$\tau_l(\cos\theta) = \frac{d}{d\theta}P_{l,1}(\cos\theta). \tag{2.99}$$

The scattering efficiency Q_s, defined as the ratio of the scattering cross-sectional area σ_s to the physical cross-sectional area πa^2, can be expressed in terms of the amplitude functions:

$$Q_s = \frac{\sigma_s}{\pi a^2} = \frac{1}{\pi x^2} \int_{4\pi} (|S_1(\theta)|^2 \cos^2\phi + |S_2(\theta)|^2 \sin^2\phi)\, d\Omega. \tag{2.100}$$

Following the integration over ϕ, Eq. (2.100) becomes

$$Q_s = \frac{1}{x^2} \int_0^\pi (|S_1(\theta)|^2 + |S_2(\theta)|^2) \sin\theta\, d\theta. \tag{2.101}$$

Likewise, the scattering anisotropy $g = \langle \cos\theta \rangle$ can be evaluated by

$$gQ_s = \frac{1}{x^2} \int_0^\pi (|S_1(\theta)|^2 + |S_2(\theta)|^2) \cos\theta \sin\theta\, d\theta. \tag{2.102}$$

The integrations over θ in Eqs. (2.101) and (2.102) can be completed using the orthogonality relations of π_l and τ_l, which yields Eqs. (2.10) and (2.11).

PROBLEMS

2.1 Show that

$$\frac{\partial j_l(kr)}{\partial r} = \frac{k}{2l+1}[l j_{l-1}(kr) - (l+1)j_{l+1}(kr)].$$

2.2 Show that

$$\frac{d\psi_l(z)}{dz} = -\frac{l}{z}\psi_l(z) + \psi_{l-1}(z).$$

2.3 Show that

$$\frac{d\zeta_l(z)}{dz} = -\frac{l}{z}\zeta_l(z) + \zeta_{l-1}(z).$$

2.4 Derive the net radiation force exerted by light on a spherical particle. (*Hint:* The photon momentum equals the photon energy divided by the speed of light.)

2.5 Plot the angular distribution of the scattered light in the Rayleigh scattering regime and calculate the scattering anisotropy.

2.6 Implement the Mie theory using an alternative program to calculate the scattering efficiency, scattering anisotropy, scattering coefficient, and reduced scattering coefficient of spherical particles suspended in a background medium.

   (a) Use the example in Section 2.5 to verify your program.

   (b) Duplicate Figure 2.2.

   (c) Summarize the asymptotic dependence of the scattering cross section on the particle size as ka varies.

2.7 Derive coefficients a_l and b_l shown in Eqs. (2.12) by completing the derivations in Appendix 2B.

2.8 **(a)** Extend the Mie theory to absorbing scatterers.

   (b) Extend the Rayleigh theory to absorbing scatterers.

READING

Van de Hulst HC (1981): *Light Scattering by Small Particles*, Dover Publications, New York. (See Sections 2.2, 2.4, 2.5, and 2.5, above.)

FURTHER READING

Bohren CF and Huffman DR (1983): *Absorption and Scattering of Light by Small Particles*, Wiley, New York.

Ishimaru A (1997): *Wave Propagation and Scattering in Random Media*, IEEE Press, New York.

CHAPTER 3
Monte Carlo Modeling of Photon Transport in Biological Tissue

3.1. INTRODUCTION

Photon transport in biological tissue can be numerically simulated by the Monte Carlo method. The trajectory of a photon is modeled as a persistent random walk, with the direction of each step depending on that of the previous step. By contrast, the directions of all the steps in a simple random walk are independent. By tracking a sufficient number of photons, we can estimate physical quantities such as diffuse reflectance.

3.2. MONTE CARLO METHOD

Although widely used in various disciplines, the Monte Carlo method defies a succinct definition. Here, we adopt the description provided by Lux and Koblinger (1991):

> In all applications of the Monte Carlo method, a stochastic model is constructed in which the expected value of a certain random variable (or of a combination of several variables) is equivalent to the value of a physical quantity to be determined. This expected value is then estimated by averaging multiple independent samples representing the random variable introduced above. For the construction of the series of independent samples, random numbers following the distribution of the variable to be estimated are used.

It is important to realize that the Monte Carlo method estimates ensemble-averaged quantities.

The Monte Carlo method offers a flexible yet rigorous approach for simulating photon transport in biological tissue. An ensemble of biological tissues is modeled for the averaged characteristics of photon transport; the ensemble consists of all instances of the tissues that are microscopically different but macroscopically identical. Rules are defined for photon propagation from the probability

---

Biomedical Optics: Principles and Imaging, by Lihong V. Wang and Hsin-i Wu
Copyright © 2007 John Wiley & Sons, Inc.

distributions of, for example, the angles of scattering and the step sizes. The statistical nature requires tracking a large number of photons, which is computationally time-consuming. Multiple physical quantities can be simultaneously estimated, however.

In this chapter, photons are treated as waves at each scattering site but as classical particles elsewhere. Coherence, polarization, and nonlinearity are neglected. Structural anisotropy—not to be confused with scattering angular anisotropy—in tissue components, such as muscle fibers or collagens, is neglected as well.

3.3. DEFINITION OF PROBLEM

The problem to be solved begins with an infinitely narrow photon beam, also referred to as a *pencil beam*, that is perpendicularly incident on a multilayered scattering medium (Figure 3.1); various physical quantities are computed as responses. The pencil beam can be represented by an impulse (Dirac delta) function of space, direction, and time; thus, the responses are termed *impulse responses* or *Green's functions*. The layers are infinitely wide and parallel to each other. Each layer is described by the following parameters: thickness d, refractive index n, absorption coefficient μ_a, scattering coefficient μ_s, and scattering anisotropy g. The top and the bottom ambient media are each described by a refractive index. Although never infinitely wide in reality, a layer can be so treated if it is much wider than the photon distribution.

Three coordinate systems are defined. A global Cartesian coordinate system (x, y, z) is used for tracking photons (Figure 3.1); the xy plane coincides with the surface of the scattering medium; the z axis is along the pencil beam.

A global cylindrical coordinate system (r, ϕ', z), which shares the z axis with the Cartesian coordinate system, is used for recording photon absorption as a function of r and z. The photon absorption distribution has cylindrical symmetry

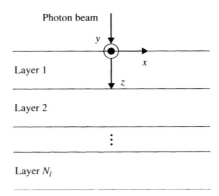

Figure 3.1. Schematic of a scattering medium with N_l layers. The y axis points out of the paper.

because of the homogeneity of each layer and the axial alignment of the pencil beam. The r coordinate is also used for recording both the diffuse reflectance and the diffuse transmittance that are functions of r as well as α, where α is the polar angle of the propagation direction of a reemitted photon with respect to the normal vector of the exit surface of the scattering medium ($-z$ axis for the top surface and $+z$ axis for the bottom surface). One can further resolve reemitted photons with the azimuthal angle ϕ'.

A local moving spherical coordinate system whose z axis is dynamically aligned with the propagation direction of the photon is used for sampling the scattering angles. Once the polar angle θ and the azimuthal angle ϕ are sampled, they are converted to direction cosines in the global Cartesian coordinate system.

The physical quantities to be computed include relative specific absorption, relative fluence, relative diffuse reflectance, and relative diffuse transmittance, all of which are relative to the incident energy. The relative specific absorption $A(r, z)$ represents the probability of photon absorption per unit volume by the medium. From $A(r, z)$, the relative fluence $F(r, z)$—which is the probability of photon flow per unit area—can be computed. The unscattered absorption from the first photon interaction events, which always take place on the z axis, is recorded separately. The relative diffuse reflectance $R_d(r, \alpha)$ for the top surface and the relative diffuse transmittance $T_d(r, \alpha)$ for the bottom surface—collectively referred to as the *relative diffuse reemittance*—are defined as the probability of photon reemission from the surfaces per unit area at r per unit solid angle around α, where a solid angle has the unit of steradians (sr). Like unscattered absorption, specularly reflected and unscattered transmitted photons are recorded separately. Physical quantities of lower dimensions can be computed from those of higher dimensions. For brevity, relative physical quantities are written as physical quantities in this chapter unless otherwise noted.

Simulated physical quantities are represented in grids on the coordinate systems. For photon absorption, a 2D homogeneous grid system is set up in the r and z directions. The grid element sizes in the r and z directions are Δr and Δz, respectively, and the total numbers of grid elements are N_r and N_z, respectively. For reemitted photons, a 1D grid system in the α direction is further set up with N_α grid elements. Since α has a range of $\pi/2$, the grid element size is $\Delta \alpha = \pi/(2N_\alpha)$. For convenience, the grid elements that should appear after a physical quantity are sometimes represented in this chapter by coordinates.

For consistency, centimeters (cm) are used as the basic unit of length throughout the simulation. For example, the thickness of each layer and the grid element sizes in the r and z directions are measured in cm. The absorption and scattering coefficients are measured in reciprocal centimeters (cm^{-1}).

3.4. PROPAGATION OF PHOTONS

This section presents the rules that govern photon propagation. Figure 3.2 shows a basic flowchart for the photon tracking part of the Monte Carlo simulation of

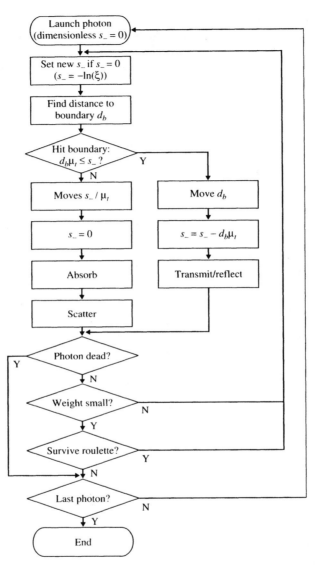

Figure 3.2. Flowchart for tracking photons in a scattering medium by the Monte Carlo method, where s_- denotes the dimensionless step size (to be discussed) and Y and N represent yes and no, respectively.

light transport in multilayered scattering media. The Monte Carlo simulation was written in ANSI (American National Standards Institute) Standard C as a software package entitled MCML (Appendix 3A). This software can be executed on any computer platform that supports ANSI Standard C. The following subsections implement many of the boxes in the flowchart.

3.4.1. Sampling of a Random Variable

The Monte Carlo method relies on the sampling of random variables from their probability distributions. The probability density function (PDF) $p(\chi)$ defines the distribution of χ over interval (a, b). The interval can also be closed or half-closed in some cases, which usually makes no practical difference. For readers unfamiliar with PDF, a brief review is given in Appendix 3B.

To sample χ, we choose a value repeatedly based on its PDF. First, a pseudorandom number ξ that is uniformly distributed between 0 and 1 is generated by computer. Then, χ is sampled by solving the following equation:

$$\int_a^\chi p(\chi)\, d\chi = \xi. \tag{3.1}$$

Since the left-hand side represents the cumulative distribution function (CDF) $P(\chi)$, we have

$$P(\chi) = \xi. \tag{3.2}$$

This equation means that if $P(\chi)$ is sampled uniformly by ξ between 0 and 1, the inverse transformation correctly samples χ as illustrated in Figure 3.3:

$$\chi = P^{-1}(\xi). \tag{3.3}$$

This sampling method, referred to as the *inverse distribution method* (IDM), is invoked repeatedly below.

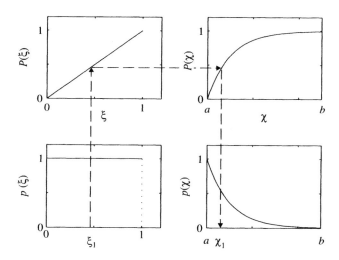

Figure 3.3. Illustration of the inverse distribution method (IDM) for sampling a random variable χ based on a random number ξ uniformly distributed between 0 and 1.

Example 3.1. Prove Eq. (3.1)

We show that when χ is sampled according to Eq. (3.1), χ follows its intended CDF $P(\chi)$.

If χ is sampled according to Eq. (3.1), the probability that a selected χ is less than χ_1 can be expressed by

$$P_{\text{IDM}}\{\chi < \chi_1\} = P\{\xi < \xi_1\}. \tag{3.4}$$

Here, $P_{\text{IDM}}\{\}$ denotes the probability in the IDM of the event in the braces; $P\{\}$ denotes the true probability of the event in the braces based on the CDF of the random variable; χ_1 is related to ξ_1 through Eq. (3.1):

$$\xi_1 = \int_a^{\chi_1} p(\chi)\, d\chi. \tag{3.5}$$

Because ξ is equidistributed between 0 and 1, we have $P\{\xi < \xi_1\} = \xi_1$. Thus, we obtain

$$P_{\text{IDM}}\{\chi < \chi_1\} = \int_a^{\chi_1} p(\chi)\, d\chi. \tag{3.6}$$

Since the right-hand side is the CDF $P(\chi)$, we have

$$P_{\text{IDM}}\{\chi < \chi_1\} = P(\chi_1), \tag{3.7}$$

which means that the sampled χ indeed follows its intended CDF.

3.4.2. Representation of a Photon Packet

A simple variance reduction technique, referred to as *implicit photon capture*, is used to improve the efficiency of the Monte Carlo simulation. This technique enables many photons to propagate as a packet of equivalent weight W along a particular trajectory.

Related parameters are structured in C to make the program easier to write, read, and modify. Thus, parameters for a photon packet are grouped into a single structure defined by

```
typedef struct {
  double x, y ,z; /* Cartesian coordinates of photon packet. */
  double ux, uy, uz;/* direction cosines of photon propagation. */
  double w; /* weight of photon packet. */
  char dead; /* 0 if photon is propagating, 1 if terminated. */
  double s_; /* dimensionless step size. */
  long scatters; /* number of scattering events experienced. */
  short layer; /* index of layer where photon packet resides. */
} PhotonStruct;
```

Structure members x, y, and z represent the coordinates of a photon packet, (x, y, z), respectively. Structure members ux, uy, and uz represent the direction cosines of the propagation direction of the photon packet, $\{\mu_x, \mu_y, \mu_z\}$, respectively. Structure member w represents the weight of the photon packet, W.

Structure member dead represents the status of the photon packet and is initialized to zero when the photon packet is launched. If the photon packet has reemitted from the scattering medium or has not survived a Russian roulette (to be discussed), this structure member is set to unity, which signals the program to stop tracking the current photon packet.

Structure member s_ represents the dimensionless step size, which is defined as the integration of extinction coefficient μ_t over the trajectory of the photon packet. In a homogeneous medium, the dimensionless step size is simply the physical path length multiplied by the extinction coefficient.

Structure member scatters stores the number of scattering events experienced by the photon packet. If this structure member is zero when the weight of a photon packet is recorded, the weight contributes to the unscattered physical quantities.

Structure member layer is the index of the layer in which the photon packet resides; it is updated only when the photon packet leaves the current layer.

3.4.3. Launching of a Photon Packet

One photon packet is launched at a time orthogonally onto the scattering medium. For each new photon packet, the coordinates (x, y, z) are initialized to $(0,0,0)$, the direction cosines $\{\mu_x, \mu_y, \mu_z\}$ are initialized to $(0,0,1)$, and the weight W is initialized to unity. Several other structure members—including dead, scatters, and layer—are also initialized.

If the upper ambient medium and the first layer have mismatched indices of refraction (n_0 and n_1, respectively), specular (Fresnel) reflection occurs. Specular reflectance in normal incidence is given by

$$R_{sp} = \left(\frac{n_0 - n_1}{n_0 + n_1}\right)^2. \tag{3.8}$$

If the first layer of refractive index n_1 is a clear medium, it causes multiple specular reflections and transmissions. If interference effect is neglected, an effective specular reflectance can be computed by

$$R_{sp} = R_{sp1} + \frac{(1 - R_{sp1})^2 R_{sp2}}{1 - R_{sp1} R_{sp2}}. \tag{3.9}$$

Here, R_{sp1} and R_{sp2} are the specular reflectances on the two boundaries of the first layer:

$$R_{sp1} = \left(\frac{n_0 - n_1}{n_0 + n_1}\right)^2, \tag{3.10}$$

$$R_{sp2} = \left(\frac{n_1 - n_2}{n_1 + n_2}\right)^2, \quad (3.11)$$

where n_2 denotes the refractive index of the second layer.

When the photon packet enters the scattering medium, the weight is decreased by R_{sp}:

$$W = 1 - R_{sp}. \quad (3.12)$$

If Eq. (3.8) is applicable, structure member layer is set to the first layer. If Eq. (3.9) is applicable, it is set to the second layer, and member z is set accordingly.

Example 3.2. Compute the specular reflectance in normal incidence between air and water, glass, and soft tissue, respectively.

For water, $n_1 = 1.33$. For an air–water interface, $R_{sp} = 2.0\%$, where $n_0 = 1$ for air.

For glass, $n_1 = 1.5$. For an air–glass interface, $R_{sp} = 4.0\%$.

For soft tissue, $n_1 = \sim 1.37$. For an interface between air and soft tissue, $R_{sp} = \sim 2.4\%$.

Example 3.3. Compute the effective specular reflectance in normal incidence for a glass slide that is sandwiched between air and water.

For air, $n_0 = 1$. For glass, $n_1 = 1.5$. For water, $n_2 = 1.33$. Thus, $R_{sp1} = 0.04$, and $R_{sp2} = 0.0035$. The effective specular reflectance $R_{sp} = 0.0432$. In this case, R_{sp} is quite close to the direct sum of R_{sp1} and R_{sp2}.

3.4.4. Step Size of a Photon Packet

The step size of a photon packet is sampled by the IDM [Eq. (3.3)], based on the PDF of the dimensional free-path length s $(0 \leq s < \infty)$ of a photon. We first consider a homogeneous medium. According to the definition of the extinction coefficient μ_t, we have

$$\mu_t = \frac{-dP\{s \geq s'\}}{P\{s \geq s'\}} \frac{1}{ds'}, \quad (3.13)$$

where $P\{\}$ denotes the probability of the event in the braces. The first fraction on the right-hand side represents the probability of interaction in interval $(s', s' + ds')$, and the second fraction represents normalization by the path length. Rearranging Eq. (3.13) yields

$$d[\ln(P\{s \geq s'\})] = -\mu_t \, ds'. \quad (3.14)$$

Integrating this equation over s' in interval $(0, s_1)$, we obtain

$$P\{s \geq s_1\} = \exp(-\mu_t s_1). \tag{3.15}$$

This is a form of the well-known Beer law.

From Eq. (3.15), the probability that an interaction occurs within s_1 is given by

$$P\{s < s_1\} = 1 - \exp(-\mu_t s_1). \tag{3.16}$$

The corresponding PDF is given by

$$p(s_1) = \frac{dP\{s < s_1\}}{ds_1} = \mu_t \exp(-\mu_t s_1). \tag{3.17}$$

According to the IDM, the CDF of s in Eq. (3.16) can be equated to ξ to yield the sampling equation for the step size

$$s_1 = -\frac{\ln(1-\xi)}{\mu_t}, \tag{3.18}$$

where $1 - \xi$ can be replaced by ξ for simplicity because ξ is uniformly distributed between 0 and 1:

$$s_1 = -\frac{\ln \xi}{\mu_t}. \tag{3.19}$$

We then consider a multilayered medium where the photon packet may experience a free path over multiple layers before an interaction occurs. In this case, the counterpart of Eq. (3.15) becomes

$$P\{s \geq s_t\} = \exp\left(-\sum_i \mu_{ti} s_i\right). \tag{3.20}$$

Here, the summation is over all the segments that the photon packet has traversed before an interaction occurs; μ_{ti} is the extinction coefficient for the ith segment, s_i is the length of the ith segment, and s_t is the total step size:

$$s_t = \sum_i s_i. \tag{3.21}$$

Equating Eq. (3.20) to ξ, we obtain the sampling equation

$$\sum_i \mu_{ti} s_i = -\ln(\xi), \tag{3.22}$$

which is a generalized form of Eq. (3.19). The left-hand side of Eq. (3.22) is the total dimensionless step size. Note that photon paths in a clear medium do not change the dimensionless step size because the extinction coefficient is zero.

Equation (3.22) is used to sample the step size in MCML, where the dimensionless step size s_- is initialized to $-\ln(\xi)$. A photon packet may travel multiple substeps of size s_i in a multilayered scattering medium before reaching an interaction site. Only when the photon packet has completed $-\ln(\xi)$ in dimensionless step size does an interaction occur. In an interaction event, the entire photon packet must experience both absorption and scattering. Since the step size is modeled, the simulation is intrinsically time-resolved.

3.4.5. Movement of a Photon Packet

Once the dimensional substep size s_i is determined, the photon packet is moved. The coordinates of the photon packet are updated by

$$x \leftarrow x + \mu_x s_i, \qquad y \leftarrow y + \mu_y s_i, \qquad z \leftarrow z + \mu_z s_i, \tag{3.23}$$

where the arrows indicate quantity substitutions. The variables on the left-hand side have the new values, and the variables on the right-hand side have the old values. In C/C++, an equal sign (=) is used for this purpose.

3.4.6. Absorption of a Photon Packet

Once the photon packet reaches an interaction site, a fraction of the weight (ΔW) is absorbed:

$$\Delta W = \frac{\mu_a}{\mu_t} W. \tag{3.24}$$

If the photon packet has not been scattered, ΔW is recorded into an array for unscattered absorption. Otherwise, it is recorded into $A(r, z)$ at the local grid element:

$$A(r, z) \leftarrow A(r, z) + \Delta W. \tag{3.25}$$

The weight must be updated by

$$W \leftarrow W - \Delta W. \tag{3.26}$$

The photon packet with the new weight then undergoes scattering at the interaction site.

3.4.7. Scattering of a Photon Packet

For scattering, the polar angle θ ($0 \leq \theta \leq \pi$) and the azimuthal angle ϕ ($0 \leq \phi < 2\pi$) are sampled by the IDM. The probability distribution of $\cos\theta$ is commonly given by the Henyey–Greenstein phase function that was originally proposed for galactic scattering:

$$p(\cos\theta) = \frac{1 - g^2}{2(1 + g^2 - 2g\cos\theta)^{3/2}}. \tag{3.27}$$

The anisotropy g, defined as $\langle\cos\theta\rangle$, has a value between -1 and 1. A value of zero indicates isotropic scattering, and a value close to unity indicates dominantly forward scattering. For most biological tissues, g is ~ 0.9. In addition to the Henyey–Greenstein phase function, the Mie theory can also provide a phase function. Note that the phase function is irrelevant to the phase of an electromagnetic wave.

Applying the IDM [Eq. (3.1)] to Eq. (3.27), we sample $\cos\theta$ as follows:

$$\cos\theta = \begin{cases} \dfrac{1}{2g}\left\{1 + g^2 - \left(\dfrac{1-g^2}{1-g+2g\xi}\right)^2\right\} & \text{if} \quad g \neq 0 \\ 2\xi - 1 & \text{if} \quad g = 0 \end{cases}. \tag{3.28}$$

The azimuthal angle ϕ, which is assumed to be uniformly distributed over interval $[0, 2\pi)$, is sampled with another independent pseudorandom number ξ:

$$\phi = 2\pi\xi. \tag{3.29}$$

Once the local polar and azimuthal angles are sampled, the new propagation direction of the photon packet can be represented in the global coordinate system as

$$\mu'_x = \frac{\sin\theta(\mu_x\mu_z\cos\phi - \mu_y\sin\phi)}{\sqrt{1-\mu_z^2}} + \mu_x\cos\theta,$$

$$\mu'_y = \frac{\sin\theta(\mu_y\mu_z\cos\phi + \mu_x\sin\phi)}{\sqrt{1-\mu_z^2}} + \mu_y\cos\theta, \tag{3.30}$$

$$\mu'_z = -\sqrt{1-\mu_z^2}\sin\theta\cos\phi + \mu_z\cos\theta.$$

If the photon direction is sufficiently close to the z axis (e.g., $|\mu_z| > 0.99999$), the following formulas are used instead so that division by a small number is avoided:

$$\mu'_x = \sin\theta\cos\phi,$$

$$\mu'_y = \sin\theta\sin\phi, \tag{3.31}$$

$$\mu'_z = \text{sgn}(\mu_z)\cos\theta,$$

where sign function $\text{sgn}(\mu_z)$ returns 1 when μ_z is positive and -1 when μ_z is negative.

Note that the direction cosines are in the global Cartesian coordinate system, whereas the polar and azimuthal angles are in the local spherical coordinate system. Since trigonometric operations are computationally intensive, alternative algebraic operations are encouraged in the program when possible.

Example 3.4. Derive Eq. (3.28) for $g \neq 0$ using the IDM [Eq. (3.1)].

Here, $\chi = \cos\theta$ and $\chi \in [-1, 1]$. Therefore, we have

$$\xi = \int_{-1}^{\chi} p(\chi)\, d\chi = \int_{-1}^{\cos\theta} p(\cos\theta')\, d\cos\theta'$$

$$= \frac{1-g^2}{2g}\left(\frac{1}{\sqrt{1+g^2-2g\cos\theta}} - \frac{1}{1+g}\right),$$

which leads to Eq. (3.28) for $g \neq 0$.

3.4.8. Boundary Crossing of a Photon Packet

During a step of dimensionless size s_-, the photon packet may hit a boundary of the current layer. Several steps are involved in boundary crossing.

Step 1. The distance d_b between the current location (x, y, z) of the photon packet and the boundary of the current layer in the photon propagation direction is computed by

$$d_b = \begin{cases} \dfrac{z_0 - z}{\mu_z} & \text{if } \mu_z < 0 \\ \infty & \text{if } \mu_z = 0, \\ \dfrac{z_1 - z}{\mu_z} & \text{if } \mu_z > 0 \end{cases} \quad (3.32)$$

where z_0 and z_1 are the z coordinates of the upper and lower boundaries of the current layer. If μ_z approaches zero, the distance approaches infinity, which is represented by constant DBL_MAX in C.

Step 2. The dimensionless step size s_- and the distance d_b are compared as follows:

$$d_b\mu_t \leq s_-, \quad (3.33)$$

where μ_t is the extinction coefficient of the current layer. If Eq. (3.33) holds, the photon packet is moved to the boundary; s_- is updated by $s_- \leftarrow s_- - d_b\mu_t$; the simulation proceeds to step 3. Otherwise, the photon packet is moved by s_-/μ_t to reach the interaction site, s_- is set to zero to signal the generation of the next dimensionless step size, and the photon packet experiences absorption and scattering. Since division by μ_t is avoided, Eq. (3.33) is applicable to clear layers ($\mu_t = 0$) as well.

Step 3. If the photon packet hits a boundary, the specular reflectance is computed. The angle of incidence α_i of the photon packet is first calculated by

$$\alpha_i = \cos^{-1}(|\mu_z|). \quad (3.34)$$

The angle of transmission α_t is then computed by Snell's law

$$n_i \sin \alpha_i = n_t \sin \alpha_t, \tag{3.35}$$

where n_i and n_t represent the refractive indices of the incident and transmitted media, respectively. If $n_i > n_t$ and α_i is greater than the critical angle $\sin^{-1}(n_t/n_i)$, the local reflectance $R_i(\alpha_i)$ equals unity. Otherwise, $R_i(\alpha_i)$ is calculated by Fresnel's formula:

$$R_i(\alpha_i) = \frac{1}{2} \left[\frac{\sin^2(\alpha_i - \alpha_t)}{\sin^2(\alpha_i + \alpha_t)} + \frac{\tan^2(\alpha_i - \alpha_t)}{\tan^2(\alpha_i + \alpha_t)} \right], \tag{3.36}$$

which is an average of the reflectances for two orthogonal linear polarization states because light is assumed to be randomly polarized.

Step 4. To determine whether the photon packet is reflected or transmitted, a pseudorandom number ξ is generated. If $\xi \leq R_i(\alpha_i)$, the photon packet is reflected; otherwise, it is transmitted. If reflected, the photon packet stays on the boundary and its direction cosines are updated by reversing the z component:

$$\{\mu_x, \mu_y, \mu_z\} \leftarrow \{\mu_x, \mu_y, -\mu_z\}. \tag{3.37}$$

If transmitted into a neighboring layer, the photon packet continues to propagate with an updated direction and step size. The new direction cosines are

$$\mu'_x = \mu_x \frac{\sin \alpha_t}{\sin \alpha_i}, \tag{3.38}$$

$$\mu'_y = \mu_y \frac{\sin \alpha_t}{\sin \alpha_i}, \tag{3.39}$$

$$\mu'_z = \operatorname{sgn}(\mu_z) \cos \alpha_t. \tag{3.40}$$

On the basis of Snell's law, Eqs. (3.38) and (3.39) can be computed more efficiently by

$$\mu'_x = \mu_x \frac{n_i}{n_t}, \tag{3.41}$$

$$\mu'_y = \mu_y \frac{n_i}{n_t}. \tag{3.42}$$

If transmitted into an ambient medium, the photon packet contributes to reemittance. If the photon packet has not been scattered, its weight is recorded into the unscattered reemittance; otherwise, its weight is recorded into either the diffuse reflectance $R_d(r, \alpha_t)$ or the diffuse transmittance $T_d(r, \alpha_t)$:

$$\begin{aligned} R_d(r, \alpha_t) &\leftarrow R_d(r, \alpha_t) + W \quad \text{if} \quad z = 0; \\ T_d(r, \alpha_t) &\leftarrow T_d(r, \alpha_t) + W \quad \text{if} \quad z \text{ is at the bottom of the medium.} \end{aligned} \tag{3.43}$$

When reemitted from the scattering medium, the photon packet completes its history (or Markov chain).

Reemission at an interface can be modeled alternatively. Rather than making the reflection of the photon packet an all-or-none event, the photon packet can be partially reflected and partially transmitted. A fraction $1 - R_i(\alpha_i)$ of the current weight of the photon packet is reemitted from the scattering medium and recorded to the local reemittance. Then, the weight is updated by $W \leftarrow W R_i(\alpha_i)$. The photon packet with the new weight is reflected and further propagated.

3.4.9. Termination of a Photon Packet

A photon packet can be terminated from the scattering medium automatically by reemission as discussed above. If the weight of a photon packet has been sufficiently decreased by many interaction events, further propagation of the photon packet yields little useful information unless interest is focused on a late stage of photon propagation. However, photon packets must be properly terminated so that energy is conserved.

A technique called *Russian roulette* is used in MCML to terminate a photon packet when the weight falls below a threshold W_{th} (e.g., $W_{th} = 0.0001$). This technique gives the photon packet one chance in m (e.g., $m = 10$) of surviving with a weight of mW. In other words, if the photon packet does not survive the Russian roulette, it is terminated with the weight set to zero; otherwise, the photon packet increases in weight from W to mW. This technique is mathematically summarized as

$$W \leftarrow \begin{cases} mW & \text{if } \xi \leq \frac{1}{m}, \\ 0 & \text{if } \xi > \frac{1}{m}, \end{cases} \quad (3.44)$$

where ξ is a uniformly distributed pseudorandom number ($0 \leq \xi \leq 1$). This method terminates photons in an unbiased manner while conserving the total energy.

3.5. PHYSICAL QUANTITIES

In this section, the process of obtaining physical quantities is discussed in detail. The units for some of the physical quantities are given at the end of their respective expressions.

Physical quantities are stored in arrays. Although photon packets propagate in infinite continuous space (limited only by computational precision), weights are recorded in finite discrete space (limited by grid element sizes). When a photon packet is recorded, its physical location may not fit into the grid system. In this case, the last grid element in the direction of the overflow collects the weight. Therefore, the last grid elements in the r and z directions do not reflect the actual values at the corresponding locations. Angle α, however, does not overflow.

As a rule of thumb for most problems, each spatial grid element should measure about 10% of the penetration depth or the transport mean free path. If the grid elements are too small, the relative errors—which are determined by the number of events occurring in each grid element—will be too large. If the grid elements are too large, the dependence of the physical quantities will not be represented with good resolution.

3.5.1. Reflectance and Transmittance

The diffuse reflectance and the diffuse transmittance are represented in MCML by two arrays, $R_{d\_r\alpha}[i_r, i_\alpha]$ and $T_{d\_r\alpha}[i_r, i_\alpha]$, respectively, where i_r and i_α are the indices of r and α ($0 \leq i_r \leq N_r - 1$ and $0 \leq i_\alpha \leq N_\alpha - 1$), respectively. The unscattered reflectance and the unscattered transmittance are stored in $R_{d\_r}[-1]$ and $T_{d\_r}[-1]$, respectively.

It can be shown that the optimal coordinates of the simulated physical quantities for the grid elements in the radial and angular directions are as follows (see Problem 3.1):

$$r(i_r) = \left[\left(i_r + \frac{1}{2}\right) + \frac{1}{12\left(i_r + \frac{1}{2}\right)}\right] \Delta r \text{ (cm)}, \tag{3.45}$$

$$\alpha(i_\alpha) = \left(i_\alpha + \frac{1}{2}\right)\Delta\alpha + \left[1 - \frac{1}{2}\Delta\alpha \cot\left(\frac{1}{2}\Delta\alpha\right)\right]\cot\left[\left(i_\alpha + \frac{1}{2}\right)\Delta\alpha\right] \text{ (rad)}. \tag{3.46}$$

The deviation of the optimized point from the center is 25% for the first radial grid element but decreases as the index of the grid element increases. Because the optimized coordinates are computed only after all photon packets are simulated, this optimization does not increase simulation time but improves accuracy.

After multiple photon packets (N) have been tracked, raw data $R_{d\_r\alpha}[i_r, i_\alpha]$ and $T_{d\_r\alpha}[i_r, i_\alpha]$ represent the total accumulated weights in each grid element. To compute the integrated weights in each direction of the 2D grid system, we sum the 2D arrays in the other dimension:

$$R_{d\_r}[i_r] = \sum_{i_\alpha=0}^{N_\alpha-1} R_{d\_r\alpha}[i_r, i_\alpha], \tag{3.47}$$

$$R_{d\_\alpha}[i_\alpha] = \sum_{i_r=0}^{N_r-1} R_{d\_r\alpha}[i_r, i_\alpha], \tag{3.48}$$

$$T_{d\_r}[i_r] = \sum_{i_\alpha=0}^{N_\alpha-1} T_{d\_r\alpha}[i_r, i_\alpha], \tag{3.49}$$

$$T_{d\_\alpha}[i_\alpha] = \sum_{i_r=0}^{N_r-1} T_{d\_r\alpha}[i_r, i_\alpha]. \tag{3.50}$$

To compute the total diffuse reflectance and the total diffuse transmittance, we sum the 1D arrays:

$$R_d = \sum_{i_r=0}^{N_r-1} R_{d\_r}[i_r], \qquad (3.51)$$

$$T_d = \sum_{i_r=0}^{N_r-1} T_{d\_r}[i_r]. \qquad (3.52)$$

These arrays describe the total weights in each grid element on the basis of N initial photon packets of unit weight. Raw data $R_{d\_r\alpha}[i_r, i_\alpha]$ and $T_{d\_r\alpha}[i_r, i_\alpha]$ are converted into probabilities of reemission per unit cross-sectional area per unit solid angle as follows:

$$R_{d\_r\alpha}[i_r, i_\alpha] \leftarrow \frac{R_{d\_r\alpha}[i_r, i_\alpha]}{\Delta a \cos\alpha \Delta\Omega N} \quad (\text{cm}^{-2}\text{sr}^{-1}), \qquad (3.53)$$

$$T_{d\_r\alpha}[i_r, i_\alpha] \leftarrow \frac{T_{d\_r\alpha}[i_r, i_\alpha]}{\Delta a \cos\alpha \Delta\Omega N} \quad (\text{cm}^{-2}\text{sr}^{-1}). \qquad (3.54)$$

The area Δa and the solid angle $\Delta\Omega$ are given by

$$\Delta a = 2\pi \left(i_r + \frac{1}{2}\right)(\Delta r)^2 \ (\text{cm}^2), \qquad (3.55)$$

$$\Delta\Omega = 4\pi \sin\left[\left(i_\alpha + \frac{1}{2}\right)\Delta\alpha\right] \sin\left(\frac{1}{2}\Delta\alpha\right) \ (\text{sr}). \qquad (3.56)$$

Raw $R_{d\_r}[i_r]$ and $T_{d\_r}[i_r]$ are converted to probabilities of reemission per unit area as follows:

$$R_{d\_r}[i_r] \leftarrow \frac{R_{d\_r}[i_r]}{N \Delta a} \quad (\text{cm}^{-2}), \qquad (3.57)$$

$$T_{d\_r}[i_r] \leftarrow \frac{T_{d\_r}[i_r]}{N \Delta a} \quad (\text{cm}^{-2}). \qquad (3.58)$$

Raw $R_{d\_r}[-1]$ and $T_{d\_r}[-1]$ are converted to total unscattered reflectance and total unscattered transmittance, respectively, by dividing them by N. Then, $R_{d\_r}[-1]$ is augmented by the specular reflectance or the effective specular reflectance.

Raw $R_{d\_\alpha}[i_\alpha]$ and $T_{d\_\alpha}[i_\alpha]$ are converted to the probabilities of reemission per unit solid angle as follows:

$$R_{d\_\alpha}[i_\alpha] \leftarrow \frac{R_{d\_\alpha}[i_\alpha]}{N \Delta\Omega} \quad (\text{sr}^{-1}), \qquad (3.59)$$

$$T_{d\_\alpha}[i_\alpha] \leftarrow \frac{T_{d\_\alpha}[i_\alpha]}{N \Delta\Omega} \quad (\text{sr}^{-1}). \qquad (3.60)$$

Raw R_d and T_d are converted to probabilities as follows:

$$R_d \leftarrow \frac{R_d}{N} \quad \text{(dimensionless)}, \tag{3.61}$$

$$T_d \leftarrow \frac{T_d}{N} \quad \text{(dimensionless)}. \tag{3.62}$$

3.5.2. Absorption and Fluence

At each interaction, the absorbed weight is stored in the specific absorption array $A_{rz}[i_r, i_z]$, where i_r and i_z are the indices of grid elements in the r and z directions, respectively ($0 \leq i_r \leq N_r - 1$ and $0 \leq i_z \leq N_z - 1$). The unscattered absorption is stored in $A_{rz}[-1, i_z]$. Whereas the optimal coordinate for i_r is shown in Eq. (3.45), the coordinate for i_z is simply

$$z(i_z) = \left(i_z + \frac{1}{2}\right) \Delta z. \tag{3.63}$$

Raw $A_{rz}[i_r, i_z]$ represents the total accumulated weight in each grid element. The total weight in each grid element in the z direction is computed by summing the 2D array in the r direction:

$$A_z[i_z] = \sum_{i_r=0}^{N_r-1} A_{rz}[i_r, i_z]. \tag{3.64}$$

Next, the total weight absorbed in each layer, $A_l[i_l]$, can be computed by

$$A_l[i_l] = \sum_{i_z} A_z[i_z], \tag{3.65}$$

where i_l is the index of a layer, and the summation includes any i_z that leads to a z coordinate in layer i_l. Further, the total weight absorbed in the scattering medium A can be computed by

$$A = \sum_{i_z=0}^{N_z-1} A_z[i_z]. \tag{3.66}$$

Then, these raw quantities are converted into the final physical quantities as follows:

$$A_{rz}[i_r, i_z] \leftarrow \frac{A_{rz}[i_r, i_z]}{N \Delta a \Delta z} \quad (\text{cm}^{-3}), \tag{3.67}$$

$$A_{rz}[-1, i_z] \leftarrow \frac{A_{rz}[-1, i_z]}{N \Delta z} \quad (\text{cm}^{-3}), \tag{3.68}$$

$$A_z[i_z] \leftarrow \frac{A_z[i_z]}{N \Delta z} \quad (\text{cm}^{-1}), \tag{3.69}$$

$$A_l[i_l] \leftarrow \frac{A_l[i_l]}{N} \quad \text{(dimensionless)}, \tag{3.70}$$

$$A \leftarrow \frac{A}{N} \quad \text{(dimensionless)}. \tag{3.71}$$

The 1D array $A_l[i_l]$ represents the probability of photon absorption in each layer. The quantity A represents the probability of photon absorption by the entire scattering medium. The 2D array $A_{rz}[i_r, i_z]$ represents the probability of photon absorption per unit volume, which can be converted into fluence $F_{rz}[i_r, i_z]$ as follows:

$$F_{rz}[i_r, i_z] = \frac{A_{rz}[i_r, i_z]}{\mu_a} \quad (\text{cm}^{-2}), \tag{3.72}$$

where μ_a denotes the local absorption coefficient. This equation breaks down in a non-absorbing medium.

The 1D array $A_z[i_z]$ represents the probability of photon absorption per unit length in the z direction, which can be converted to a dimensionless quantity $F_z[i_z]$ as follows:

$$F_z[i_z] = \frac{A_z[i_z]}{\mu_a} \quad \text{(dimensionless)}. \tag{3.73}$$

$F_z[i_z]$ represents the internal fluence as a function of z apart from a constant factor. This equation also breaks down in a non-absorbing medium.

Example 3.5. Show the equivalence of Eq. (3.73) to the convolution over an infinitely wide uniform beam (see Chapter 4).

According to Eqs. (3.64), (3.67) and (3.69), the final converted $A_z[i_z]$ and $A_{rz}[i_r, i_z]$ have the following relationship:

$$A_z[i_z] = \sum_{i_r=0}^{N_r-1} A_{rz}[i_r, i_z]\Delta a(i_r), \tag{3.74}$$

where $\Delta a(i_r)$ is computed by Eq. (3.55).

Employing Eqs. (3.72) and (3.73), we convert Eq. (3.74) to

$$F_z[i_z] = \sum_{i_r=0}^{N_r-1} F_{rz}[i_r, i_z]\Delta a(i_r). \tag{3.75}$$

This equation is a discrete representation of the following integral:

$$F_z(z) = \int_0^\infty F_{rz}(r, z) 2\pi r \, dr. \tag{3.76}$$

This integration is essentially the convolution over an infinitely wide uniform beam of unit fluence. Therefore, $F_z[i_z]$ represents the fluence distribution along the z axis in response to such a beam.

3.6. COMPUTATIONAL EXAMPLES

As computational examples, MCML simulation results are compared with results from other theories and other investigators' Monte Carlo simulations. These comparisons partially validate the MCML.

3.6.1. Total Diffuse Reflectance and Total Transmittance

The total diffuse reflectance R_d and the total transmittance T_t (sum of both the unscattered and the scattered transmittances) are computed for a slab of scattering medium with the following optical properties: relative refractive index $n_{rel} = 1$, absorption coefficient $\mu_a = 10 \text{ cm}^{-1}$, scattering coefficient $\mu_s = 90 \text{ cm}^{-1}$, anisotropy $g = 0.75$, and thickness $d = 0.02$ cm. The relative refractive index is defined as the ratio of the refractive index of the scattering medium to that of the ambient medium. If $n_{rel} = 1$, the boundaries are termed *refractive-index-matched*. After 10 Monte Carlo simulations of 50,000 photon packets each are completed, the averages and the standard errors (standard deviations of the averages) of the total diffuse reflectance and total transmittance are computed and compared with the data from van de Hulst's (1980) table and Prahl et al.'s (1989) Monte Carlo simulations (Table 3.1). Because the unscattered transmittance is $\exp[-(\mu_a + \mu_s)d] = e^{-2} = 0.13534$, the total diffuse transmittance equals $0.66096 - 0.13534 = 0.52562$. All results are in good agreement. It is worth noting that standard errors are expected to decrease proportionally with the square root of the number of photon packets tracked owing to the central limit theorem.

For a semiinfinite scattering medium that has a refractive-index-mismatched boundary ($n_{rel} \neq 1$), the total reflectance is computed similarly and compared in Table 3.2 with the data from Giovanelli (1955) and Prahl et al.'s (1989) Monte Carlo simulations. The scattering medium has the following optical properties: $n_{rel} = 1.5$, $\mu_a = 10 \text{ cm}^{-1}$, $\mu_s = 90 \text{ cm}^{-1}$, $g = 0$ (isotropic scattering). Then, 10 Monte Carlo simulations of 5000 photon packets each are completed to compute the average and the standard error of the total diffuse reflectance.

TABLE 3.1. Comparison of Results from MCML with van de Hulst's Table and the Monte Carlo Data of Prahl et al. (1989)[a]

Source	R_d Average	R_d Error	T_t Average	T_t Error
MCML	0.09734	0.00035	0.66096	0.00020
van de Hust (1980)	0.09739	—	0.66096	—
Prahl et al. (1989)	0.09711	0.00033	0.66159	0.00049

[a] "R_d average" and "R_d error" list the averages and the standard errors of the total diffuse reflectance, respectively. Columns "T_t average" and "T_t error" list the averages and the standard errors of the total transmittance, respectively.

TABLE 3.2. Comparison of Total Reflectance[a] from MCML with Data from Giovanelli (1955) and Prahl et al. (1989)

Source	R_d Average	R_d Error
MCML	0.25907	0.00170
Giovanelli 1995	0.2600	—
Prahl et al. (1989)	0.26079	0.00079

[a]The total reflectance includes both the specular reflectance (0.04) and the diffuse reflectance

3.6.2. Angularly Resolved Diffuse Reflectance and Transmittance

The angularly resolved diffuse reflectance and transmittance are also computed for a slab of scattering medium with the following optical properties: $n_{\text{rel}} = 1$, $\mu_a = 10 \text{ cm}^{-1}$, $\mu_s = 90 \text{ cm}^{-1}$, $g = 0.75$, and $d = 0.02$ cm. In this simulation, 500,000 photon packets are tracked, and the number of angular grid elements is 30. The results from MCML are compared in Figure 3.4 with the data from van de Hulst's (1980) table. Because van de Hulst used a different definition of reflectance and transmittance and also used an incident flux of π, his data are multiplied by $\cos \alpha$ and then divided by π before the comparison.

3.6.3. Depth-Resolved Fluence

As an example, the depth-resolved internal fluence $F_z[i_z]$ is simulated by MCML for two semiinfinite scattering media with refractive-index-matched and refractive-index-mismatched boundaries, respectively (Figure 3.5). The optical parameters are $n_{\text{rel}} = 1.0$ or 1.37, $\mu_a = 0.1 \text{ cm}^{-1}$, $\mu_s = 100 \text{ cm}^{-1}$, and $g = 0.9$. One million photon packets are tracked in each simulation. The grid element size and the number of grid elements in the z direction are 0.005 cm and 200, respectively.

As can be seen, the fluence near the surface is greater than unity because scattered light augments the fluence. Furthermore, the internal fluence in the scattering medium with a refractive-index-mismatched boundary is greater than that in the medium with a refractive-index-matched boundary, because photons can be bounced back into the scattering medium by the mismatched boundary.

When z is greater than the transport mean free path l'_t, diffusion theory predicts that the internal fluence distribution is

$$F(z) = K F_0 \exp\left(-\frac{z}{\delta}\right). \tag{3.77}$$

Here, K is a scaling factor that depends on the relative index of refraction, F_0 is the incident irradiance (unity in MCML), and δ is the penetration depth. Consequently, the two curves in the diffusive regime should be separated by a constant factor only, which means that the tails of the two curves should be

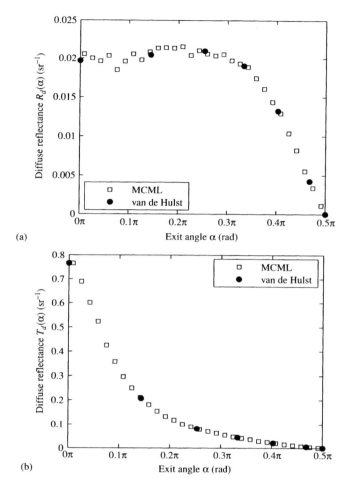

Figure 3.4. Angularly resolved (a) diffuse reflectance $R_d(\alpha)$ and (b) diffuse transmittance $T_d(\alpha)$.

parallel in a log-linear plot (Figure 3.5). The two curves shown here are parallel when $z > l'_t = [\mu_a + \mu_s(1-g)]^{-1} \approx 0.1$ cm.

We fit the parallel portions of the two curves with exponential functions. The decay constants for the curves are approximately 0.578 cm for the refractive-index-matched boundary and 0.575 cm for the refractive-index-mismatched boundary. Both values are close to the one predicted by the diffusion theory

$$\delta = \frac{1}{\sqrt{3\mu_a[\mu_a + \mu_s(1-g)]}} = 0.574 \text{ cm}, \qquad (3.78)$$

which is independent of the relative index of refraction.

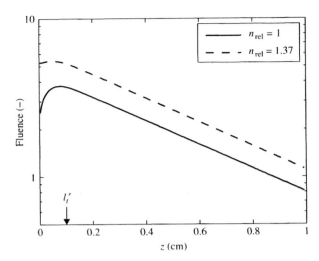

Figure 3.5. Comparison of the internal fluence as a function of depth z for two semiinfinite scattering media with a refractive-index-matched boundary and a refractive-index-mismatched boundary, respectively. Minus ($-$) represents a dimensionless unit.

APPENDIX 3A. SUMMARY OF MCML

The entire source code for MCML can be found on the Web at ftp://ftp.wiley.com/public/sci_tech_med/biomedical_optics. The whole program for MCML is divided into several files. Header file mcml.h defines the data structures and some constants. File mcmlmain.c contains the main function and the status-reporting function. File mcmlio.c deals with reading and writing data. File mcmlgo.c contains most of the photon-tracking code. File mcmlnr.c contains several functions for dynamical data allocations and error reports. Readers should read the main function first.

In MCML, 1D and 2D physical quantities are stored in 1D or 2D arrays, respectively. These arrays are dynamically allocated so that users are allowed to specify the array sizes at runtime without wasting computer memory, an advantage that static arrays do not provide.

The following list is generated by command cflow -d3 -n --omit-arguments --omit-symbol-names mcml*.c, which shows the structure of the program (MCML 1.2.2) with the nesting depth limited to 3:

```
1  main() <int () at MCMLMAIN.C:186>:
2      ShowVersion() <void () at MCMLIO.C:71>:
3          CtrPuts() <void () at MCMLIO.C:48>:
4              printf()
5              puts()
6      GetFnameFromArgv() <void () at MCMLMAIN.C:130>:
```

```
 7          strcpy()
 8      GetFile() <FILE * () at MCMLIO.C:111>:
 9          printf()
10          scanf()
11          strlen()
12          exit()
13          fopen()
14      CheckParm() <void () at MCMLIO.C:531>:
15          ReadNumRuns() <short () at MCMLIO.C:222>:
16          printf()
17          ReadParm() <void () at MCMLIO.C:442>:
18          FnameTaken() <Boolean () at MCMLIO.C:504>:
19          sprintf()
20          free()
21          nrerror() <void () at MCMLNR.C:17>:
22          FreeFnameList() <void () at MCMLIO.C:517>:
23          rewind()
24      ReadNumRuns() <short () at MCMLIO.C:222>:
25          FindDataLine() <char * () at MCMLIO.C:201>:
26          strcpy()
27          nrerror() <void () at MCMLNR.C:17>:
28          sscanf()
29      ReadParm() <void () at MCMLIO.C:442>:
30          ReadFnameFormat() <void () at MCMLIO.C:242>:
31          ReadNumPhotons() <void () at MCMLIO.C:260>:
32          ReadDzDr() <void () at MCMLIO.C:277>:
33          ReadNzNrNa() <void () at MCMLIO.C:293>:
34          ReadNumLayers() <void () at MCMLIO.C:316>:
35          ReadLayerSpecs() <void () at MCMLIO.C:390>:
36          CriticalAngle() <void () at MCMLIO.C:421>:
37      DoOneRun() <void () at MCMLMAIN.C:145>:
38          InitProfile()
39          cecho2file()
40          InitOutputData() <void () at MCMLIO.C:562>:
41          Rspecular() <double () at MCMLGO.C:116>:
42          PunchTime() <time_t () at MCMLMAIN.C:59>:
43          printf()
44          PredictDoneTime() <void () at MCMLMAIN.C:94>:
45          LaunchPhoton() <void () at MCMLGO.C:140>:
46          HopDropSpin() <void () at MCMLGO.C:734>:
47          exit()
48          ReportResult() <void () at MCMLMAIN.C:115>:
49          FreeData() <void () at MCMLIO.C:598>:
50      fclose()
```

APPENDIX 3B. PROBABILITY DENSITY FUNCTION

The probability density function (PDF), expressed as $p(x)$, is a function that gives the probability that random variable x assumes a value between x_1 and x_2 as follows:

$$P\{x_1 \le x \le x_2\} = \int_{x_1}^{x_2} p(x)\,dx. \tag{3.79}$$

The PDF has the following properties:

$$\int_{-\infty}^{+\infty} p(x)\,dx = 1 \tag{3.80}$$

and

$$p(x) \ge 0 \quad \text{for} \quad x \in (-\infty, +\infty). \tag{3.81}$$

The relationship between the PDF and the cumulative distribution function (CDF) $P(x)$ is

$$P(x) = \int_{-\infty}^{x} p(x')\,dx' \tag{3.82}$$

or

$$p(x) = \frac{d}{dx} P(x). \tag{3.83}$$

PROBLEMS

3.1 Derive Eqs. (3.45) and (3.46).

3.2 Derive Eq. (3.30). Note that the formula depends on the choice of the local moving coordinate system. The Monte Carlo simulation, however, leads to the same result.

3.3 Find the effective specular reflectance in normal incidence from a water layer ($n = 1.33$) placed between air and biological tissue ($n = 1.37$).

3.4 Show that the number of scattering events occurring in path length s follows the Poisson distribution. Neglect absorption.

3.5 Show that the mean and standard deviation of the free path between scattering events are both equal to $1/\mu_s$. Neglect absorption.

3.6 An alternative to Eq. (3.28) when $g \ne 0$ is

$$\cos\theta = \frac{1}{2g}\left[1 + g^2 - \left(\frac{1-g^2}{1+g-2g\xi}\right)^2\right].$$

Why?

3.7 An absorbing slab of thickness d has absorption coefficient μ_a with negligible scattering. A collimated laser beam of intensity I_{in} is normally incident on the slab.

(a) Assume that the mismatch between the refractive indices of the ambient media and the absorbing medium is negligible. Calculate the transmitted light intensity I_{out}.

(b) Assume that the mismatch between the refractive indices of the ambient media and the absorbing medium causes a specular reflection R on each surface of the slab. Recalculate the transmitted light intensity I_{out}.

3.8 A pencil beam is incident on a semiinfinite reference scattering medium. The spatially and temporally resolved diffuse reflectance $R_0(r, t)$ is known for the following optical parameters: absorption coefficient μ_{a0}, scattering coefficient μ_{s0}, and scattering anisotropy g_0, where r denotes the radial distance between the observation point and the point of incidence, and t denotes time. The speed of light in the medium is c.

(a) Write the expression for the new diffuse reflectance $R_1(r, t)$ assuming that the absorption coefficient is changed from μ_{a0} to μ_{a1} but the other optical parameters are unchanged.

(b) Write the expression for the new diffuse reflectance $R_2(r, t)$ where both the absorption coefficient and the scattering coefficient are scaled by the same factor C and the other optical parameter is unchanged: $\mu_{a2} = C\mu_{a0}$ and $\mu_{s2} = C\mu_{s0}$.

3.9 Consider a Gaussian beam with a radius of R, a total energy of 1 mJ, and a radial energy density distribution of $S(r) = (2/\pi R^2)\exp(-2r^2/R^2)$. Derive the sampling expression for the radius r based on the random number ξ that is uniformly distributed between 0 and 1.

3.10 Implement a Monte Carlo simulation of photon transport in a semiinfinite scattering medium in C/C++ or another programming language. The Henyey–Greenstein function is assumed to be the phase function. Inputs include n_{rel}, μ_a, μ_s, and g. Outputs from the program include the total reflectance R and the depth-resolved fluence.

(a) Use the results in Table 3.2 to verify your program.

(b) Calculate for $n_{rel} = 1.37$, $\mu_a = 0.1$ cm^{-1}, $\mu_s = 100$ cm^{-1}, and $g = 0.9$.

(c) Reproduce Figure 3.5.

3.11 Use the original Monte Carlo code written for Problem 3.10 and a modified version to verify that the following two algorithms are statistically identical. Compare the total diffuse reflectances and the total absorptions using a statistical test. Use $n_{rel} = 1$, $\mu_a = 0.1$ cm^{-1}, $\mu_s = 100$ cm^{-1}, and $g = 0.9$.

(a) Sample the step size with $s = -\ln(\xi)/\mu_t$ and calculate the weight loss at each step by $\Delta W = \frac{\mu_a}{\mu_t} W$.

(b) Sample the step size with $s = -\ln(\xi)/\mu_s$ and calculate the weight loss at each step by $\Delta W = W[1 - \exp(-\mu_a s)]$.

READING

Giovanelli RG (1955): Reflection by semi-infinite diffusers, *Optica Acta* **2**: 153–162. (See Section 3.6, above.)

Lux I and Koblinger L (1991): *Monte Carlo Particle Transport Methods: Neutron and Photon Calculations*, CRC Press, Boca Raton. (See Section 3.2, above.)

Marquez G, Wang LHV, Lin SP, Schwartz JA, and Thomsen SL (1998): Anisotropy in the absorption and scattering spectra of chicken breast tissue, *Appl. Opt.* **37**(4): 798–804. (See Section 3.2, above.)

Prahl SA, Keijzer M, Jacques SL, and Welch AJ (1989): A Monte Carlo model of light propagation in tissue, in *Dosimetry of Laser Radiation in Medicine and Biology*, Muller GJ and Sliney DH, eds., SPIE Press, **IS5**: 102–111. (See Section 3.6, above.)

van de Hulst HC (1980): *Multiple Light Scattering: Tables, Formulas, and Applications*, Academic Press, New York. (See Section 3.6, above.)

Wang LHV, Jacques SL, and Zheng LQ (1995): MCML—Monte Carlo modeling of light transport in multi-layered tissues, *Comput. Meth. Prog. Biomed.* **47**: 131–146. (All sections in this chapter.)

FURTHER READING

Ahrens JH and Dieter U (1972): Computer methods for sampling for the exponential and normal distributions, *Commun. ACM* **15**: 873–882.

Baranoski GVG, Krishnaswamy A, and Kimmel B (2004): An investigation on the use of data-driven scattering profiles in Monte Carlo simulations of ultraviolet light propagation in skin tissues, *Phys. Med. Biol.* **49**(20): 4799–4809.

Bartel S and Hielscher AH (2000): Monte Carlo simulations of the diffuse backscattering Mueller matrix for highly scattering media, *Appl. Opt.* **39**(10): 1580–1588.

Boas DA, Culver JP, Stott JJ, and Dunn AK (2002): Three dimensional Monte Carlo code for photon migration through complex heterogeneous media including the adult human head, *Opt. Express* **10**(3): 159–170.

Carter LL and Cashwell ED (1975): *Particle-Transport Simulation with the Monte Carlo Method* USERDA Technical Information Center, Oak Ridge, TN.

Cashwell ED and Everett CJ (1959): *A Practical Manual on the Monte Carlo Method for Random Walk Problems*, Pergamon Press, New York.

Churmakov DY, Meglinski IV, Piletsky SA, and Greenhalgh DA (2003): Analysis of skin tissues spatial fluorescence distribution by the Monte Carlo simulation, *J. Phys. D—Appl. Phys.* **36**(14): 1722–1728.

Cote D and Vitkin IA (2005): Robust concentration determination of optically active molecules in turbid media with validated three-dimensional polarization sensitive Monte Carlo calculations, *Opt. Express* **13**(1): 148–163.

Ding L, Splinter R, and Knisley SB (2001): Quantifying spatial localization of optical mapping using Monte Carlo simulations, *IEEE Trans. Biomed. Eng.* **48**(10): 1098–1107.

Flock ST, Patterson MS, Wilson BC, and Wyman DR (1989): Monte-Carlo modeling of light-propagation in highly scattering tissues. 1. Model predictions and comparison with diffusion-theory, *IEEE Trans. Biomed. Eng.* **36**(12): 1162–1168.

Flock ST, Wilson BC, and Patterson MS (1989): Monte-Carlo modeling of light-propagation in highly scattering tissues. 2. Comparison with measurements in phantoms, *IEEE Trans. Biomed. Eng.* **36**(12): 1169–1173.

Gardner CM and Welch AJ (1994): Monte-Carlo simulation of light transport in tissue—unscattered absorption events, *Appl. Opt.* **33**(13): 2743–2745.

Hendricks JS and Booth TE (1985): MCNP variance reduction overview, *Lecture Notes Phys.* (AXV64) **240**: 83–92.

Henniger J, Minet O, Dang HT, and Beuthan J (2003): Monte Carlo simulations in complex geometries: Modeling laser light transport in real anatomy of rheumatoid arthritis, *Laser Phys.* **13**(5): 796–803.

Igarashi M, Gono K, Obi T, Yamaguchi M, and Ohyama N (2004): Monte Carlo simulations of reflected spectra derived from tissue phantom with double-peak particle size distribution, *Opt. Rev.* **11**(2): 61–67.

Jacques SL, Later CA, and Prahl SA (1987): Angular dependence of HeNe laser light scattering by human dermis, *Lasers Life Sci.* **1**: 309–333.

Jacques SL (1989): Time resolved propagation of ultrashort laser-pulses within turbid tissues, *Appl. Opt.* **28**(12): 2223–2229.

Jacques SL (1989): Time-resolved reflectance spectroscopy in turbid tissues, *IEEE Trans. Biomed. Eng.* **36**(12): 1155–1161.

Jaillon F and Saint-Jalmes H (2003): Description and time reduction of a Monte Carlo code to simulate propagation of polarized light through scattering media, *Appl. Opt.* **42**(16): 3290–3296.

Kahn H (1950): Random sampling Monte Carlo techniques in neutron attenuation problems I, *Nucleonics* **6**: 27–37.

Kalos MH and Whitlock PA (1986): *Monte Carlo Methods*, Wiley, New York.

Keijzer M, Jacques SL, Prahl SA, and Welch AJ (1989): Light distributions in artery tissue—monte-carlo simulations for finite-diameter laser-beams, *Lasers Surg. Med.* **9**(2): 148–154.

Keijzer M, Pickering JW, and Gemert MJCv (1991): Laser beam diameter for port wine stain treatment, *Lasers Surg. Med.* **11**: 601–605.

Kienle A and Patterson MS (1996): Determination of the optical properties of turbid media from a single Monte Carlo simulation, *Phys. Med. Biol.* **41**(10): 2221–2227.

Li H, Tian J, Zhu FP, Cong WX, Wang LHV, Hoffman EA, and Wang G (2004): A mouse optical simulation environment (MOSE) to investigate bioluminescent phenomena in the living mouse with the Monte Carlo method, *Acad. Radiol.* **11**(9): 1029–1038.

Lu Q, Gan XS, Gu M, and Luo QM (2004): Monte Carlo modeling of optical coherence tomography imaging through turbid media, *Appl. Opt.* **43**(8): 1628–1637.

MacLaren MD, Marsaglia G, and Bray AT (1964): A fast procedure for generating exponential random variables, *Commun. ACM* **7**: 298–300.

Marsaglia G (1961): Generating exponential random variables, *Ann. Math. Stat.* **32**: 899–900.

McShane MJ, Rastegar S, Pishko M, and Cote GL (2000): Monte carlo modeling for implantable fluorescent analyte sensors, *IEEE Trans. Biomed. Eng.* **47**(5): 624–632.

Meglinski IV (2001): Monte Carlo simulation of reflection spectra of random multilayer media strongly scattering and absorbing light, *Quantum Electron.* **31**(12): 1101–1107.

Meglinski IV and Matcher SD (2001): Analysis of the spatial distribution of detector sensitivity in a multilayer randomly inhomogeneous medium with strong light scattering and absorption by the Monte Carlo method, *Opt. Spectrosc.* **91**(4): 654–659.

Mroczka J and Szczepanowski R (2005): Modeling of light transmittance measurement in a finite layer of whole blood—a collimated transmittance problem in Monte Carlo simulation and diffusion model, *Optica Applicata* **35**(2): 311–331.

Nishidate I, Aizu Y, and Mishina H (2004): Estimation of melanin and hemoglobin in skin tissue using multiple regression analysis aided by Monte Carlo simulation, *J. Biomed. Opt.* **9**(4): 700–710.

Patwardhan SV, Dhawan AP, and Relue PA (2005): Monte Carlo simulation of light-tissue interaction: Three-dimensional simulation for trans-illumination-based imaging of skin lesions, *IEEE Trans. Biomed. Eng.* **52**(7): 1227–1236.

Plauger PJ and Brodie J (1989): *Standard C*, Microsoft Press, Redmond, WA.

Plauger PJ, Brodie J, and Plauger PJ (1992): *ANSI and ISO Standard C : Programmer's Reference*, Microsoft Press, Redmond, WA.

Qian ZY, Victor SS, Gu YQ, Giller CA, and Liu HL (2003): "Look-ahead distance" of a fiber probe used to assist neurosurgery: Phantom and Monte Carlo study, *Opt. Express* **11**(16): 1844–1855.

Ramella-Roman JC, Prahl SA, and Jacques SL (2005): Three Monte Carlo programs of polarized light transport into scattering media: Part I, *Opt. Express* **13**(12): 4420–4438.

Sharma SK and Banerjee S (2003): Role of approximate phase functions in Monte Carlo simulation of light propagation in tissues, *J. Opt. A* **5**(3): 294–302.

Swartling J, Pifferi A, Enejder AMK, and Andersson-Engels S (2003): Accelerated Monte Carlo models to simulate fluorescence spectra from layered tissues, *J. Opt. Soc. Am. A* **20**(4): 714–727.

Tycho A, Jorgensen TM, Yura HT, and Andersen PE (2002): Derivation of a Monte Carlo method for modeling heterodyne detection in optical coherence tomography systems, *Appl. Opt.* **41**(31): 6676–6691.

van de Hulst HC (1980): *Multiple Light Scattering: Tables, Formulas, and Applications*, Academic Press, New York.

Wang LHV and Jacques SL (1993): Hybrid model of Monte Carlo simulation diffusion theory for light reflectance by turbid media, *J. Opt. Soc. Am. A* **10**: 1746–1752.

Wang LHV and Jacques SL (1994): Optimized radial and angular positions in Monte Carlo modeling, *Med. Phys.* **21**: 1081–1083.

Wang LHV, Nordquist RE, and Chen WR (1997): Optimal beam size for light delivery to absorption-enhanced tumors buried in biological tissues and effect of multiple-beam delivery: A Monte Carlo study, *Appl. Opt.* **36**(31): 8286–8291.

Wang LHV (2001): Mechanisms of ultrasonic modulation of multiply scattered coherent light: A Monte Carlo model, *Opt. Lett.* **26**(15): 1191–1193.

Wang RKK (2002): Signal degradation by coherence tomography multiple scattering in optical of dense tissue: A Monte Carlo study towards optical clearing of biotissues, *Phys. Med. Biol.* **47**(13): 2281–2299.

Wang XD, Yao G, and Wang LHV (2002): Monte Carlo model and single-scattering approximation of the propagation of polarized light in turbid media containing glucose, *Appl. Opt.* **41**(4): 792–801.

Wang XD, Wang LHV, Sun CW, and Yang CC (2003): Polarized light propagation through scattering media: Time-resolved Monte Carlo simulations and experiments, *J. Biomed. Opt.* **8**(4): 608–617.

Wang XY, Zhang CP, Zhang LS, Qi SW, Xu T, and Tian JG (2003): Reconstruction of optical coherence tomography image based on Monte Carlo method, *J. Infrared Millim. Waves* **22**(1): 68–70.

Wilson BC and Adam GA (1983): Monte Carlo model for the absorption and flux distributions of light in tissue, *Med. Phys.* **10**: 824–830.

Wilson BC and Jacques SL (1990): Optical reflectance and transmittance of tissues: Principles and applications, *IEEE J. Quantum Electron.* **26**: 2186–2199.

Wong BT and Menguc MP (2002): Comparison of Monte Carlo techniques to predict the propagation of a collimated beam in participating media, *Num. Heat Transfer B—Fundamentals* **42**(2): 119–140.

Xiong GL, Xue P, Wu JG, Miao Q, Wang R, and Ji L (2005): Particle-fixed Monte Carlo model for optical coherence tomography, *Opt. Express* **13**(6): 2182–2195.

Yadavalli VK, Russell RJ, Pishko MV, McShane MJ, and Cote GL (2005): A Monte Carlo simulation of photon propagation in fluorescent poly(ethylene glycol) hydrogel microsensors, *Sensors Actuators B—Chemical* **105**(2): 365–377.

Yang Y, Soyemi OO, Landry MR, and Soller BR (2005): Influence of a fat layer on the near infrared spectra of human muscle: Quantitative analysis based on two-layered Monte Carlo simulations and phantom experiments, *Opt. Express* **13**(5): 1570–1579.

Yao G and Wang LHV (1999): Monte Carlo simulation of an optical coherence tomography signal in homogeneous turbid media, *Phys. Med. Biol.* **44**(9): 2307–2320.

Yao G and Haidekker MA (2005): Transillumination optical tomography of tissue-engineered blood vessels: A Monte Carlo simulation, *Appl. Opt.* **44**(20): 4265–4271.

CHAPTER 4

Convolution for Broadbeam Responses

4.1. INTRODUCTION

The Monte Carlo program MCML, introduced in Chapter 3, computes responses to a pencil beam normally incident on a multilayered scattering medium. These responses are referred to as *Green's functions* or *impulse responses*. When a collimated photon beam is of finite width, the Monte Carlo method is still able to compute the responses by distributing the incident positions over the cross section of the beam. Each broad beam, however, requires a new time-consuming Monte Carlo simulation, even if the other parameters are unchanged. Convolution of Green's functions for the same multilayered scattering medium, however, can efficiently compute the responses to a broad beam. Such convolution was implemented in a program named CONV (Appendix 4A). Like MCML, CONV is written in ANSI Standard C and hence can be executed on various computer platforms. Although convolution is applicable to collimated beams of any intensity distribution, only Gaussian and top-hat (flat-top) beams are considered in CONV version 1.

4.2. GENERAL FORMULATION OF CONVOLUTION

Convolution is applicable to a system that is stationary (time-invariant), linear, and translation-invariant. The system here consists of horizontal layers of homogeneous scattering media that have stationary properties (see Figure 3.1 in Chapter 3). The input to the system is a collimated photon beam perpendicularly incident on the surface of the scattering medium. The responses can be any observable physical quantities, such as specific absorption, fluence, reflectance, or transmittance. The linearity implies that (1) the responses increase by the same factor if the input intensity increases by a constant factor and (2) any response to two photon beams together is the sum of the two responses to each photon beam alone. The translation invariance in space here means that if the photon beam is

---

Biomedical Optics: Principles and Imaging, by Lihong V. Wang and Hsin-i Wu
Copyright © 2007 John Wiley & Sons, Inc.

shifted in any horizontal direction by any distance, the responses are also shifted in the same direction by the same distance. The translation invariance in time indicates that if the photon beam is delayed by a given time, the responses are also shifted by the same delay. Therefore, responses to spatially and temporally broad beams can be computed using the convolution of the impulse responses; only spatial convolution is described in this chapter, however.

Impulse responses to a normally incident pencil beam are first computed using MCML, where a Cartesian coordinate system is set up as described in Chapter 3. The origin of the coordinate system is the incident point of the pencil beam on the surface of the scattering medium, and the z axis is along the pencil beam; hence, the xy plane is on the surface of the scattering medium.

We denote a particular response to a collimated broad photon beam as $C(x, y, z)$ and denote the corresponding impulse response as $G(x, y, z)$. If the broad collimated light source has intensity profile $S(x, y)$, the response to this broad beam can be obtained through the following convolution

$$C(x, y, z) = \int_{-\infty}^{\infty} \int_{-\infty}^{\infty} G(x - x', y - y', z) S(x', y') \, dx' \, dy', \quad (4.1)$$

which can be reformulated with a change of variables $x'' = x - x'$ and $y'' = y - y'$:

$$C(x, y, z) = \int_{-\infty}^{\infty} \int_{-\infty}^{\infty} G(x'', y'', z) S(x - x'', y - y'') \, dx'' \, dy''. \quad (4.2)$$

Because the multilayered structure has planar symmetry and the photon beam is perpendicularly incident on the surface of the scattering medium, $G(x, y, z)$ possesses cylindrical symmetry. Consequently, the Green function in Eq. (4.1) depends only on the distance r_{os} between the source point (x', y') and the observation point (x, y), rather than on their absolute locations:

$$r_{os} = \sqrt{(x - x')^2 + (y - y')^2}. \quad (4.3)$$

If $S(x', y')$ also has cylindrical symmetry about the origin, it becomes a function of only the radius r':

$$r' = \sqrt{x'^2 + y'^2}. \quad (4.4)$$

On the basis of these symmetries, we reformulate Eqs. (4.1) and (4.2) to

$$C(x, y, z) = \int_{-\infty}^{\infty} \int_{-\infty}^{\infty} G\left(\sqrt{(x - x')^2 + (y - y')^2}, z\right) S\left(\sqrt{x'^2 + y'^2}\right) dx' \, dy', \quad (4.5)$$

$$C(x, y, z) = \int_{-\infty}^{\infty} \int_{-\infty}^{\infty} G\left(\sqrt{x''^2 + y''^2}, z\right) S\left(\sqrt{(x - x'')^2 + (y - y'')^2}\right) dx'' \, dy''. \quad (4.6)$$

Because $C(x, y, z)$ has the same cylindrical symmetry, Eqs. (4.5) and (4.6) can be rewritten in cylindrical coordinates (r, ϕ):

$$C(r, z) = \int_0^\infty S(r')r' \left[\int_0^{2\pi} G\left(\sqrt{r^2 + r'^2 - 2rr'\cos\phi'}, z\right) d\phi' \right] dr', \quad (4.7)$$

$$C(r, z) = \int_0^\infty G(r'', z)r'' \left[\int_0^{2\pi} S\left(\sqrt{r^2 + r''^2 - 2rr''\cos\phi''}\right) d\phi'' \right] dr''. \quad (4.8)$$

In Eq. (4.8), the integration over ϕ'' is independent of z and hence needs to be computed only once for all z values. In some cases, the integration over ϕ'' can be solved analytically; thus, the 2D integral in Eq. (4.8) is reduced to a computationally more efficient 1D integral. Therefore, Eq. (4.8) is more advantageous computationally than Eq. (4.7).

Example 4.1. Derive Eq. (4.7) from Eq. (4.5).

The differential area element can be changed from $dx\,dy$ to $r\,dr\,d\phi$, and the corresponding limits of the integrations are from 0 to $+\infty$ for r and from 0 to 2π for ϕ. In the polar coordinates aligned with (x, y), (x, y) is represented by $(r, 0)$ and (x', y') by (r', ϕ'). We convert Eq. (4.3) into $r_{os} = \sqrt{r^2 + r'^2 - 2rr'\cos\phi'}$. Thus, Eq. (4.7) can be obtained from Eq. (4.5).

4.3. CONVOLUTION OVER A GAUSSIAN BEAM

For a Gaussian beam, the convolution can be further simplified. The intensity profile of the beam is given by

$$S(r') = S_0 \exp\left[-2\left(\frac{r'}{R}\right)^2\right]. \quad (4.9)$$

Here, R denotes the $1/e^2$ radius of the beam; S_0 denotes the intensity at the center of the beam, and S_0 is related to the total power P_0 by

$$S_0 = \frac{2P_0}{\pi R^2}. \quad (4.10)$$

Substituting Eq. (4.9) into Eq. (4.8), we obtain

$$C(r, z) = S(r) \int_0^\infty G(r'', z) \exp\left[-2\left(\frac{r''}{R}\right)^2\right]$$
$$\times \left[\int_0^{2\pi} \exp\left(\frac{4rr''\cos\phi''}{R^2}\right) d\phi'' \right] r''\,dr''. \quad (4.11)$$

The inner integral in the square brackets resembles the integral representation of the zeroth-order modified Bessel function, which is defined by

$$I_0(x) = \frac{1}{2\pi} \int_0^{2\pi} \exp(x \sin \phi)\, d\phi \qquad (4.12)$$

or

$$I_0(x) = \frac{1}{2\pi} \int_0^{2\pi} \exp(x \cos \phi)\, d\phi. \qquad (4.13)$$

By using Eq. (4.13), we can rewrite Eq. (4.11) as

$$C(r, z) = 2\pi S(r) \int_0^\infty G(r'', z) \exp\left[-2\left(\frac{r''}{R}\right)^2\right] I_0\left(\frac{4rr''}{R^2}\right) r''\, dr''. \qquad (4.14)$$

Example 4.2. Derive Eq. (4.10).

The total power

$$\begin{aligned} P_0 &= \int_0^\infty S(r')2\pi r'\, dr' = 2\pi S_0 \int_0^\infty \exp\left[-2\left(\frac{r'}{R}\right)^2\right] r'\, dr' \\ &= -\frac{\pi R^2}{2} S_0 \exp\left[-2\left(\frac{r'}{R}\right)^2\right]\bigg|_0^\infty = \frac{\pi R^2}{2} S_0, \end{aligned} \qquad (4.15)$$

which leads to Eq. (4.10).

Example 4.3. Show that Eqs. (4.12) and (4.13) are equivalent.

Letting $\phi = \phi' + (\pi/2)$ and splitting the integral in Eq. (4.12) into two parts, we obtain

$$\begin{aligned} \int_0^{2\pi} \exp(x \sin \phi)\, d\phi &= \int_{-\pi/2}^{3\pi/2} \exp(x \cos \phi')\, d\phi' \\ &= \int_{-\pi/2}^0 \exp(x \cos \phi')\, d\phi' + \int_0^{3\pi/2} \exp(x \cos \phi')\, d\phi'. \end{aligned} \qquad (4.16)$$

Letting $\phi' = \phi'' + 2\pi$ in the first integral on the right-hand side gives

$$\int_{-\pi/2}^0 \exp(x \cos \phi')\, d\phi' = \int_{3\pi/2}^{2\pi} \exp(x \cos \phi'')\, d\phi''. \qquad (4.17)$$

Substituting Eq. (4.17) into Eq. (4.16) and merging the two integrals on the right-hand side of Eq. (4.16), we obtain

$$\int_0^{2\pi} \exp(x \sin \phi)\, d\phi = \int_{3\pi/2}^{2\pi} \exp(x \cos \phi'')\, d\phi'' + \int_0^{3\pi/2} \exp(x \cos \phi')\, d\phi'$$
$$= \int_0^{2\pi} \exp(x \cos \phi)\, d\phi. \tag{4.18}$$

Since both ϕ' and ϕ'' are dummy variables, they are both replaced with ϕ above.

4.4. CONVOLUTION OVER A TOP-HAT BEAM

For a top-hat beam of radius R, the source function becomes

$$S(r') = \begin{cases} S_0 & \text{if } r' \leq R \\ 0 & \text{if } r' > R \end{cases}, \tag{4.19}$$

where S_0 denotes the intensity inside the beam. We have

$$S_0 = \frac{P_0}{\pi R^2}, \tag{4.20}$$

where P_0 denotes the total power of the beam.

Substituting Eq. (4.19) into Eq. (4.8), we obtain

$$C(r, z) = 2\pi S_0 \int_0^\infty G(r'', z) I_\phi(r, r'') r''\, dr'', \tag{4.21}$$

where

$$I_\phi(r, r'') = \begin{cases} 1 & \text{if } R \geq r + r'' \\ \frac{1}{\pi} \cos^{-1}\left(\frac{r^2 + r''^2 - R^2}{2rr''}\right) & \text{if } |r - r''| \leq R < r + r'' \\ 0 & \text{if } R < |r - r''| \end{cases}. \tag{4.22}$$

From Eq. (4.22), the limits of integration in Eq. (4.21) can be changed to a finite range

$$C(r, z) = 2\pi S_0 \int_a^{r+R} G(r'', z) I_\phi(r, r'') r''\, dr'', \tag{4.23}$$

where

$$a = \max(0, r - R). \tag{4.24}$$

Function max( ) takes on the greater of the two arguments.

If R tends to infinity, Eq. (4.23) becomes

$$C(r, z) = 2\pi S_0 \int_0^\infty G(r'', z) r'' \, dr''. \quad (4.25)$$

This equation implies that in MCML, if the absorbed weights in all r grid elements are summed and then divided by the total number of tracked photon packets, the result represents the specific absorption as a function of z for an infinitely wide beam of unit intensity.

4.5. NUMERICAL SOLUTION TO CONVOLUTION

As described in Chapter 3, a grid system is used in the Monte Carlo model. A 2D homogeneous grid system is set up in the r and z directions. The grid element sizes are Δr and Δz in the r and z directions, respectively; the total numbers of grid elements in the r and z directions are N_r and N_z, respectively.

When the photon beam is Gaussian or top-hat, the 2D convolution becomes 1D. Because the Monte Carlo simulation assigns physical quantities to discrete grid elements, an appropriate integration algorithm is based on the extended trapezoidal rule. This algorithm is ideal for a nonsmooth integrand that is linearly interpolated between available data points; it is implemented in C as a function named trapzd(), which is called by another function named qtrap().

Another method of integration is to evaluate the integrand at the original grid points. This approach, however, does not offer any control over the integration accuracy. For a top-hat beam, for example, N_r is 50, and R is about $5\Delta r$. If $C(0, z)$ is computed from Eq. (4.23), the integration interval $[0, R]$ covers only $5\Delta r$. Thus, only five function evaluations contribute to the integration and may yield unacceptable accuracy. By contrast, the extended trapezoidal rule continues to perform function evaluations until a user-specified accuracy is reached.

The sequence of integrand evaluations in the extended trapezoidal rule is shown in Figure 4.1a. Subsequent calls to trapzd() incorporate the previous evaluations and evaluate the integrand only at the new points. To integrate $f(x)$ over interval $[a, b]$, we evaluate $f(a)$ and $f(b)$ in the first step as noted by 1 and 2 in Figure 4.1a. To refine the grid, we evaluate $f\left(\frac{1}{2}(a + b)\right)$ in the second step as noted by 3. This process is repeated until the integral evaluation reaches a specified accuracy. The bottom line shows all function evaluations after four calls.

4.5.1. Interpolation and Extrapolation of Physical Quantities

The physical quantities under discussion have been computed using MCML over a grid system. As discussed in Chapter 3, the optimal r coordinate is

$$r(i_r) = \left[\left(i_r + \frac{1}{2}\right) + \frac{1}{12(i_r + \frac{1}{2})}\right] \Delta r, \quad (4.26)$$

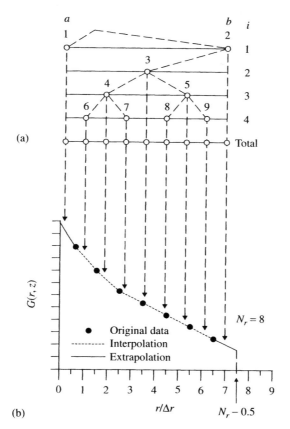

Figure 4.1. (a) Sequence of integrand evaluations in the extended trapezoidal rule of integration; (b) interpolation and extrapolation of the physical quantities. In this example, $N_r = 8$. Symbols a and b denote the integration limits, and i denotes the iteration index. Arrows point to where the integrand is evaluated. Solid circles represent the original data points. Dashed and solid lines represent linear interpolation and extrapolation, respectively.

where i_r is the index of the grid element ($0 \leq i_r \leq N_r - 1$). For $i_r = 0, r(0)$ is $\frac{2}{3}\Delta r$ instead of $\frac{1}{2}\Delta r$. The offset between the optimized and the centered coordinates in each grid element decreases as i_r increases.

In qtrap(), the integrand—of which G is only a part—is evaluated at points that may not fall on the original grid as illustrated in Figure 4.1. Linear interpolations are used for those points that fall between two original grid points, and linear extrapolations are used for those points that are located beyond the original grid system (Figure 4.1b). Extrapolation is extended only up to $(N_r - 0.5)\Delta r$ because further extrapolation is unreliable. In MCML, the last cells in the r direction are used to collect contributions from photon packets that do not fit into the grid system and thus do not represent the true local physical quantities. Therefore, the upper limit for extrapolation is $(N_r - 0.5)\Delta r$ instead of $(N_r + 0.5)\Delta r$, and the physical quantity beyond $(N_r - 0.5)\Delta r$ is set to zero. We denote

$$r_{\max} = (N_r - 0.5)\Delta r. \quad (4.27)$$

4.5.2. Integrand Evaluation for a Gaussian Beam

Although the integration in Eq. (4.14) must converge for physical reasons, it may cause overflow in a computer because the modified Bessel function increases rapidly as the argument increases. Therefore, a proper reformulation of the integrand is necessary.

We note that the modified Bessel function has the following asymptotic approximation for large x values:

$$I_0(x) \approx \frac{\exp(x)}{\sqrt{2\pi x}}. \quad (4.28)$$

We define the following new function on the basis of I_0

$$I_{0e}(x) = I_0(x)\exp(-x) \quad (4.29)$$

or

$$I_0(x) = I_{0e}(x)\exp(x), \quad (4.30)$$

where I_{0e} is always well bounded. Substituting Eqs. (4.9) and (4.30) into Eq. (4.14), we obtain

$$C(r, z) = 2\pi S_0 \int_0^\infty G(r'', z) \exp\left[-2\left(\frac{r'' - r}{R}\right)^2\right] I_{0e}\left(\frac{4rr''}{R^2}\right) r'' dr''. \quad (4.31)$$

Because both the exponential and the I_{0e} terms are well bounded, the integrand can be computed without overflow. Since Eq. (4.28) is not actually used in the computation, Eq. (4.31) does not carry any asymptotic approximation.

Computation speed is another issue. The evaluation of $\exp(\ )I_{0e}(\ )$ in Eq. (4.31) is a major part of the computation, which can take up to 90% of the total time. For multi-dimensional physical quantities (e.g., the fluence as a function of r and z), the convolution may repeatedly evaluate $\exp(\ )I_{0e}(\ )$ at the same r coordinate as the integration is computed for different z coordinates. Therefore, if the values of $\exp(\ )I_{0e}(\ )$ are stored during the convolution for one z coordinate, computation time can be reduced. Because the number of function evaluations is unknown in advance, the function evaluations should be saved with dynamic data allocation. Since the evaluation sequence in qtrap() resembles a binary tree as shown in Figure 4.1a, a binary tree can be used to store the function evaluations. Although the first two nodes are out of balance, the subtree below node 3 is perfectly balanced.

4.5.3. Integration Limits for a Gaussian Beam

Although the upper integration limit for a Gaussian beam in Eq. (4.31) is infinity, it can be converted to a finite value by a change of variables; then, the integration can be computed by a function called midexp(). This approach, however, is not computationally efficient. Therefore, qtrap() is preferred; the upper integration limit, however, must be reduced to a finite value. To this end, the integrand is nonzero if

$$|r'' - r| \leq KR \tag{4.32}$$

or

$$r - KR \leq r'' \leq r + KR, \tag{4.33}$$

where K is a constant that can be set in CONV. For example, if K is 4, the exponential term in Eq. (4.31) is $\exp(-32) \approx 10^{-14}$.

The computation of G covers only interval $[0, r_{max}]$, where r_{max} is as given by Eq. (4.27). Combining this limit and Eq. (4.33), we rewrite Eq. (4.31) as

$$C(r, z) = 2\pi S_0 \int_a^b G(r'', z) \exp\left[-2\left(\frac{r'' - r}{R}\right)^2\right] I_{0e}\left(\frac{4rr''}{R^2}\right) r'' \, dr'', \tag{4.34}$$

where

$$a = \max(0, r - KR), \tag{4.35}$$

$$b = \min(r_{max}, r + KR). \tag{4.36}$$

Functions max( ) and min( ) take on the greater and the lesser of the two arguments, respectively.

4.5.4. Integration for a Top-Hat Beam

The integrand for a top-hat beam can be evaluated more easily than that for a Gaussian beam, because the integration limits are finite and the integrand causes no overflow. However, evaluation of I_ϕ in Eq. (4.23) is also time-consuming. As in the integrand evaluation for Gaussian beams, the evaluated I_ϕ values for qtrap() are stored in a binary tree for computational efficiency.

Since the physical quantities are computed only in interval $[0, r_{max}]$, Eq. (4.23) can be expressed as

$$C(r, z) = 2\pi S_0 \int_a^b G(r'', z) I_\phi(r, r'') r'' \, dr'', \tag{4.37}$$

where

$$a = \max(0, r - R), \tag{4.38}$$

$$b = \min(r_{max}, r + R). \tag{4.39}$$

4.5.5. First Interactions

In MCML, absorption from the first photon–tissue interactions is recorded separately. The first interactions always occur on the z axis and hence contribute to the specific absorption or related physical quantities as a delta function. The total impulse response can be expressed in two parts

$$G(r, z) = G_1(0, z)\frac{\delta(r)}{2\pi r} + G_2(r, z), \tag{4.40}$$

where the first term results from the first interactions and the second, from subsequent interactions.

For a Gaussian beam, substituting Eq. (4.40) into Eq. (4.34) yields

$$C(r, z) = G_1(0, z)S(r) + 2\pi S_0 \int_a^b G_2(r'', z)$$
$$\times \exp\left[-2\left(\frac{r'' - r}{R}\right)^2\right] I_{0e}\left(\frac{4rr''}{R^2}\right) r'' dr''. \tag{4.41}$$

For a top-hat beam, substituting Eq. (4.40) into Eq. (4.37) yields

$$C(r, z) = G_1(0, z)S(r) + 2\pi S_0 \int_a^b G_2(r'', z) I_\phi(r, r'') r'' dr''. \tag{4.42}$$

The numerical results obtained with and without separately recording the first interactions are compared in the next section.

4.5.6. Truncation Error in Convolution

As shown in Eqs. (4.36) and (4.39), the upper integration limits may be bounded by r_{max}. For a top-hat beam, if

$$r \leq r_{max} - R, \tag{4.43}$$

the limited grid coverage in the r direction does not affect the convolution; otherwise, it truncates the convolution and leads to error in the convolution for $r > r_{max} - R$. Thus, to convolve reliably for physical quantities at r in response to a top-hat beam, we must ensure that r_{max} in the Monte Carlo simulation is large enough that Eq. (4.43) holds.

For a Gaussian beam, no simple formula similar to Eq. (4.43) exists because a Gaussian beam theoretically extends to infinity. At $r \gg R$, a Gaussian beam and a top-hat beam of the same R and S_0 have comparable convolution results. Therefore, Eq. (4.43) can be used approximately for Gaussian beams as well.

4.6. COMPUTATIONAL EXAMPLES

In this section, the error that is caused by not recording the first photon–matter interactions separately is illustrated, and a numerical example of convolution is presented. The impulse responses are computed by MCML for a scattering medium described in Table 4.1. The grid element sizes in the r and z directions are both 0.01 cm. The numbers of grid elements in the r and z directions are 50 and 40, respectively. One million photon packets are tracked.

The impulse fluence near the surface of the scattering medium ($z = 0.005$ cm) is shown in Figure 4.2a, where the first interactions are recorded separately. If they were recorded into the first r grid element instead, it would augment the fluence in the first grid element by 1.95×10^3 cm^{-2}, which is significantly greater than the current value of 1.34×10^3 cm^{-2}. For comparison, the impulse response is convolved over a top-hat beam of 1-nJ energy and 0.01-cm radius both with, and without, recording the first interactions separately (Figure 4.2b). The convolved results differ at $r = 0.015$ cm by as much as 120%.

The impulse response is also convolved over a Gaussian beam (1-nJ total energy, 0.1-cm radius), where the convolution error is set to 0.01. The contour lines of the fluence distribution before and after the convolution are shown in Figure 4.3.

APPENDIX 4A. SUMMARY OF CONV

The entire source code of CONV can be found on the Web at ftp://ftp.wiley.com/public/sci_tech_med/biomedical_optics. The program is divided into several files. Header file conv.h defines the data structures and some constants. File convmain.c contains primarily the main function. File convi.c handles data reading. File convo.c handles data writing. File convconv.c implements the

TABLE 4.1. Optical Properties and Structure of a Three-Layered Scattering Medium[a]

Layer	n	μ_a (cm^{-1})	μ_s (cm^{-1})	g	Thickness (cm)
1	1.37	1.0	100.0	0.9	0.1
2	1.37	1.0	10.0	0	0.1
3	1.37	2.0	10.0	0.7	0.2

[a] Refractive indices for top and bottom ambient media are both 1.0.

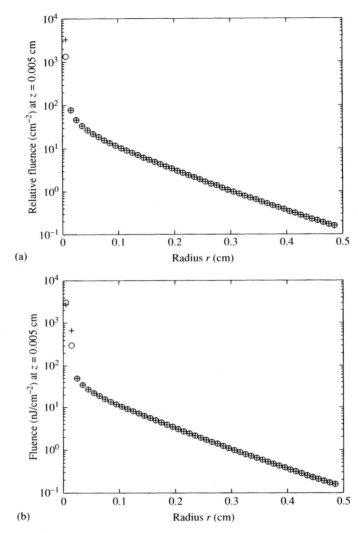

Figure 4.2. (a) Relative fluence at $z = 0.005$ cm in response to a pencil beam computed by MCML; (b) fluence at $z = 0.005$ cm in response to a top-hat beam computed by CONV. Circles and crosses represent data with and without, respectively, the first interactions scored separately.

actual convolution. File `conviso.c` handles calculation of contours. File `convnr.c` contains several functions for dynamical data allocations and error reports. Readers should read the main function first.

The following list is generated by command `cflow -d4 -n --omit-arguments --omit-symbol-names conv*.c`, which shows the structure of the program with

APPENDIX 4A. SUMMARY OF CONV

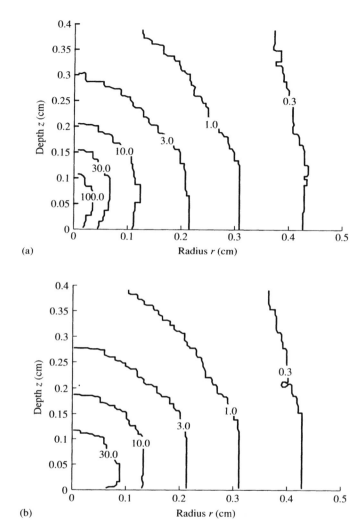

Figure 4.3. (a) Relative fluence distribution in response to a pencil beam computed by MCML; (b) fluence distribution in response to a Gaussian beam computed by CONV.

the nesting depth limited to 4:

```
1 main() <int () at CONVMAIN.C:143>:
2     ShowVersion() <void () at CONVO.C:37>:
3         puts()
4         CenterStr() <char * () at CONVO.C:11>:
5             strlen()
6             strcpy()
```

```
 7              strcat()
 8     printf()
 9     gets()
10     strlen()
11     BranchMainCmd() <void () at CONVMAIN.C:122>:
12         strlen()
13         BranchMainCmd1() <void () at CONVMAIN.C:63>:
14             toupper()
15             ReadMcoFile() <void () at CONVI.C:568>:
16             LaserBeam() <void () at CONVCONV.C:92>:
17             ConvResolution() <void () at CONVCONV.C:127>:
18             ConvError() <void () at CONVCONV.C:156>:
19             ShowMainMenu() <void () at CONVMAIN.C:25>:
20             QuitProgram() <void () at CONVMAIN.C:44>:
21             puts()
22         BranchMainCmd2() <void () at CONVMAIN.C:92>:
23             toupper()
24             OutputOrigData() <void () at CONVO.C:784>:
25             OutputConvData() <void () at CONVCONV.C:1017>:
26             ContourOrigData() <void () at CONVO.C:893>:
27             ContourConvData() <void () at CONVCONV.C:1137>:
28             ScanOrigData() <void () at CONVO.C:1211>:
29             ScanConvData() <void () at CONVCONV.C:1506>:
30             puts()
31         puts()
```

PROBLEMS

4.1 Derive Eq. (4.11).

4.2 Derive Eq. (4.21).

4.3 Derive Eq. (4.23).

4.4 Derive Eq. (4.25).

4.5 Derive Eq. (4.31).

4.6 Derive Eqs. (4.41) and (4.42).

4.7 Write a computer program that can convolve the impulse responses over a flat beam.

4.8 Take the Fourier transformation of Eqs. (4.1) and (4.2) with respect to x and y.

4.9 Write the time-domain counterparts of Eqs. (4.1) and (4.2) for an arbitrary pulse profile $S(t)$ of the incident pencil beam. In this case, an impulse response $G(x, y, z, t)$ is defined as a time-resolved response to a temporally infinitely short $\delta(t)$ photon beam. Assuming the response to $S(t)$ to be experimentally measured, explain how to recover the impulse response $G(x, y, z, t)$ through deconvolution.

4.10 Take the Fourier transformation of the time-domain counterparts of Eqs. (4.1) and (4.2) with respect to t.

4.11 Assume that the incident photon beam is finite in both time and space and can be represented by $S(x, y, z, t)$. Write the convolution over this beam.

4.12 Although Eqs. (4.9) and (4.19) are related to the total power of the lightbeam, explain why the total energy can be used when, for example, the lightbeam is infinitely short-pulsed.

READING

Wang LHV, Jacques SL, and Zheng LQ (1997): CONV—convolution for responses to a finite diameter photon beam incident on multi-layered tissues, *Comput. Meth. Prog. Biomed.* **54**(3): 141–150. (All sections in this chapter.)

FURTHER READING

Amelink A and Sterenborg H (2004): Measurement of the local optical properties of turbid media by differential path-length spectroscopy, *Appl. Opt.* **43**(15): 3048–3054.

Carp SA, Prahl SA, and Venugopalan V (2004): Radiative transport in the delta-P-1 Approximation: Accuracy of fluence rate and optical penetration depth predictions in turbid semi-infinite media, *J. Biomed. Opt.* **9**(3): 632–647.

Choi B, Majaron B, and Nelson JS (2004): Computational model to evaluate port wine stain depth profiling using pulsed photothermal radiometry, *J. Biomed. Opt.* **9**(2): 299–307.

Diaz SH, Aguilar G, Lavernia EJ, and Wong BJF (2001): Modeling the thermal response of porcine cartilage to laser irradiation, *IEEE J. Select. Topics Quantum Electron.* **7**(6): 944–951.

Ding L, Splinter R, and Knisley SB (2001): Quantifying spatial localization of optical mapping using Monte Carlo simulations, *IEEE Trans. Biomed. Eng.* **48**(10): 1098–1107.

Fried NM, Hung VC, and Walsh JT (1999): Laser tissue welding: Laser spot size and beam profile studies, *IEEE J. Select. Topics Quantum Electron.* **5**(4): 1004–1012.

Gardner CM and Welch AJ (1994): Monte-Carlo simulation of light transport in tissue—unscattered absorption events, *Appl. Opt.* **33**(13): 2743–2745.

Garofalakis A, Zacharakis G, Filippidis G, Sanidas E, Tsiftsis D, Ntziachristos V, Papazoglou TG, and Ripoll J (2004): Characterization of the reduced scattering coefficient for optically thin samples: theory and experiments, *J. Opt. A* **6**(7): 725–735.

Giller CA, Liu RL, Gurnani P, Victor S, Yazdani U, and German DC (2003): Validation of a near-infrared probe for detection of thin intracranial white matter structures, *J. Neurosurg.* **98**(6): 1299–1306.

Hidovic-Rowe D and Claridge E (2005): Modelling and validation of spectral reflectance for the colon, *Phys. Med. Biol.* **50**(6): 1071–1093.

Johns M, Giller CA, and Liu HL (2001): Determination of hemoglobin oxygen saturation from turbid media using reflectance spectroscopy with small source-detector separations, *Appl. Spectrosc.* **55**(12): 1686–1694.

Johns M, Giller CA, German DC, and Liu HL (2005): Determination of reduced scattering coefficient of biological tissue from a needle-like probe, *Opt. Express* **13**(13): 4828–4842.

Klavuhn KG and Green D (2002): Importance of cutaneous cooling during photothermal epilation: Theoretical and practical considerations, *Lasers Surg. Med.* **31**(2): 97–105.

Laufer JG, Beard PC, Walker SP, and Mills TN (2001): Photothermal determination of optical coefficients of tissue phantoms using an optical fibre probe, *Phys. Med. Biol.* **46**(10): 2515–2530.

Lee JH, Kim S, and Kim YT (2004): Diffuse-diffuse photon coupling via nonscattering void in the presence of refractive index mismatch on the void boundary, *Med. Phys.* **31**(8): 2237–2248.

Marquez G, Wang LHV, Lin SP, Schwartz JA, and Thomsen SL (1998): Anisotropy in the absorption and scattering spectra of chicken breast tissue, *Appl. Opt.* **37**(4): 798–804.

McShane MJ, Rastegar S, Pishko M, and Cote GL (2000): Monte carlo modeling for implantable fluorescent analyte sensors, *IEEE Trans. Biomed. Eng.* **47**(5): 624–632.

Prahl SA, Keijzer M, Jacques SL, and Welch AJ (1989): A Monte Carlo model of light propagation in tissue, in *Dosimetry of Laser Radiation in Medicine and Biology*, Muller GJ and Sliney DH, eds., SPIE Press, **IS5**: 102–111.

Press WH, Flannery BP, Teukolsky SA, and Vetterling WT (1992): *Numerical Recipes in C*, Cambridge Univ. Press.

Reuss JL (2005): Multilayer modeling of reflectance pulse oximetry, *IEEE Trans. Biomed. Eng.* **52**(2): 153–159.

Shah RK, Nemati B, Wang LHV, and Shapshay SM (2001): Optical-thermal simulation of tonsillar tissue irradiation, *Lasers Surg. Med.* **28**(4): 313–319.

Wang LHV and Jacques SL (1994): Optimized radial and angular positions in Monte Carlo modeling, *Med. Phys.* **21**: 1081–1083.

Wang LHV, Jacques SL, and Zheng LQ (1995): MCML—Monte Carlo modeling of photon transport in multi-layered tissues, *Comput. Meth. Prog. Biomed.* **47**: 131–146.

CHAPTER 5

Radiative Transfer Equation and Diffusion Theory

5.1. INTRODUCTION

Photon transport in biological tissue can be modeled analytically by the radiative transfer equation (RTE), which is considered equivalent to the numerical Monte Carlo method covered in Chapter 3. Because the RTE is difficult to solve, it is often approximated to a diffusion equation, which provides solutions that are more computationally efficient but less accurate than those provided by the Monte Carlo method.

5.2. DEFINITIONS OF PHYSICAL QUANTITIES

Spectral radiance L_ν, the most general physical quantity discussed in this chapter, is defined as the energy flow per unit normal area per unit solid angle per unit time per unit temporal frequency (temporofrequency) bandwidth, where the normal area is perpendicular to the flow direction. Radiance L is defined as the spectral radiance integrated over a narrow frequency range $[\nu, \nu + \Delta\nu]$:

$$L(\vec{r}, \hat{s}, t) = L_\nu(\vec{r}, \hat{s}, t)\Delta\nu \quad (\text{W m}^{-2}\text{sr}^{-1}). \quad (5.1)$$

Here, $\vec{r}$ denotes position, $\hat{s}$ denotes unit direction vector, t denotes time, and the parentheses enclose the unit of the physical quantity on the left-hand side of the equation. The amount of radiant energy dE that is transported across differential area element dA within differential solid angle element $d\Omega$ during differential time element dt (Figure 5.1) is given by

$$dE = L(\vec{r}, \hat{s}, t)(\hat{s} \cdot \hat{n})\, dA\, d\Omega\, dt \quad (\text{J}). \quad (5.2)$$

Here, $\hat{n}$ denotes the unit outward normal vector of dA; $\hat{s} \cdot \hat{n}$ denotes the dot product of the two unit vectors, which equals the cosine of the angle between them. The radiance is the dependent variable in the RTE (to be derived). Several additional physical quantities can be derived from the radiance.

Biomedical Optics: Principles and Imaging, by Lihong V. Wang and Hsin-i Wu
Copyright © 2007 John Wiley & Sons, Inc.

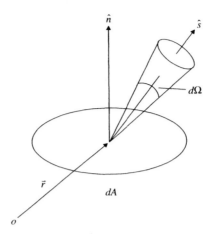

Figure 5.1. Schematic of energy flow through a differential area element dA within a differential solid angle element $d\Omega$.

Fluence rate (or intensity) Φ is defined as the energy flow per unit area per unit time regardless of the flow direction; it is expressed as the radiance integrated over the entire 4π solid angle:

$$\Phi(\vec{r},t) = \int_{4\pi} L(\vec{r},\hat{s},t)\, d\Omega \quad (\text{W/m}^2). \tag{5.3}$$

An infinitesimal sphere of surface area dS receives power in the amount of $\Phi(\vec{r},t)\,dS$. In spherical coordinates, we have

$$\Phi(\vec{r},t) = \int_{\theta=0}^{\pi} \int_{\phi=0}^{2\pi} L(\vec{r},\hat{s},t) \sin\theta\, d\phi\, d\theta \tag{5.4}$$

and

$$\hat{s} = (\sin\theta\cos\phi,\ \sin\theta\sin\phi,\ \cos\theta), \tag{5.5}$$

where θ and ϕ denote the polar and azimuthal angles, respectively.

Fluence F is defined as the time-integrated fluence rate:

$$F(\vec{r}) = \int_{-\infty}^{+\infty} \Phi(\vec{r},t)\, dt \quad (\text{J/m}^2). \tag{5.6}$$

Current density $\vec{J}$ is defined as the net energy flow per unit area per unit time; it can be expressed as

$$\vec{J}(\vec{r},t) = \int_{4\pi} \hat{s} L(\vec{r},\hat{s},t)\, d\Omega \quad (\text{W/m}^2), \tag{5.7}$$

which is the vector counterpart of Eq. (5.3). Current density points to the direction of the prevalent flow since flows in opposite directions partially offset each other. Current density is also referred to as *energy flux*; the term *flux*, however, can also refer to a vector quantity integrated over a given area.

Energy density u_e is defined as the energy of the propagating electromagnetic wave per unit volume; it can be obtained by

$$u_e = \frac{\Phi}{c} \quad (\text{J/m}^3), \tag{5.8}$$

where c is the speed of light in the medium.

Photon density U is defined as the number of propagating photons per unit volume; for monochromatic light, it can be expressed as

$$U = \frac{u_e}{h\nu} = \frac{\Phi}{ch\nu} \quad (\text{m}^{-3}), \tag{5.9}$$

where h is the Planck constant and $h\nu$ is the energy of a single photon.

Specific power deposition (or specific absorption rate) A_p is defined as the optical energy absorbed by the medium per unit volume per unit time; it can be expressed as

$$A_p = \mu_a \Phi \quad (\text{W/m}^3), \tag{5.10}$$

where μ_a is the absorption coefficient of the medium.

Specific energy deposition (or specific absorption) A_e is defined as the time-integrated specific power deposition:

$$A_e(\vec{r}) = \int_{-\infty}^{+\infty} A_p(\vec{r}, t) \, dt \quad (\text{J/m}^3). \tag{5.11}$$

5.3. DERIVATION OF RADIATIVE TRANSPORT EQUATION

With the quantities defined above, we now heuristically derive the RTE from the principle of conservation of energy, where coherence, polarization, and nonlinearity are neglected. The optical properties—including refractive index n, absorption coefficient μ_a, scattering coefficient μ_s, and scattering anisotropy g—are assumed to be time-invariant but space-variant. Only elastic scattering is considered in this chapter.

Consider a stationary differential cylindrical volume element as shown in Figure 5.2. Here, ds is the differential length element of the cylinder along photon propagation direction $\hat{s}$; dA is the differential area element perpendicular to direction $\hat{s}$. Below, we consider all possible contributions to the energy change in this volume element within differential solid angle element $d\Omega$ around direction $\hat{s}$. In addition, $d\Omega'$ is a differential solid angle element around direction $\hat{s}'$.

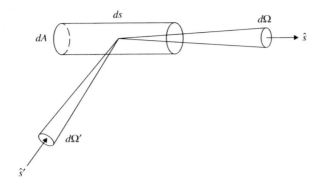

Figure 5.2. Schematic of a stationary differential cylindrical volume element.

5.3.1. Contribution 1: Divergence

If the local photon beam is not collimated, divergence is nonzero. Energy diverging out of the volume element or the solid angle element per unit time is given by

$$dP_{\text{div}} = \frac{\partial L(\vec{r}, \hat{s}, t)}{\partial s} ds \, dA \, d\Omega = \frac{\partial L(\vec{r}, \hat{s}, t)}{\partial s} d\Omega \, dV, \quad (5.12)$$

where $dV = dA \, ds$. This contribution is positive for actual divergence and negative for actual convergence.

In divergence form, Eq. (5.12) becomes

$$dP_{\text{div}} = \hat{s} \cdot \nabla L(\vec{r}, \hat{s}, t) \, d\Omega \, dV = \nabla \cdot [L(\vec{r}, \hat{s}, t)\hat{s}] \, d\Omega \, dV. \quad (5.13)$$

This contribution is due to local "noninteractive" beam propagation; thus, it can exist even in a nonscattering medium. Scattering elsewhere, however, can affect the local divergence. It can be seen later that this contribution still exists when the absorption and scattering coefficients are set to zero in the radiative transfer equation.

5.3.2. Contribution 2: Extinction

Energy loss per unit time in the volume element within the solid angle element due to absorption and scattering is given by

$$dP_{\text{ext}} = (\mu_t \, ds)[L(\vec{r}, \hat{s}, t) \, dA \, d\Omega], \quad (5.14)$$

where $\mu_t \, ds$ represents the probability of extinction—by either absorption or scattering—in ds. Light scattered from all directions into solid angle element $d\Omega$ is considered in the next subsection.

5.3.3. Contribution 3: Scattering

Energy incident on the volume element from any direction $\hat{s}'$ and scattered into $d\Omega$ around direction $\hat{s}$ per unit time is given by

$$dP_{\text{sca}} = (N_s \, dV) \left[\int_{4\pi} L(\vec{r}, \hat{s}', t) P(\hat{s}', \hat{s}) \sigma_s \, d\Omega' \right] d\Omega. \tag{5.15}$$

Here, N_s denotes the number density of the scatterers and σ_s denotes the scattering cross section of a scatterer. Thus, $N_s \, dV$ denotes the number of scatterers in the volume element; $L(\vec{r}, \hat{s}', t) \sigma_s \, d\Omega'$ denotes the energy intercepted by a single scatterer within solid angle $d\Omega'$ per unit time. The phase function $P(\hat{s}', \hat{s})$ is a PDF:

$$\int_{4\pi} P(\hat{s}', \hat{s}) \, d\Omega = 1. \tag{5.16}$$

The product $P(\hat{s}', \hat{s}) \, d\Omega$ represents the probability of light with propagation direction $\hat{s}'$ being scattered into $d\Omega$ around direction $\hat{s}$. Often, the phase function depends only on the angle between the scattered and incident directions, that is

$$P(\hat{s}', \hat{s}) = P(\hat{s}' \cdot \hat{s}), \tag{5.17}$$

where $\hat{s}' \cdot \hat{s}$ equals the cosine of the angle between the two unit vectors. We limit our consideration to this case. The scattering anisotropy can be expressed as

$$g = \int_{4\pi} (\hat{s}' \cdot \hat{s}) P(\hat{s}' \cdot \hat{s}) \, d\Omega. \tag{5.18}$$

From $\mu_s = N_s \sigma_s$ and Eq. (5.17), Eq. (5.15) can be rewritten as

$$dP_{\text{sca}} = (\mu_s \, dV) \left[\int_{4\pi} L(\vec{r}, \hat{s}', t) P(\hat{s}' \cdot \hat{s}) \, d\Omega' \right] d\Omega. \tag{5.19}$$

5.3.4. Contribution 4: Source

Energy produced by a source in the volume element within the solid angle element per unit time is given by

$$dP_{\text{src}} = S(\vec{r}, \hat{s}, t) \, dV \, d\Omega, \tag{5.20}$$

where S carries the unit of W/(m^3 sr).

5.3.5. Conservation of Energy

The change in energy in the volume element within the solid angle element per unit time is given by

$$dP = \frac{\partial L(\vec{r}, \hat{s}, t)/c}{\partial t} dV \, d\Omega, \quad (5.21)$$

where L/c represents the propagating energy per unit volume per unit solid angle. This rate of change is a result of the balance among the two negative and two positive contributions described above. The principle of conservation of energy requires

$$dP = -dP_{\text{div}} - dP_{\text{ext}} + dP_{\text{sca}} + dP_{\text{src}}. \quad (5.22)$$

Substituting Eqs. (5.13), (5.14), and (5.19)–(5.21) into Eq. (5.22), we obtain

$$\frac{\partial L(\vec{r}, \hat{s}, t)/c}{\partial t} = -\hat{s} \cdot \nabla L(\vec{r}, \hat{s}, t) - \mu_t L(\vec{r}, \hat{s}, t) \\ + \mu_s \int_{4\pi} L(\vec{r}, \hat{s}', t) P(\hat{s}' \cdot \hat{s}) \, d\Omega' + S(\vec{r}, \hat{s}, t), \quad (5.23)$$

which is the well-known RTE (or the Boltzmann equation).

For time-independent responses, the left-hand side of Eq. (5.23) is zero:

$$\frac{\partial L(\vec{r}, \hat{s}, t)}{\partial t} = 0. \quad (5.24)$$

To reach a time-independent state requires the use of a time-invariant light source, that is, a constant-power continuous-wave lightbeam. For a pulsed light source, time-independent responses are still applicable to time-integrated physical quantities such as specific energy deposition.

5.4. DIFFUSION THEORY

The RTE is difficult to solve since it has six independent variables $(x, y, z, \theta, \phi, t)$. Usually, the RTE is simplified in the diffusion approximation. The diffusion approximation assumes that the radiance in a high-albedo ($\mu_a \ll \mu_s$) scattering medium is nearly isotropic after sufficient scattering.

5.4.1. Diffusion Expansion of Radiance

Spherical harmonics $Y_{n,m}$ form a basis set, on which the radiance can be expanded. In the diffusion approximation, the radiance is expanded to the first order

$$L(\vec{r}, \hat{s}, t) \approx \sum_{n=0}^{1} \sum_{m=-n}^{n} L_{n,m}(\vec{r}, t) Y_{n,m}(\hat{s}), \qquad (5.25)$$

where $L_{n,m}$ are the expansion coefficients. The term for $n = 0$ and $m = 0$ on the right-hand side represents the isotropic component, whereas the terms for $n = 1$ and $m = 0, \pm 1$ represent the anisotropic component.

In terms of the associated Legendre polynomials $P_{n,m}$ and a periodic function of ϕ, we have

$$Y_{n,m}(\hat{s}) = Y_{n,m}(\theta, \phi) = (-1)^m \sqrt{\frac{(2n+1)(n-m)!}{4\pi(n+m)!}} P_{n,m}(\cos\theta) e^{im\phi}, \qquad (5.26)$$

where

$$P_{n,m}(x) = \frac{(1-x^2)^{m/2}}{2^n n!} \frac{d^{m+n}}{dx^{m+n}} (x^2 - 1)^n. \qquad (5.27)$$

When $m = 0$, $P_{n,m}$ reduces to the (unassociated) Legendre polynomials P_n. If the expansion of L in Eq. (5.25) continues to $n = N$, the approximation is known as the P_N approximation. Hence, the diffusion approximation is also known as the P_1 approximation.

The spherical harmonics for $n = 1$ are

$$\begin{aligned} Y_{0,0}(\theta, \phi) &= \frac{1}{\sqrt{4\pi}}, \\ Y_{1,-1}(\theta, \phi) &= \sqrt{\frac{3}{8\pi}} \sin\theta e^{-i\phi}, \\ Y_{1,0}(\theta, \phi) &= \sqrt{\frac{3}{4\pi}} \cos\theta, \\ Y_{1,1}(\theta, \phi) &= -\sqrt{\frac{3}{8\pi}} \sin\theta e^{i\phi}. \end{aligned} \qquad (5.28)$$

The following symmetry and orthogonality (or orthonormality) exist:

$$Y_{n,-m}(\theta, \phi) = (-1)^m Y^*_{n,m}(\theta, \phi), \qquad (5.29)$$

$$\int_{4\pi} Y_{n,m}(\hat{s}) Y^*_{n',m'}(\hat{s}) \, d\Omega = \delta_{nn',mm'}. \qquad (5.30)$$

Here, $*$ denotes complex conjugation; $\delta_{nn',mm'}$ denotes the Krönecker delta function, which equals 1 if both $n = n'$ and $m = m'$ hold and 0 otherwise. The integral in Eq. (5.30) is referred to as the *inner product*, analogous to the dot product of two vectors.

Substituting Eq. (5.25) into Eq. (5.3), we obtain

$$\Phi(\vec{r}, t) = 4\pi L_{0,0}(\vec{r}, t) Y_{0,0}(\hat{s}) \tag{5.31}$$

or

$$L_{0,0}(\vec{r}, t) Y_{0,0}(\hat{s}) = \frac{\Phi(\vec{r}, t)}{4\pi}, \tag{5.32}$$

which means that the isotropic term in Eq. (5.25) is equal to the fluence rate divided by the entire 4π solid angle.

Unit vector $\hat{s}$ can also be expressed in terms of the spherical harmonics:

$$\hat{s} = (\sin\theta\cos\phi, \quad \sin\theta\sin\phi, \quad \cos\theta)$$

$$= \sqrt{\frac{2\pi}{3}} \left(Y_{1,-1}(\hat{s}) - Y_{1,1}(\hat{s}), \quad i[Y_{1,-1}(\hat{s}) + Y_{1,1}(\hat{s})], \quad \sqrt{2} Y_{1,0}(\hat{s}) \right). \tag{5.33}$$

Multiplying Eq. (5.25) by $\hat{s}$ and substituting the result into Eq. (5.7), we obtain

$$\vec{J}(\vec{r}, t) \cdot \hat{s} = \frac{4\pi}{3} \sum_{m=-1}^{1} L_{1,m}(\vec{r}, t) Y_{1,m}(\hat{s}) \tag{5.34}$$

or

$$\sum_{m=-1}^{1} L_{1,m}(\vec{r}, t) Y_{1,m}(\hat{s}) = \frac{3}{4\pi} \vec{J}(\vec{r}, t) \cdot \hat{s}. \tag{5.35}$$

Note that $\vec{J}(\vec{r}, t) \cdot \hat{s} = |\vec{J}(\vec{r}, t)| \cos\alpha$, where α denotes the angle between $\vec{J}(\vec{r}, t)$ and $\hat{s}$. Therefore, the anisotropic term in Eq. (5.25) is proportional to the projection of $\vec{J}(\vec{r}, t)$ onto $\hat{s}$.

Substituting both Eqs. (5.32) and (5.35) into Eq. (5.25), we obtain

$$L(\vec{r}, \hat{s}, t) = \frac{1}{4\pi} \Phi(\vec{r}, t) + \frac{3}{4\pi} \vec{J}(\vec{r}, t) \cdot \hat{s}, \tag{5.36}$$

which is illustrated in Figure 5.3.

Example 5.1. Derive Eq. (5.31).

Substituting Eq. (5.25) into Eq. (5.3), we obtain

$$\Phi(\vec{r}, t) = \int_{4\pi} L_{0,0}(\vec{r}, t) Y_{0,0}(\hat{s}) \, d\Omega + \int_{4\pi} L_{1,-1}(\vec{r}, t) Y_{1,-1}(\hat{s}) \, d\Omega \tag{5.37}$$

$$+ \int_{4\pi} L_{1,0}(\vec{r}, t) Y_{1,0}(\hat{s}) \, d\Omega + \int_{4\pi} L_{1,1}(\vec{r}, t) Y_{1,1}(\hat{s}) \, d\Omega.$$

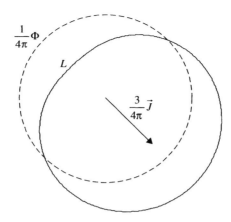

Figure 5.3. Illustration of the effect of current density $\vec{J}$ on the radiance. Solid line represents the isotropic term; dashed line represents the total radiance.

Noting that $L_{n,m}$ is independent of $\hat{s}$ and using Eq. (5.28), we have

$$\int_{4\pi} L_{0,0}(\vec{r},t) Y_{0,0}(\hat{s}) \, d\Omega = L_{0,0}(\vec{r},t) Y_{0,0}(\hat{s}) \int_{4\pi} d\Omega = 4\pi L_{0,0}(\vec{r},t) Y_{0,0}(\hat{s}), \tag{5.38}$$

$$\int_{4\pi} L_{1,-1}(\vec{r},t) Y_{1,-1}(\hat{s}) \, d\Omega = \sqrt{\frac{3}{8\pi}} L_{1,-1}(\vec{r},t) \int_0^{2\pi} e^{-i\phi} d\phi \int_0^{\pi} \sin^2\theta \, d\theta = 0, \tag{5.39}$$

$$\int_{4\pi} L_{1,0}(\vec{r},t) Y_{1,0}(\hat{s}) \, d\Omega = 2\pi \sqrt{\frac{3}{4\pi}} L_{1,0}(\vec{r},t) \int_0^{\pi} \cos\theta \sin\theta \, d\theta = 0, \tag{5.40}$$

$$\int_{4\pi} L_{1,1}(\vec{r},t) Y_{1,1}(\hat{s}) \, d\Omega = -\sqrt{\frac{3}{8\pi}} L_{1,1}(\vec{r},t) \int_0^{2\pi} e^{i\phi} d\phi \int_0^{\pi} \sin^2\theta \, d\theta = 0. \tag{5.41}$$

Finally, we obtain

$$\Phi(\vec{r},t) = 4\pi L_{0,0}(\vec{r},t) Y_{0,0}(\hat{s}). \tag{5.42}$$

Example 5.2. Derive Eq. (5.34).

Substituting Eqs. (5.33) and (5.25) into Eq. (5.7) and using Eqs. (5.29) and (5.30), we obtain

$$\vec{J}(\vec{r},t) = \sqrt{\frac{2\pi}{3}} \left(-L_{1,1} + L_{1,-1}, -i(L_{1,1} + L_{1,-1}), \sqrt{2} L_{1,0} \right). \tag{5.43}$$

92 RADIATIVE TRANSFER EQUATION AND DIFFUSION THEORY

From Eqs. (5.33) and (5.43), we obtain

$$\hat{s} \cdot \vec{J}(\vec{r}, t) = \frac{4\pi}{3} \sum_{m=-1}^{1} L_{1,m}(\vec{r}, t) Y_{1,m}(\hat{s}). \tag{5.44}$$

5.4.2. Source

The source S is assumed to be isotropic; that is, $S(\vec{r}, \hat{s}, t)$ is independent of $\hat{s}$:

$$S(\vec{r}, \hat{s}, t) = \frac{S(\vec{r}, t)}{4\pi}. \tag{5.45}$$

A collimated source can be approximately converted into an isotropic source (to be discussed).

5.4.3. Scalar Differential Equation

Substituting the diffusion expansion of $L(\vec{r}, \hat{s}, t)$ [Eq. (5.36)] into the RTE [Eq. (5.23)] and integrating over the full 4π solid angle, we obtain the following scalar differential equation:

$$\frac{\partial \Phi(\vec{r}, t)}{c \partial t} + \mu_a \Phi(\vec{r}, t) + \nabla \cdot \vec{J}(\vec{r}, t) = S(\vec{r}, t). \tag{5.46}$$

Example 5.3. Derive Eq. (5.46).

We substitute Eq. (5.36) into Eq. (5.23), integrate over the full 4π solid angle (scalar sum), and then evaluate each term as follows:

1. For the left-hand side, on the basis of Eq. (5.3), we obtain

$$\int_{4\pi} \frac{\partial L(\vec{r}, \hat{s}, t)/c}{\partial t} d\Omega = \frac{\partial \Phi(\vec{r}, t)}{c \partial t}. \tag{5.47}$$

2. For the first term on the right-hand side, on the basis of the vector identities $\hat{s} \cdot \nabla L = \nabla \cdot (\hat{s} L) - L \nabla \cdot \hat{s}$ and $\nabla \cdot \hat{s} = 0$ and then Eq. (5.7), we obtain

$$-\int_{4\pi} \hat{s} \cdot \nabla L(\vec{r}, \hat{s}, t) d\Omega = -\int_{4\pi} \nabla \cdot [\hat{s} L(\vec{r}, \hat{s}, t)] d\Omega$$

$$= -\nabla \cdot \int_{4\pi} \hat{s} L(\vec{r}, \hat{s}, t) d\Omega = -\nabla \cdot \vec{J}(\vec{r}, t). \tag{5.48}$$

3. For the second term, from Eq. (5.3), we obtain

$$-\mu_t \int_{4\pi} L(\vec{r}, \hat{s}, t) d\Omega = -\mu_t \Phi(\vec{r}, t). \tag{5.49}$$

4. For the third term, we have

$$\mu_s \int_{4\pi} \left[\int_{4\pi} L(\vec{r}, \hat{s}', t) P(\hat{s}' \cdot \hat{s}) d\Omega' \right] d\Omega$$

$$= \frac{\mu_s}{4\pi} \int_{4\pi} \int_{4\pi} [\Phi(\vec{r}, t) + 3\vec{J}(\vec{r}, t) \cdot \hat{s}'] P(\hat{s}' \cdot \hat{s}) d\Omega' d\Omega. \quad (5.50)$$

We first evaluate the following two integrals:

$$\int_{4\pi} \left[\int_{4\pi} \Phi(\vec{r}, t) P(\hat{s}' \cdot \hat{s}) d\Omega' \right] d\Omega = \Phi(\vec{r}, t) \int_{4\pi} \left[\int_{4\pi} P(\hat{s}' \cdot \hat{s}) d\Omega' \right] d\Omega$$

$$= \Phi(\vec{r}, t) \int_{4\pi} d\Omega = 4\pi \Phi(\vec{r}, t)$$

$$(5.51)$$

and

$$\int_{4\pi} \int_{4\pi} [\vec{J}(\vec{r}, t) \cdot \hat{s}'] P(\hat{s}' \cdot \hat{s}) d\Omega' d\Omega$$

$$= |\vec{J}(\vec{r}, t)| \int_{4\pi} \left[\int_{4\pi} P(\hat{s}' \cdot \hat{s}) d\Omega \right] \cos\theta' d\Omega' \quad (5.52)$$

$$= |\vec{J}(\vec{r}, t)| \int_{4\pi} \cos\theta' d\Omega'$$

$$= 0.$$

Here, $\vec{J}$ is aligned with the z' axis, and $d\Omega' = \sin\theta' d\theta' d\phi'$. Therefore

$$\mu_s \int_{4\pi} \left[\int_{4\pi} L(\vec{r}, \hat{s}', t) P(\hat{s}' \cdot \hat{s}) d\Omega' \right] d\Omega = \mu_s \Phi(\vec{r}, t). \quad (5.53)$$

5. For the last term, using Eq. (5.45), we obtain

$$\int_{4\pi} S(\vec{r}, \hat{s}, t) d\Omega = \frac{1}{4\pi} \int_{4\pi} S(\vec{r}, t) d\Omega = S(\vec{r}, t). \quad (5.54)$$

Combining these five parts completes the proof, where $\mu_t - \mu_s = \mu_a$ is used.

5.4.4. Vector Differential Equation

Substituting the diffusion expansion of $L(\vec{r}, \hat{s}, t)$ [Eq. (5.36)] into the RTE [Eq. (5.23)], multiplying both sides by $\hat{s}$, and integrating over the full 4π solid angle, we obtain the following vector differential equation:

$$\frac{\partial \vec{J}(\vec{r}, t)}{c \partial t} + (\mu_a + \mu_s')\vec{J}(\vec{r}, t) + \frac{1}{3}\nabla \Phi(\vec{r}, t) = 0, \quad (5.55)$$

where

$$\mu'_s = \mu_s(1-g) \tag{5.56}$$

is referred to as the *transport* (or *reduced*) *scattering coefficient*. The sum $\mu_a + \mu'_s$ is referred to as the *transport* (or *reduced*) *interaction coefficient* μ'_t:

$$\mu'_t = \mu_a + \mu'_s. \tag{5.57}$$

The reciprocal of μ'_t is referred to as the *transport mean free path* l'_t:

$$l'_t = \frac{1}{\mu'_t}. \tag{5.58}$$

Example 5.4. Derive Eq. (5.55).

We substitute Eq. (5.36) into Eq. (5.23), multiply both sides by $\hat{s}$, integrate over the full 4π solid angle (vector sum), and then evaluate each term as follows.

1. On the left-hand side, on the basis of Eq. (5.7), we obtain

$$\int_{4\pi} \hat{s} \frac{\partial L(\vec{r}, \hat{s}, t)/c}{\partial t} d\Omega = \frac{\partial \vec{J}(\vec{r}, t)}{c \partial t}. \tag{5.59}$$

2. For the first-term on the right-hand side, we have

$$\int_{4\pi} \hat{s}(\hat{s} \cdot \nabla L) d\Omega$$
$$= \frac{1}{4\pi} \int_{4\pi} \hat{s}(\hat{s} \cdot \nabla \Phi) d\Omega + \frac{3}{4\pi} \int_{4\pi} \hat{s}[\hat{s} \cdot \nabla(\vec{J} \cdot \hat{s})] d\Omega. \tag{5.60}$$

It can be shown that the two integrals on the right-hand side here have the following results (see Problems 5.1 and 5.2):

$$\int_{4\pi} \hat{s}(\hat{s} \cdot \nabla \Phi) d\Omega = \frac{4\pi}{3} \nabla \Phi, \tag{5.61}$$

$$\int_{4\pi} \hat{s}[\hat{s} \cdot \nabla(\vec{J} \cdot \hat{s})] d\Omega = 0. \tag{5.62}$$

3. For the second-term on the right-hand side, from Eq. (5.7), we obtain

$$\mu_t \int_{4\pi} \hat{s} L(\vec{r}, \hat{s}, t) d\Omega = \mu_t \vec{J}(\vec{r}, \hat{s}, t). \tag{5.63}$$

4. For the third-term on the right-hand side, we have

$$\int_{4\pi}\int_{4\pi}\hat{s}[L(\vec{r},\hat{s}',t)P(\hat{s}'\cdot\hat{s})\,d\Omega']\,d\Omega$$
$$=\frac{1}{4\pi}\int_{4\pi}\int_{4\pi}\hat{s}[\Phi(\vec{r},t)P(\hat{s}'\cdot\hat{s})\,d\Omega']\,d\Omega$$
$$+\frac{3}{4\pi}\int_{4\pi}\hat{s}\left\{\int_{4\pi}[\vec{J}(\vec{r},t)\cdot\hat{s}']P(\hat{s}'\cdot\hat{s})\,d\Omega'\right\}\,d\Omega. \tag{5.64}$$

For the first integral, we have

$$\int_{4\pi}\int_{4\pi}\hat{s}[\Phi(\vec{r},t)P(\hat{s}'\cdot\hat{s})\,d\Omega']\,d\Omega$$
$$=\Phi(\vec{r},t)\int_{4\pi}\hat{s}\left[\int_{4\pi}P(\hat{s}'\cdot\hat{s})\,d\Omega'\right]d\Omega = \Phi(\vec{r},t)\int_{4\pi}\hat{s}\,d\Omega = 0. \tag{5.65}$$

For the second integral, we have

$$\int_{4\pi}\hat{s}\left\{\int_{4\pi}[\vec{J}(\vec{r},t)\cdot\hat{s}']P(\hat{s}'\cdot\hat{s})\,d\Omega'\right\}d\Omega$$
$$=\int_{4\pi}\left[\int_{4\pi}\hat{s}P(\hat{s}'\cdot\hat{s})\,d\Omega\right][\vec{J}(\vec{r},t)\cdot\hat{s}']\,d\Omega'. \tag{5.66}$$

On the basis of the identity

$$\hat{s}=\hat{s}'(\hat{s}\cdot\hat{s}')+\hat{s}'\times(\hat{s}\times\hat{s}'), \tag{5.67}$$

the inner integral in Eq. (5.66) is split into two integrals. The first one is

$$\int_{4\pi}\hat{s}'(\hat{s}'\cdot\hat{s})P(\hat{s}'\cdot\hat{s})\,d\Omega=\hat{s}'g. \tag{5.68}$$

The second one is

$$\int_{4\pi}\hat{s}'\times(\hat{s}\times\hat{s}')P(\hat{s}'\cdot\hat{s})\,d\Omega=\hat{s}'\times\left[\left(\int_{4\pi}\hat{s}P(\hat{s}'\cdot\hat{s})\,d\Omega\right)\times\hat{s}'\right]. \tag{5.69}$$

Since $P(\hat{s}'\cdot\hat{s})$ is azimuthally symmetric about $\hat{s}$, $\int_{4\pi}\hat{s}P(\hat{s}'\cdot\hat{s})\,d\Omega$ is parallel with $\hat{s}'$; hence, its cross-product with $\hat{s}'$ is zero. Therefore, Eq. (5.66) becomes

$$\int_{4\pi} \hat{s} \left\{ \int_{4\pi} [\vec{J}(\vec{r},t) \cdot \hat{s}'] P(\hat{s}' \cdot \hat{s}) \, d\Omega' \right\} d\Omega$$
$$= g \int_{4\pi} \hat{s}' [\vec{J}(\vec{r},t) \cdot \hat{s}'] \, d\Omega' \qquad (5.70)$$
$$= \frac{4\pi}{3} g \vec{J}(\vec{r},t),$$

where the last step is based on Problem 5.1.

5. For the last term on the right-hand side, from Eq. (5.45), we obtain

$$\int_{4\pi} \hat{s} S(\vec{r}, \hat{s}, t) \, d\Omega = \frac{S(\vec{r},t)}{4\pi} \int_{4\pi} \hat{s} \, d\Omega = 0. \qquad (5.71)$$

Combining these five parts completes the proof.

5.4.5. Diffusion Equation

We notice that Eqs. (5.46) and (5.55) do not contain $\hat{s}$ as does Eq. (5.23) but contain two physical quantities $\vec{J}(\vec{r},t)$ and $\Phi(\vec{r},t)$. We now aim to obtain a single differential equation containing $\Phi(\vec{r},t)$ only.

We further assume that the fractional change in $\vec{J}(\vec{r},t)$ within l'_t is small, specifically

$$\left(\frac{l'_t}{c} \right) \left(\frac{1}{|\vec{J}(\vec{r},t)|} \left| \frac{\partial \vec{J}(\vec{r},t)}{\partial t} \right| \right) \ll 1, \qquad (5.72)$$

where the first pair of parentheses contains the time duration for photons to traverse l'_t (which may be referred to as the *transport mean free time*) and the second pair of parentheses contains the fractional change in the current density per unit time. Equation (5.72) can be rewritten as

$$\left| \frac{\partial \vec{J}(\vec{r},t)}{c \partial t} \right| \ll (\mu_a + \mu'_s) |\vec{J}(\vec{r},t)|. \qquad (5.73)$$

Under this condition, the time-dependent term in Eq. (5.55) is negligible, leading to

$$\vec{J}(\vec{r},t) = -D \nabla \Phi(\vec{r},t), \qquad (5.74)$$

which is referred to as *Fick's law*. A negative sign appears above because diffusion current density is always along the negative gradient. The constant D is referred to as the diffusion coefficient:

$$D = \frac{1}{3(\mu_a + \mu'_s)}. \tag{5.75}$$

Fick's law describes the diffusion of photons in a scattering medium. In fact, Fick's law can describe diffusion in many other forms such as pollutant diffusion in air, ink diffusion in water, and heat diffusion in metal. It is not, however, applicable to propagations driven by external forces, such as electron drift in an external electrical field and particle drift under external pressure.

Substituting Eq. (5.74) into Eq. (5.36), we obtain

$$L(\vec{r}, \hat{s}, t) = \frac{1}{4\pi}\Phi(\vec{r}, t) - \frac{3}{4\pi}D\nabla\Phi(\vec{r}, t) \cdot \hat{s}, \tag{5.76}$$

which expresses the radiance in terms of the fluence rate alone.

Substituting Eq. (5.74) into Eq. (5.46), we obtain

$$\frac{\partial \Phi(\vec{r}, t)}{c\partial t} + \mu_a \Phi(\vec{r}, t) - \nabla \cdot [D\nabla\Phi(\vec{r}, t)] = S(\vec{r}, t), \tag{5.77}$$

which is referred to as the *diffusion equation*. If the absorption coefficient is zero, this diffusion equation reduces to the heat diffusion equation. If the diffusion coefficient is space-invariant, we have a simpler version:

$$\frac{\partial \Phi(\vec{r}, t)}{c\partial t} + \mu_a \Phi(\vec{r}, t) - D\nabla^2\Phi(\vec{r}, t) = S(\vec{r}, t). \tag{5.78}$$

The diffusion equation does not depend on vector $\hat{s}$ and hence has 4 instead of 6 degrees of freedom; it can be used to solve for the fluence rate instead of the radiance. Note that the diffusion equation does not depend on μ_s and g independently but on their combination μ'_s. This degeneracy is referred to as the *similarity relation*, which is valid in the context of the diffusion approximation.

Two approximations are made in the derivation of the diffusion equation from the RTE: (1) the expansion of the radiance is limited to the first-order spherical harmonics and (2) the fractional change in the current density in one transport mean free path is much less than unity. The interpretation of the first approximation is that the radiance is nearly isotropic (omnidirectional) owing to directional broadening. The interpretation of the second approximation is that the photon current is temporally broadened relative to the transport mean free time. Both broadenings are caused by multiple scattering events. Consequently, these two approximations can be translated into a single condition $\mu'_s \gg \mu_a$, because all of the diffuse photons must have sustained a sufficient number of scattering events before being absorbed. In addition, we also require that the observation point be sufficiently far from sources and boundaries. However, boundary conditions can be applied to improve accuracy.

5.4.6. Impulse Responses in an Infinite Scattering Medium

For an infinitely short-pulsed point source, $S(\vec{r}, t) = \delta(\vec{r}, t)$, the solution to the diffusion equation [Eq. (5.78)] for $t > 0$ is

$$\Phi(\vec{r}, t) = \frac{c}{(4\pi Dct)^{3/2}} \exp\left(-\frac{r^2}{4Dct} - \mu_a ct\right), \quad (5.79)$$

which is an impulse response, also referred to as a *Green function*, in an infinite homogeneous scattering medium. The exponential decay $\exp(-\mu_a ct)$ actually represents a form of Beer's law with respect to time due to absorption, whereas the other terms represent broadening due to scattering. Note that Eq. (5.79) incorrectly predicts a nonzero fluence rate anywhere in space at time 0^+, which violates causality.

For an arbitrary infinitely short-pulsed point source located at $\vec{r}'$ and peaked at t', Eq. (5.78) becomes

$$\frac{\partial \Phi(\vec{r}, t; \vec{r}', t')}{c\partial t} + \mu_a \Phi(\vec{r}, t; \vec{r}', t') - D\nabla^2 \Phi(\vec{r}, t; \vec{r}', t') = \delta(\vec{r} - \vec{r}')\delta(t - t'), \quad (5.80)$$

which yields a new form of Green's function for $t > t'$:

$$\Phi(\vec{r}, t; \vec{r}', t') = \frac{c}{[4\pi Dc(t-t')]^{3/2}} \exp\left[-\frac{|\vec{r} - \vec{r}'|^2}{4Dc(t-t')} - \mu_a c(t-t')\right]. \quad (5.81)$$

We should note that this solution depends on the distance between source point $\vec{r}'$ and observation point $\vec{r}$ but that it is independent of the roles of the source and the detector, which indicates reciprocity. In other words, if the source and the observation points are exchanged, the solution remains the same. This is the well-known principle of reciprocity, which is applicable to many wave phenomena.

From the Green function, Green's theorem provides a solution for any arbitrary source in space and time, $S(\vec{r}', t')$:

$$\Phi(\vec{r}, t) = \int_0^t \int_0^\infty \Phi(\vec{r}, t; \vec{r}', t') S(\vec{r}', t') \, d\vec{r}' \, dt'. \quad (5.82)$$

The integral represents a superposition of impulse responses weighed by the source distribution; it is actually a convolution here because the Green function is translation-invariant.

In a time-independent state, Eq. (5.78) becomes

$$\Phi(\vec{r}) - \frac{1}{\mu_{\text{eff}}^2} \nabla^2 \Phi(\vec{r}) = \frac{S(\vec{r})}{\mu_a}. \quad (5.83)$$

Here, μ_{eff} denotes the effective attenuation coefficient:

$$\mu_{\text{eff}} = \sqrt{\frac{\mu_a}{D}} = \sqrt{3\mu_a(\mu_a + \mu_s')}. \quad (5.84)$$

For a time-independent point source, $S(\vec{r}) = \delta(\vec{r})$; the solution to Eq. (5.83) is

$$\Phi(\vec{r}) = \frac{1}{4\pi D r} \exp(-\mu_{\text{eff}} r), \tag{5.85}$$

which is a time-independent 3D impulse response or a Green function in an infinite homogeneous scattering medium.

For an infinitely broad isotropic source in a 1D case, $S(z) = \delta(z)$; Eq. (5.83) reduces to

$$\Phi_{1D}(z) - \frac{1}{\mu_{\text{eff}}^2} \frac{d^2 \Phi_{1D}(z)}{dz^2} = \frac{\delta(z)}{\mu_a}, \tag{5.86}$$

which yields the following solution in an infinite homogeneous scattering medium:

$$\Phi_{1D}(z) = \frac{\mu_{\text{eff}}}{2\mu_a} \exp(-\mu_{\text{eff}} |z|). \tag{5.87}$$

The $1/e$ decay constant in this equation is the penetration depth δ:

$$\delta = 1/\mu_{\text{eff}}. \tag{5.88}$$

Comparing Eq. (5.87) with the following Beer's law for a nonscattering medium

$$\Phi_{1D}(z) = \Phi(0) \exp(-\mu_a |z|), \tag{5.89}$$

we observe that the ratio $\mu_{\text{eff}}/\mu_a = \sqrt{3(\mu_a + \mu_s')/\mu_a}$ can be interpreted as the ratio of the "mean" photon path length in the scattering medium to the depth. However, if $\mu_s = 0$, we have $\mu_{\text{eff}} = \sqrt{3}\mu_a$, which implies erroneously that the "mean" photon path is greater than the depth even in a nonscattering medium. This breakdown of the diffusion theory occurs because condition $\mu_s' \gg \mu_a$ is not satisfied.

Example 5.5. Derive Eq. (5.85).

Solution 1. The following Fourier transformation pair is used:

$$\Psi(\vec{k}) = \int \Phi(\vec{r}) \exp(-i\vec{k} \cdot \vec{r}) \, d\vec{r}, \tag{5.90}$$

$$\Phi(\vec{r}) = \frac{1}{(2\pi)^3} \int \Psi(\vec{k}) \exp(i\vec{k} \cdot \vec{r}) \, d\vec{k}. \tag{5.91}$$

Since $S(\vec{r}) = \delta(\vec{r})$, Eq. (5.83) becomes

$$\Phi(\vec{r}) - \frac{1}{\mu_{\text{eff}}^2} \nabla^2 \Phi(\vec{r}) = \frac{\delta(\vec{r})}{\mu_a}. \tag{5.92}$$

Taking the Fourier transformation of this equation, we obtain

$$\Psi(\vec{k}) = \frac{1}{\mu_a(1 + k^2/\mu_{\text{eff}}^2)}. \tag{5.93}$$

Taking the inverse Fourier transformation of this equation yields Eq. (5.85):

$$\begin{aligned}
\Phi(\vec{r}) &= \frac{1}{(2\pi)^3 \mu_a} \int \frac{\exp(i\vec{k}\cdot\vec{r})}{1 + k^2/\mu_{\text{eff}}^2} d\vec{k} \\
&= \frac{1}{(2\pi)^2 \mu_a} \int_0^\infty \frac{k^2\, dk}{1 + k^2/\mu_{\text{eff}}^2} \int_0^\pi \exp(ikr\cos\theta) \sin\theta\, d\theta \\
&= \frac{-1}{(2\pi)^2 i\mu_a r} \int_0^\infty \frac{k\, dk}{1 + k^2/\mu_{\text{eff}}^2} \exp(ikr\cos\theta)\big|_0^\pi \\
&= \frac{2}{(2\pi)^2 \mu_a r} \int_0^\infty \frac{k\sin(kr)\, dk}{1 + k^2/\mu_{\text{eff}}^2} \\
&= \frac{\mu_{\text{eff}}^2 \exp(-\mu_{\text{eff}} r)}{4\pi \mu_a r} \\
&= \frac{1}{4\pi D r} \exp(-\mu_{\text{eff}} r).
\end{aligned} \tag{5.94}$$

Solution 2. The general solution to Eq. (5.92) is from the following homogeneous equation:

$$\Phi(\vec{r}) - \frac{1}{\mu_{\text{eff}}^2} \nabla^2 \Phi(\vec{r}) = 0. \tag{5.95}$$

Since $\Phi(\vec{r})$ is independent of θ and ϕ, Eq. (5.95) can be rewritten as

$$r^2 \frac{d^2\Phi(r)}{dr^2} + 2r\frac{d\Phi(r)}{dr} - \mu_{\text{eff}}^2 r^2 \Phi(r) = 0. \tag{5.96}$$

This is a transformed Bessel equation that has the following solution after the imaginary part of the solution is discarded

$$\Phi(r) = \frac{C}{\mu_{\text{eff}} r} \exp(-\mu_{\text{eff}} r), \tag{5.97}$$

where C is a constant. Substituting Eq. (5.97) into Eq. (5.92) yields $C = \mu_{\text{eff}}/(4\pi D)$, and thus Eq. (5.85) is derived.

Example 5.6. Derive Eq. (5.87).

Solution 1. The general solution of Eq. (5.86) is $\Phi_{1D}(z) = C \exp(-\mu_{eff}|z|)$. Integrating Eq. (5.86) with respect to z from $-\infty$ to ∞ gives $\int_{-\infty}^{\infty} \Phi_{1D}(z)\,dz = 1/\mu_a$ since

$$\int_{-\infty}^{\infty} \frac{d^2 \Phi_{1D}(z)}{dz^2}\,dz = 0$$

and $\int_{-\infty}^{\infty} \delta(z)\,dz = 1$. Integrating the general solution gives

$$\int_{-\infty}^{\infty} \Phi_{1D}(z)\,dz = \frac{2C}{\mu_{eff}}.$$

Therefore, $C = (\mu_{eff}/2\mu_a)$ and Eq. (5.87) is derived.

Solution 2. Integrating the 3D solution given by Eq. (5.85) over the source plane yields

$$\Phi_{1D}(z) = \int_0^{\infty} \Phi(r) 2\pi \rho\,d\rho = \int_z^{\infty} \Phi(r) 2\pi r\,dr, \tag{5.98}$$

where $z^2 + \rho^2 = r^2$. Completing the integration yields Eq. (5.87).

5.5. BOUNDARY CONDITIONS

5.5.1. Refractive-Index-Matched Boundary

If a nonscattering ambient medium and a scattering medium have the same index of refraction, the interface between them is referred to as a *refractive-index-matched boundary*. For example, an interface between water and soft tissue is approximately refractive-index-matched. At this kind of boundary, no light propagates into the scattering medium from the ambient medium (Figure 5.4). This boundary condition is mathematically expressed as

$$L(\vec{r}, \hat{s}, t) = 0 \quad \text{for} \quad \hat{s} \cdot \hat{n} > 0, \tag{5.99}$$

where $\vec{r}$ denotes a point on the boundary and $\hat{n}$ denotes the unit normal vector of the interface pointing into the scattering medium. If the z axis is defined along $\hat{n}$, we have $\hat{s} \cdot \hat{n} = \cos\theta$, where θ is the polar angle of $\hat{s}$. Because the radiance is nonnegative, an equivalent boundary condition can be expressed as

$$\int_{\hat{s}\cdot\hat{n}>0} L(\vec{r}, \hat{s}, t) \hat{s} \cdot \hat{n}\,d\Omega = 0, \tag{5.100}$$

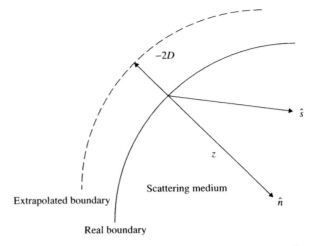

Figure 5.4. Schematic representation of the boundary condition.

which means that the direction-integrated radiance toward the scattering medium is zero.

In the diffusion approximation, the boundary condition becomes

$$\Phi(\vec{r}, t) + 2\vec{J}(\vec{r}, t) \cdot \hat{n} = 0. \tag{5.101}$$

Substituting Fick's law [Eq. (5.74)] into Eq. (5.101), we obtain

$$\Phi(\vec{r}, t) - 2D\nabla\Phi(\vec{r}, t) \cdot \hat{n} = 0 \tag{5.102}$$

or

$$\Phi(\vec{r}, t) - 2D\frac{\partial \Phi(\vec{r}, t)}{\partial z} = 0. \tag{5.103}$$

This boundary condition mathematically falls into the category of homogeneous Cauchy boundary conditions because a linear combination of the fluence rate and its normal derivative on the boundary is zero.

Using the Taylor expansion to the first order, we obtain

$$\Phi(z = -2D, t) = \Phi(z = 0, t) - 2D \left. \frac{\partial \Phi(\vec{r}, t)}{\partial z} \right|_{z=0} = 0, \tag{5.104}$$

which means that the fluence rate at $z = -2D$ is approximately zero. On an extrapolated boundary at

$$z_b = -2D, \tag{5.105}$$

the fluence rate is approximately zero. This boundary condition mathematically falls into the category of homogeneous Dirichlet boundary conditions because the fluence rate on the boundary is zero.

Example 5.7. Derive Eq. (5.101).

Substituting the diffusion expansion of the radiance [Eq. (5.36)] into Eq. (5.100), we obtain

$$\int_{\hat{s}\cdot\hat{n}>0} L(\vec{r},\hat{s},t)\hat{s}\cdot\hat{n}\,d\Omega = \int_{\hat{s}\cdot\hat{n}>0}\left[\frac{1}{4\pi}\Phi(\vec{r},t) + \frac{3}{4\pi}\vec{J}(\vec{r},t)\cdot\hat{s}\right]\hat{s}\cdot\hat{n}\,d\Omega$$

$$= \frac{1}{4\pi}\Phi(\vec{r},t)\int_{\hat{s}\cdot\hat{n}>0}\hat{s}\cdot\hat{n}\,d\Omega + \frac{3}{4\pi} \quad (5.106)$$

$$\times \int_{\hat{s}\cdot\hat{n}>0}[\vec{J}(\vec{r},t)\cdot\hat{s}]\hat{s}\cdot\hat{n}\,d\Omega = 0.$$

For a smooth boundary, the first integral on the right-hand side equals π. The second integral can be evaluated as follows:

$$\int_{\hat{s}\cdot\hat{n}>0}[\vec{J}(\vec{r},t)\cdot\hat{s}]\hat{s}\cdot\hat{n}\,d\Omega = \int_0^{\pi/2}\int_0^{2\pi}[J_x(\vec{r},t)\sin\theta\cos\phi + J_y(\vec{r},t)\sin\theta\sin\phi$$

$$+ J_z(\vec{r},t)\cos\theta]\sin\theta\cos\theta\,d\phi\,d\theta \quad (5.107)$$

$$= 2\pi\int_0^{\pi/2}J_z(\vec{r},t)\cos^2\theta\sin\theta\,d\theta$$

$$= \frac{2\pi}{3}J_z(\vec{r},t) = \frac{2\pi}{3}\vec{J}(\vec{r},t)\cdot\hat{n}.$$

Therefore

$$\int_{\hat{s}\cdot\hat{n}>0} L(\vec{r},\hat{s},t)\hat{s}\cdot\hat{n}\,d\Omega = \frac{1}{4}\Phi(\vec{r},t) + \frac{1}{2}\vec{J}(\vec{r},t)\cdot\hat{n} = 0, \quad (5.108)$$

which leads to Eq. (5.101).

5.5.2. Refractive-Index-Mismatched Boundary

When the ambient and scattering media have different indices of refraction, the interface between them is referred to as a *refractive-index-mismatched boundary*. For example, an interface between air and soft tissue is refractive-index-mismatched. In this case, the boundary condition is modified as follows owing to the Fresnel reflections

$$\Phi(\vec{r},t) - 2C_R D\nabla\Phi(\vec{r},t)\cdot\hat{n} = 0 \quad (5.109)$$

or

$$\Phi(\vec{r}, t) - 2C_R D \frac{\partial \Phi(\vec{r}, t)}{\partial z} = 0, \qquad (5.110)$$

where

$$C_R = \frac{1 + R_{\text{eff}}}{1 - R_{\text{eff}}}. \qquad (5.111)$$

The effective reflection coefficient R_{eff} represents the percentage of the outgoing radiance integrated over all directions pointing toward the ambient medium that is converted to incoming radiance integrated over all directions pointing toward the scattering medium. R_{eff} can be calculated as follows:

$$R_{\text{eff}} = \frac{R_\Phi + R_J}{2 - R_\Phi + R_J}, \qquad (5.112)$$

where

$$R_\Phi = \int_0^{\pi/2} 2 \sin\theta \cos\theta R_F(\cos\theta) \, d\theta, \qquad (5.113)$$

$$R_J = \int_0^{\pi/2} 3 \sin\theta (\cos\theta)^2 R_F(\cos\theta) \, d\theta, \qquad (5.114)$$

$$R_F(\cos\theta) = \frac{1}{2}\left(\frac{n_{\text{rel}}\cos\theta' - \cos\theta}{n_{\text{rel}}\cos\theta' + \cos\theta}\right)^2 + \frac{1}{2}\left(\frac{n_{\text{rel}}\cos\theta - \cos\theta'}{n_{\text{rel}}\cos\theta + \cos\theta'}\right)^2 \text{ for } 0 \le \theta < \theta_c, \qquad (5.115)$$

$$R_F(\cos\theta) = 1 \text{ for } \theta_c \le \theta \le \frac{\pi}{2}. \qquad (5.116)$$

The angle of incidence is determined by

$$\theta = \cos^{-1}(\hat{s} \cdot \hat{n}). \qquad (5.117)$$

The angle of refraction is determined by Snell's law as

$$\theta' = \sin^{-1}(n_{\text{rel}} \sin\theta), \qquad (5.118)$$

where the relative refractive index n_{rel} is the ratio of the refractive index of the scattering medium to that of the ambient medium. The critical angle is given by

$$\theta_c = \sin^{-1} \frac{1}{n_{\text{rel}}}. \qquad (5.119)$$

Likewise, the distance between the extrapolated boundary and the actual boundary is modified to

$$z_b = -2C_R D. \tag{5.120}$$

Example 5.8. Derive Eq. (5.109).

For a refractive-index-mismatched boundary, we have

$$\int_{\hat{s}\cdot\hat{n}>0} L(\vec{r},\hat{s},t)\hat{s}\cdot\hat{n}\,d\Omega = \int_{\hat{s}\cdot\hat{n}<0} R_F(\hat{s}\cdot\hat{n})L(\vec{r},\hat{s},t)\hat{s}\cdot\hat{n}\,d\Omega, \tag{5.121}$$

where $\hat{s}\cdot\hat{n} = \cos\theta$ and the Fresnel reflection R_F of the light—presumed to be unpolarized—at the boundary is given by Eqs. (5.115) and (5.116).

We define an effective reflection coefficient as

$$R_{\text{eff}} = \frac{\int_{\hat{s}\cdot\hat{n}<0} R_F(\hat{s}\cdot\hat{n})L(\vec{r},\hat{s},t)\hat{s}\cdot\hat{n}\,d\Omega}{\int_{\hat{s}\cdot\hat{n}<0} L(\vec{r},\hat{s},t)\hat{s}\cdot\hat{n}\,d\Omega}. \tag{5.122}$$

As in Example 5.7, R_{eff} is evaluated in the diffusive regime using the diffusion expansion of radiance [Eq. (5.36)]

$$R_{\text{eff}} = \frac{\frac{1}{4}R_\Phi \Phi(\vec{r},t) - R_J \frac{1}{2}\vec{J}(\vec{r},t)\cdot\hat{n}}{\frac{1}{4}\Phi(\vec{r},t) - \frac{1}{2}\vec{J}(\vec{r},t)\cdot\hat{n}}, \tag{5.123}$$

where

$$R_\Phi = \int_0^{\pi/2} 2\sin\theta\cos\theta R_F(\cos\theta)\,d\theta, \tag{5.124}$$

$$R_J = \int_0^{\pi/2} 3\sin\theta(\cos\theta)^2 R_F(\cos\theta)\,d\theta. \tag{5.125}$$

Similarly, boundary condition Eq. (5.121) leads to

$$\frac{1}{4}\Phi(\vec{r},t) + \frac{1}{2}\vec{J}(\vec{r},t)\cdot\hat{n} = \frac{1}{4}R_\Phi\Phi(\vec{r},t) - R_J\frac{1}{2}\vec{J}(\vec{r},t)\cdot\hat{n} \tag{5.126}$$

in the diffusive regime.

Merging Eq. (5.126) and Eq. (5.123) yields

$$R_{\text{eff}} = \frac{R_\Phi + R_J}{2 - R_\Phi + R_J}, \tag{5.127}$$

which can be solved numerically. Fitting this equation can provide an empirical formula for R_{eff}.

Substituting Eq. (5.122) into Eq. (5.121) yields

$$\int_{\hat{s}\cdot\hat{n}>0} L(\vec{r},\hat{s},t)\hat{s}\cdot\hat{n}\,d\Omega = R_{\text{eff}}\int_{\hat{s}\cdot\hat{n}<0} L(\vec{r},\hat{s},t)\hat{s}\cdot\hat{n}\,d\Omega. \qquad (5.128)$$

In the diffusive regime, this equation can be rewritten as

$$\frac{1}{4}\Phi(\vec{r},t) + \frac{1}{2}\vec{J}(\vec{r},t)\cdot\hat{n} = \frac{1}{4}R_{\text{eff}}\Phi(\vec{r},t) - R_{\text{eff}}\frac{1}{2}\vec{J}(\vec{r},t)\cdot\hat{n}. \qquad (5.129)$$

Substituting Fick's law [Eq. (5.74)] into this equation yields boundary condition Eq. (5.109).

5.6. DIFFUSE REFLECTANCE

Measured diffuse reflectance can be used, for example, to determine the optical properties of biological tissue noninvasively. The *relative diffuse reflectance* (or simply *diffuse reflectance*) is defined here as the probability of photon reemission per unit surface area from a scattering medium. Although the Monte Carlo method can predict diffuse reflectance accurately, it is computationally intensive. In particular, when the absorption coefficient is much less than the scattering coefficient, photons may propagate over long distances before being absorbed. Fortunately, the diffusion theory offers an alternative rapid approach although it is inaccurate near the light source.

The task here is to compute the diffuse reflectance in response to an infinitely narrow photon beam (a pencil beam) normally incident on a semiinfinite homogeneous scattering medium that has a refractive-index-matched boundary (Figure 5.5a). The problem is solved below by the diffusion theory along with the boundary condition. Key factors that affect the accuracy of the diffusion theory are also discussed.

5.6.1. Steps of Approximation

Three steps of approximation are involved in the solution (Figure 5.5): (1) the anisotropically scattering medium (Figure 5.5a) is converted into an isotropically scattering medium (Figure 5.5b), based on the similarity relation; (2) the unit-power pencil beam is converted into an equivalent isotropic point source at $z = l'_t$ with a power equal to transport albedo a' (Figure 5.5c); see also Problem 5.3; and (3) the surface of the scattering medium is removed after an image source is added above the surface at $z = -(l'_t + 2z_b)$ to satisfy the boundary condition.

An image point source is mirror-symmetric with the original point source about the extrapolated boundary at $z = -z_b$ [Eq. (5.105)]; it is added to satisfy the boundary condition so that the original single source in a semiinfinite medium can be converted into double sources in an infinite medium. The response to a single source in a semiinfinite medium (Figure 5.5c) can be approximated by a

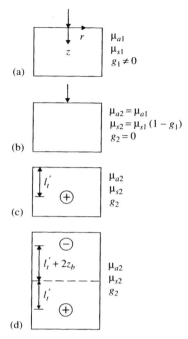

Figure 5.5. Illustrations of the steps of approximation (the boxes represent the scattering media): (a) a pencil beam incident on the original scattering medium with absorption coefficient μ_{a1}, scattering coefficient μ_{s1}, and nonzero anisotropy g_1; (b) a pencil beam incident on an isotropically scattering medium with $\mu_{a2} = \mu_{a1}$, $\mu_{s2} = \mu_{s1}(1 - g_1)$, and $g_2 = 0$; (c) an isotropic point source under the surface of the isotropically scattering medium; (d) an image point source added to approximately satisfy the boundary condition—with this addition, the physical boundary (dashed line) is removed (circled signs indicate the polarities of the sources).

superposition of the two responses to each of the double sources in an infinite medium (Figure 5.5d). The latter problem can be solved easily because it is free of boundaries. This approach is akin to the common practice of solving electrostatic problems with a zero-potential conducting boundary. Therefore, instead of dealing with a pencil beam incident on a semiinfinite anisotropically scattering medium (Figure 5.5a), we deal with two isotropic point sources in an infinite scattering medium (Figure 5.5d).

5.6.2. Formulation

A cylindrical coordinate system (r, θ, z) is set up. The origin of the coordinate system is the point of light incidence on the surface of the scattering medium, and the z axis is along the pencil beam.

The fluence rate that is generated by a unit-power point source in an infinite scattering medium is described by Eq. (5.85) and is rewritten as follows in the cylindrical coordinates

$$\Phi_\infty(r, \theta, z; r', \theta', z') = \frac{1}{4\pi D\rho} \exp(-\mu_{\text{eff}}\rho), \tag{5.130}$$

where ρ is the distance between observation point (r, θ, z) and source point (r', θ', z'):

$$\rho = \sqrt{r^2 + r'^2 - 2rr'\cos(\theta - \theta') + (z - z')^2}. \tag{5.131}$$

A linear combination of the solutions for each of the two isotropic sources in Figure 5.5d, according to Eq. (5.130), yields approximately the fluence rate in response to the original isotropic point source in the original semiinfinite scattering medium:

$$\Phi(r, \theta, z; r', \theta', z') = a'\Phi_\infty(r, \theta, z; r', \theta', z') - a'\Phi_\infty(r, \theta, z; r', \theta', -z' - 2z_b), \tag{5.132}$$

where $z' = l'_t$ and a' denotes the transport albedo.

According to Fick's law, the diffuse reflectance from the semiinfinite scattering medium is approximately the current density projected to the surface normal:

$$R_d(r) = D \left.\frac{\partial \Phi}{\partial z}\right|_{z=0}. \tag{5.133}$$

Substituting Eq. (5.132) into Eq. (5.133), we obtain

$$R_d(r) = a'\frac{z'(1 + \mu_{\text{eff}}\rho_1)\exp(-\mu_{\text{eff}}\rho_1)}{4\pi\rho_1^3} + a'\frac{(z' + 4D)(1 + \mu_{\text{eff}}\rho_2)\exp(-\mu_{\text{eff}}\rho_2)}{4\pi\rho_2^3}. \tag{5.134}$$

Here, ρ_1 is the distance between observation point $(r, 0, 0)$ and original source point $(0, 0, z')$ and ρ_2 is the distance between observation point $(r, 0, 0)$ and image source point $(0, 0, -z' - 2z_b)$.

Example 5.9. Derive Eq. (5.134).

From Eqs. (5.130) and (5.131), we derive

$$\frac{\partial \Phi_\infty}{\partial \rho} = -\frac{1}{4\pi D}\frac{1 + \mu_{\text{eff}}\rho}{\rho^2}\exp(-\mu_{\text{eff}}\rho), \tag{5.135}$$

$$\frac{\partial \rho}{\partial z} = \frac{z - z'}{\rho}. \tag{5.136}$$

Therefore

$$\left.\frac{\partial \Phi_\infty(r, \theta, z; r', \theta', z')}{\partial z}\right|_{z=0} = \left.\frac{\partial \Phi_\infty}{\partial \rho}\frac{\partial \rho}{\partial z}\right|_{z=0} = \frac{z'}{4\pi D}\frac{1 + \mu_{\text{eff}}\rho_1}{\rho_1^3}\exp(-\mu_{\text{eff}}\rho_1). \tag{5.137}$$

Likewise, we have

$$\left.\frac{\partial \Phi_\infty(r, \theta, z; r', \theta', -z' - 2z_b)}{\partial z}\right|_{z=0} = -\frac{z' + 4D}{4\pi D} \frac{1 + \mu_{\text{eff}}\rho_2}{\rho_2^3} \exp(-\mu_{\text{eff}}\rho_2). \tag{5.138}$$

Combining Eqs. (5.137) and (5.138) leads to Eq. (5.134).

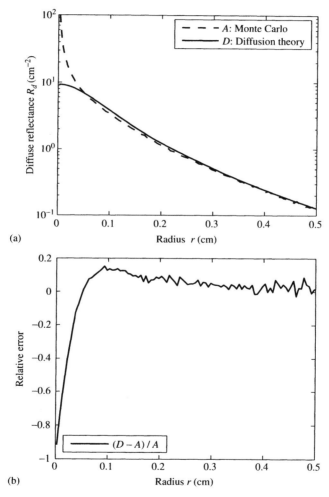

Figure 5.6. (a) Diffuse reflectance in response to a pencil beam incident on a semi-infinite scattering medium. Curve A is from the Monte Carlo simulation for the case in Figure 5.5a. Curve D is from the diffusion theory for the case in Figure 5.5d. (b) Relative error between the two curves in part (a), which is the difference between curves D and A divided by curve A point-by-point.

5.6.3. Validation of Diffusion Theory

In this section, we evaluate each step of the approximation described above using the accurate Monte Carlo method. The following optical properties are used: $n_{rel} = 1$, $\mu_{a1} = 0.1$ cm^{-1}, $\mu_{s1} = 100$ cm^{-1}, and $g_1 = 0.9$. As shown in Figure 5.6, the diffuse reflectance $R_d(r)$ from the diffusion theory is accurate only when r is greater than $\sim l'_t$ ($l'_t = 0.1$ cm here).

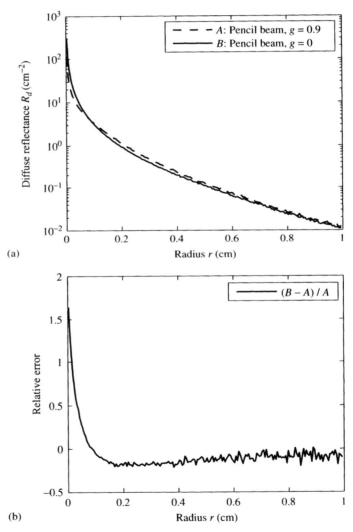

Figure 5.7. (a) Comparison between the diffuse reflectance distributions from the anisotropically (see Figure 5.5a) and isotropically (see Figure 5.5b) scattering media calculated using the Monte Carlo method; (b) relative error versus r.

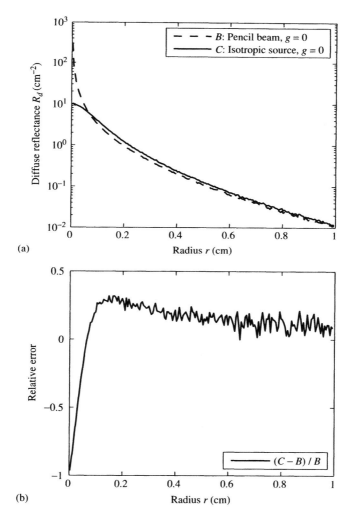

Figure 5.8. (a) Comparison between the diffuse reflectance distributions from an isotropically scattering medium in response to a pencil beam (see Figure 5.5b) and an isotropic point source (see Figure 5.5c) calculated using the Monte Carlo method; (b) relative error versus r.

Deviations caused by each step of the approximation are illustrated in Figures 5.7–5.9. Curves A, B, and C are from the Monte Carlo method, whereas curve D is from the diffusion theory; curves A–D are associated with parts (a)–(d) in Figure 5.5.

The error due to the approximation of Figure 5.5a with 5.5b is shown in Figure 5.7. The scattering anisotropy is converted from $g = 0.9$ to $g = 0$ while

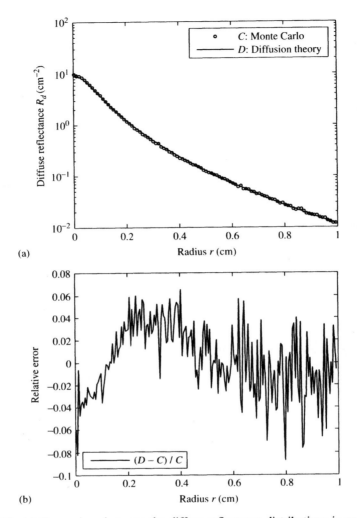

Figure 5.9. (a) Comparison between the diffuse reflectance distributions in response to an isotropic point source in a semiinfinite scattering medium (see Figure 5.5c) and a pair of isotropic point sources in an infinite scattering medium (see Figure 5.5d). Curves C and D are from the Monte Carlo method and the diffusion theory, respectively. (b) Relative error versus r.

μ'_s is held constant. The relative error decreases with increasing r; it is >100% near $r = 0$ and ~20% near $r = 2l'_t = 0.2$ cm.

The error due to the approximation of Figure 5.5b with 5.5c is shown in Figure 5.8. This pencil beam is converted to a single isotropic point source at $z = l'_t = 0.1$ cm. Such a conversion causes a severe underestimation of $R_d(r)$ near $r = 0$.

The error due to the approximation of Figure 5.5c to 5.5d is shown in Figure 5.9. Curves C and D are calculated by the Monte Carlo method and the diffusion theory, respectively; they show relatively small systematic differences.

Although the diffusion theory is acceptable when the isotropic point source is far from the surface of the scattering medium as demonstrated in Figure 5.9, it becomes less accurate as the source approaches the surface (Figure 5.10). To demonstrate this point, we compare the results from the Monte Carlo method and

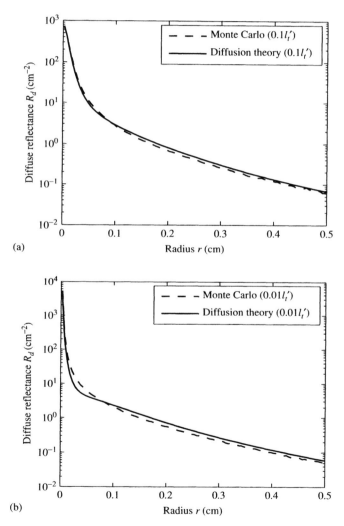

Figure 5.10. Comparison between the diffuse reflectance distributions from the Monte Carlo method and the diffusion theory. An isotropic light source is placed at (a) $z = 0.1 l_t'$ and then (b) $z = 0.01 l_t'$ in an isotropically scattering semiinfinite medium.

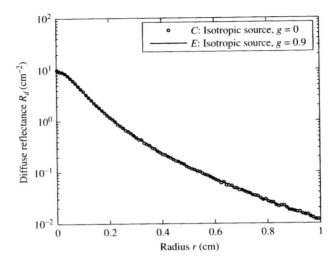

Figure 5.11. Comparison between the diffuse reflectance distributions from the Monte Carlo method in response to an isotropic point source placed at $z = l'_t$ in two scattering media whose optical properties are related by the similarity relation. For curve C, $g = 0$; for curve E, $g = 0.9$; for both, $n_{\text{rel}} = 1$, $\mu_a = 0.1$ cm^{-1}, and $\mu'_s = 10$ cm^{-1}.

the diffusion theory for the configurations in Figures 5.5c and 5.5d, respectively. The point source, however, is placed at $z = 0.1 l'_t$ and then at $z = 0.01 l'_t$ instead of at $z = l'_t$. As expected from the diffusion theory, data for $z = 0.01 l'_t$ are less accurate than those for $z = 0.1 l'_t$.

Although the conversion from Figure 5.5a to 5.5b introduces considerable error in $R_d(r)$ near the source as shown in Figure 5.7, it is acceptable if the photons originate isotropically deep inside the scattering medium, as demonstrated below. In response to an isotropic point source at $z = l'_t$, the diffuse reflectance distributions from an isotropically scattering medium (as in Figure 5.5c) and an anisotropically scattering medium are computed by the Monte Carlo method; they are approximately equal to each other (Figure 5.11).

5.7. PHOTON PROPAGATION REGIMES

The cumulative effect of photon scattering by a medium can be loosely classified into four regimes. The term *ballistic regime* refers to photons that have undergone no scattering; *quasiballistic regime* refers to photons that have sustained a few scattering events but retain a strong memory of the original incidence direction. The term *quasidiffusive regime* refers to photons that have sustained many scattering events and retain only a weak memory of the original incidence direction; *diffusive regime* refers to photons that have suffered many scattering events that they have almost completely lost their memory of the original incidence direction.

In addition, *nonballistic photons* are those that have deviated from the ballistic path. The propagation regimes can be approximately related to the propagation time t through the mean free path l_t and the transport mean free path l'_t.

On the basis of Beer's law, the probability of no scattering for a photon decays with time t is as follows:

$$P(ct) = \exp\left(-\frac{ct}{l_t}\right). \tag{5.139}$$

Accordingly, we define the *ballistic regime* to cover $ct \leq l_t$, within which the probability of no scattering is $P(ct) \geq \exp(-1) = 37\%$. We define the *quasiballistic regime* to cover $l_t < ct \leq l'_t$, within which the probability of no scattering falls between $\exp(-l'_t/l_t)$ and $\exp(-1) : \exp(-1) > P(ct) \geq \exp(-l'_t/l_t)$.

When a pencil beam is incident within an infinite scattering medium, the photons spread into a photon cloud. From cumulant expansion, it is found that the center of the photon cloud approaches l'_t according to

$$l'_t - z_c = l'_t \exp\left(-\frac{ct}{l'_t}\right), \tag{5.140}$$

where z_c is the distance between the weighted center of the photon cloud and the point of incidence. We define a new constant as

$$\varepsilon_r = \frac{l'_t - z_c}{l'_t}. \tag{5.141}$$

Thus, we have

$$\varepsilon_r = \exp\left(-\frac{ct}{l'_t}\right). \tag{5.142}$$

Accordingly, we define the *quasidiffusive regime* to cover $l'_t < ct \leq 10 l'_t$, within which we have $\exp(-1) > \varepsilon_r \geq \exp(-10) = 4.5 \times 10^{-5}$. We define the *diffusive regime* to cover $ct > 10 l'_t$, within which we have $\varepsilon_r \leq \exp(-10)$. If $l_t = 0.1$ mm and $l'_t = 1$ mm, the four scattering regimes are divided at path lengths of 0.1, 1, and 10 mm. The classification holds in scattering dominant media.

If $\mu_a \ll \mu'_s$, the mean number of scattering events that photons experience within the dividing path lengths (N_s) can be estimated. Within l_t (the ballistic regime), $N_s \leq 1$ holds. Within l'_t (the quasiballistic regime), we have

$$N_s \leq \frac{l'_t}{l_t} \approx \frac{1}{1-g}, \tag{5.143}$$

which equals 10, for example, if $g = 0.9$. Likewise, within $10 l'_t$ (the quasidiffusive regime), we have

$$N_s \leq \frac{10 l'_t}{l_t} \approx \frac{10}{1-g}, \tag{5.144}$$

which equals 100, for example, if $g = 0.9$.

PROBLEMS

5.1 Show that $\int_{4\pi} \hat{s}(\hat{s} \cdot \vec{A}) d\Omega = (4\pi/3)\vec{A}$, where $\vec{A}$ is independent of $\hat{s}$.

5.2 Show that $\int_{4\pi} \hat{s}[\hat{s} \cdot \nabla(\vec{J} \cdot \hat{s})] d\Omega = 0$.

5.3 Show that a pencil beam normally incident on a semiinfinite medium can be approximated by an isotropic source placed one transport mean free path below the surface. Extend to a 1D case, namely, an infinitely broad beam normally incident on a semiinfinite medium. Explain why this is an approximation.

5.4 Verify Eq. (5.85) using a Monte Carlo simulation for $\mu_a = 0.1$ cm^{-1}, $\mu_s = 100$ cm^{-1}, and $g = 0.9$. (*Hint*: Use spherical coordinates to record photon absorption.)

5.5 Duplicate Figures 5.6–5.11.

5.6 Derive the average number of scattering events in one transport mean free path given $\mu_a = 0$.

5.7 Derive Eq. (5.10).

5.8 Show that $P(\hat{s}' \cdot \hat{s}) = p(\cos\theta)/2\pi$, where $p(\cos\theta)$ is defined as in Chapter 3.

5.9 Plot Eq. (5.36) in polar coordinates as a function of α, where $\vec{J}(\vec{r}, t) \cdot \hat{s} = |\vec{J}(\vec{r}, t)|\cos\alpha$. Set $\Phi(\vec{r}, t) = 1$ and plot for $3|\vec{J}(\vec{r}, t)| = 3, 1, 0.3, 0.1$, and 0.03.

5.10 Plot the Henyey–Greenstein phase function in polar coordinates as a function of θ for $g = 0, 0.1, 0.5, 0.9$, and 0.99.

5.11 (a) Using the Monte Carlo method, compute $L(\vec{r}, \hat{s}, t)$ integrated over time t as a function of the polar angle θ at $z = (0.1, 0.5, 1.0, 2.0)l_t'$ below a pencil beam in an infinite medium, where $\mu_a = 0.1$ cm^{-1}, $\mu_s = 100$ cm^{-1}, and $g = 0.9$. Plot the result in polar coordinates. (b) Use the least-squares fitting algorithm available in MATLAB to fit the derived distributions to $a + b\cos\theta$. List b/a versus z in a table.

5.12 Modify the Monte Carlo code written for Chapter 3 to compute and plot the specific absorption distributions on the z axis in response to a pencil beam in two infinite scattering media of $g = 0.9$ and $g = 0$. Both media have $\mu_a = 0.1$ cm^{-1} and $\mu_s' = 10$ cm^{-1}. The range of z should cover several transport mean free paths.

5.13 Integrate Eq. (5.79) over the entire space and explain the result. Then, set $\mu_a = 0$ and explain the result.

5.14 Integrate Eq. (5.79) over time from 0 to $+\infty$ and explain the result.

5.15 Using a Monte Carlo program, compute and plot the 1D depth-resolved fluence rate as a function of z in response to a pencil beam incident normally on a semiinfinite scattering medium. Fit the curve for μ_{eff} and compare with the value predicted by the diffusion theory. Compare the depth of the peak fluence rate with l'_t.

5.16 Derive Eq. (5.85) from Eq. (5.83) using an alternative method.

5.17 Derive Eq. (5.79) using the Fourier transformation.

5.18 Assuming that the absorption coefficient is zero, from Eq. (5.85), derive the current density and explain the conservation of energy.

5.19 (a) The phase function $P(\hat{s}' \cdot \hat{s})$ is highly forward-directed in biological tissue. Explain why it is not expanded in spherical harmonics in the derivation of the diffusion theory. (b) Explain that since $P(\hat{s}' \cdot \hat{s})$ is azimuthally symmetric about $\hat{s}$, $\int_{4\pi} \hat{s} P(\hat{s}' \cdot \hat{s}) d\Omega$ is parallel with $\hat{s}'$.

5.20 One approximation in the diffusion theory is that the fractional change in the current density in one transport mean free path is much less than unity. Explain why this approximation can be translated to the statement that the reduced scattering coefficient must be much greater than the absorption coefficient.

5.21 The diffusion equation derived in this chapter does not conform to the postulate of causality. If a second-order temporal wave equation term is added, this problem can be corrected. The new equation is referred to as the telegraphy equation:

$$\frac{\partial \Phi(\vec{r}, t)}{c \partial t} + \mu_a \Phi(\vec{r}, t) - \nabla \cdot [D \nabla \Phi(\vec{r}, t)] + 3D \frac{\partial \Phi^2(\vec{r}, t)}{c^2 \partial t^2} = S(\vec{r}, t).$$

Derive this equation.

5.22 Derive the RTE by considering a differential area that moves along photon propagation direction $\hat{s}$. $\{Hint: (dL/ds) = (\partial L/\partial s) + [(\partial L/\partial t)(\partial t/\partial s)].\}$

READING

Boas DA (1996): *Diffuse Photon Probes of Structural and Dynamical Properties of Scattering Media*, Ph.D. dissertation, Univ. Pennsylvania, Philadelphia. (See Sections 5.2–5.5, above.)

Cai W, Lax M, and Alfano RR (2000): Cumulant solution of the elastic Boltzmann transport equation in an infinite uniform medium, *Phys. Rev. E* **61**(4): 3871–3876. (See Section 5.7, above.)

Haskell RC, Svaasand LO, Tsay TT, Feng TC, and Mcadams MS (1994): Boundary-conditions for the diffusion equation in radiative-transfer, *J. Opt. Soc. Am. A* **11**(10): 2727–2741. (See Section 5.5, above.)

Ishimaru A (1978): *Wave Propagation and Scattering in Random Media*, Academic Press, New York. (See Sections 5.2–5.5, above.)

Wang LHV and Jacques SL (2000): Source of error in calculation of optical diffuse reflectance from turbid media using diffusion theory, *Comput. Meth. Prog. Biomed.* **61**(3): 163–170. (See Section 5.6, above.)

FURTHER READING

Aronson R (1995): Boundary-conditions for diffusion of light, *J. Opt. Soc. Am. A* **12**(11): 2532–2539.

Case KM and Zweifel PF (1967): *Linear Transport Theory*, Addison-Wesley, Reading, MA.

Chandrasekhar S (1960): *Radiative Transfer*, Dover Publications, New York.

Cheong WF, Prahl SA, and Welch AJ (1990): A review of the optical-properties of biological tissues, *IEEE J. Quantum Electron.* **26**(12): 2166–2185.

Faris GW (2002): Diffusion equation boundary conditions for the interface between turbid media: A comment, *J. Opt. Soc. Am. A* **19**(3): 519–520.

Farrell TJ, Patterson MS, and Wilson B (1992): A diffusion-theory model of spatially resolved, steady-state diffuse reflectance for the noninvasive determination of tissue optical properties in vivo, *Med. Phys.* **19**(4): 879–888.

Flock ST, Wilson BC, and Patterson MS (1989): Monte-Carlo modeling of light-propagation in highly scattering tissues. 2. comparison with measurements in phantoms, *IEEE Trans. Biomed. Eng.* **36**(12): 1169–1173.

Groenhuis RAJ, Ferwerda HA, and Tenbosch JJ (1983): Scattering and absorption of turbid materials determined from reflection measurements. 1. Theory, *Appl. Opt.* **22**(16): 2456–2462.

Keijzer M, Star WM, and Storchi PRM (1988): Optical diffusion in layered media, *Appl. Opt.* **27**(9): 1820–1824.

Kienle A, Patterson MS, Dognitz N, Bays R, Wagnieres G, and van den Bergh H (1998): Noninvasive determination of the optical properties of two-layered turbid media, *Appl. Opt.* **37**(4): 779–791.

Markel VA and Schotland JC (2002): Inverse problem in optical diffusion tomography. II. Role of boundary conditions, *J. Opt. Soc. Am. A* **19**(3): 558–566.

Shafirstein G, Baumler W, Lapidoth M, Ferguson S, North PE, and Waner M (2004): A new mathematical approach to the diffusion approximation theory for selective photothermolysis modeling and its implication in laser treatment of port-wine stains, *Lasers Surg. Med.* **34**(4): 335–347.

Wyman DR, Patterson MS, and Wilson BC (1989): Similarity relations for the interaction parameters in radiation transport, *Appl. Opt.* **28**(24): 5243–5249.

You JS, Hayakawa CK, and Venugopalan V (2005): Frequency domain photon migration in the delta-P-1 approximation: Analysis of ballistic, transport, and diffuse regimes, *Phys. Rev. E* **72**(2): 021903.

CHAPTER 6

Hybrid Model of Monte Carlo Method and Diffusion Theory

6.1. INTRODUCTION

The Monte Carlo method and the diffusion theory have complementary attributes for modeling photon transport in a scattering medium. The Monte Carlo method is accurate but computationally inefficient, whereas the diffusion theory is inaccurate but computationally efficient. A hybrid of the two approaches, however, is constructed to combine the advantages of both. The hybrid model computes as much as 100 times faster than the Monte Carlo method yet improves the accuracy of the diffusion theory.

6.2. DEFINITION OF PROBLEM

A pencil beam is normally incident on a slab of homogeneous scattering medium. The geometric and optical properties of the slab are described by thickness d, relative refractive index n_{rel}, absorption coefficient μ_a, scattering coefficient μ_s, and scattering anisotropy g, where n_{rel} is the ratio of the refractive index of the scattering medium to that of the ambient medium. The Henyey–Greenstein phase function is assumed. Cylindrical coordinates (r, ϕ, z) are used; the origin is the point of incidence of the pencil beam on the top surface of the slab; the z axis points along the pencil beam. The diffuse reflectance and the diffuse transmittance versus r are computed.

6.3. DIFFUSION THEORY

The diffusion theory for a scattering slab with refractive-index-mismatched ($n_{rel} \neq 1$) boundaries is an extension of the theory for a semiinfinite scattering medium with a refractive-index-matched ($n_{rel} = 1$) boundary that was covered in Chapter 5. The fluence rate Φ at observation point (r, ϕ, z) in response

---

Biomedical Optics: Principles and Imaging, by Lihong V. Wang and Hsin-i Wu
Copyright © 2007 John Wiley & Sons, Inc.

to an isotropic point source of unit power at (r', ϕ', z') in an infinite scattering medium is

$$\Phi_\infty(r, \phi, z; r', \phi', z') = \frac{1}{4\pi D} \frac{\exp(-\mu_{\text{eff}}\rho)}{\rho}. \tag{6.1}$$

Here, ρ is the distance between the observation point and the source point, D is the diffusion coefficient, and μ_{eff} is the effective attenuation coefficient:

$$\rho = \sqrt{r^2 + r'^2 - 2rr'\cos(\phi - \phi') + (z - z')^2}, \tag{6.2}$$

$$D = \frac{1}{3[\mu_a + \mu_s(1-g)]}, \tag{6.3}$$

$$\mu_{\text{eff}} = \sqrt{\mu_a/D}. \tag{6.4}$$

To compute the fluence rate in response to an isotropic point source in a scattering slab on the basis of Eq. (6.1), we first theoretically convert the slab into an infinite medium by satisfying the boundary conditions with an array of image sources, akin to the infinite array of images seen by a person standing between two parallel mirrors. Two extrapolated boundaries, at which the fluence rate is approximately zero, are used; they are separated from the slab surfaces by a distance of z_b (Figure 6.1)

$$z_b = 2C_R D, \tag{6.5}$$

where C_R is related to the effective reflection coefficient R_{eff}. If $n_{\text{rel}} = 1$, then $C_R = 1$. Otherwise, C_R is estimated by

$$C_R = \frac{1 + R_{\text{eff}}}{1 - R_{\text{eff}}}. \tag{6.6}$$

Here, the following empirical formula is used (the exact formula can be found in Chapter 5):

$$R_{\text{eff}} = -1.440 n_{\text{rel}}^{-2} + 0.710 n_{\text{rel}}^{-1} + 0.668 + 0.0636 n_{\text{rel}}. \tag{6.7}$$

An original isotropic point source at (r', θ', z') and its images are shown in Figure 6.1. The images are caused by reflections from the two extrapolated boundaries, where each reflection alternates the polarity of the point source. The z coordinates of the ith source pair are given by

$$z_{i\pm} = -z_b + 2i(d + 2z_b) \pm (z' + z_b), \tag{6.8}$$

where $i = 0, \pm 1, \pm 2, \ldots$. The source pair at $z_{0\pm}$ (the original and its image) straddles the top boundary of the slab. The source pair at $z_{1\pm}$ is the image of the

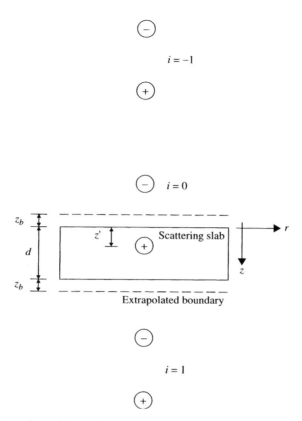

Figure 6.1. Illustration of the original point source inside the scattering slab and the image point sources outside the slab. Circled signs indicate the polarities of the sources.

pair at $z_{0\pm}$ with respect to the bottom extrapolated boundary. The source pair at $z_{-1\pm}$ is the image of the pair at $z_{1\pm}$ with respect to the top extrapolated boundary. Although infinite, the image series can be truncated after several source pairs.

With these image sources, the boundary conditions are satisfied; hence, the true boundaries can be removed. Consequently, the original point source in the scattering slab is converted to an array of isotropic sources in an infinite homogeneous medium. The fluence rate from the original source in the slab is approximated by

$$\Phi(r, \phi, z; r', \phi', z') = \sum_{i=i_{\min}}^{i_{\max}} [\Phi_\infty(r, \phi, z; r', \phi', z'_{i+}) - \Phi_\infty(r, \phi, z; r', \phi', z'_{i-})], \quad (6.9)$$

where $i_{\min}$ and $i_{\max}$ are the lower and upper indices, respectively, of the truncated source pair series. The diffuse reflectance and the diffuse transmittance from the slab are given by

$$R(r, \phi, 0; r', \phi', z') = \sum_{i=i_{\min}}^{i_{\max}} [R_\infty(r, \phi, 0; r', \phi', z'_{i+})$$
$$- R_\infty(r, \phi, 0; r', \phi', z'_{i-})], \qquad (6.10)$$

$$T(r, \phi, d; r', \phi', z') = \sum_{i=i_{\min}}^{i_{\max}} [T_\infty(r, \phi, d; r', \phi', z'_{i+})$$
$$- T_\infty(r, \phi, d; r', \phi', z'_{i-})], \qquad (6.11)$$

where

$$R_\infty(r, \phi, 0; r', \phi', z') = D \frac{\partial \Phi_\infty}{\partial z}\bigg|_{z=0} = \frac{z'(1 + \mu_{\text{eff}}\rho)\exp(-\mu_{\text{eff}}\rho)}{4\pi\rho^3}, \qquad (6.12)$$

$$T_\infty(r, \phi, d; r', \phi', z') = -D \frac{\partial \Phi_\infty}{\partial z}\bigg|_{z=d} = \frac{(d - z')(1 + \mu_{\text{eff}}\rho)\exp(-\mu_{\text{eff}}\rho)}{4\pi\rho^3}.$$
$$(6.13)$$

6.4. HYBRID MODEL

Accurate conversion of the incident pencil beam into an isotropically emitting light source deep in the scattering medium can improve the accuracy of the diffusion theory. Such a conversion can be provided by the Monte Carlo method. The combination of the Monte Carlo method and the diffusion theory is referred to as a *hybrid model*.

In the Monte Carlo step, the incident pencil beam is converted into a distributed isotropic source while reemitted photons are recorded. Since the diffusion theory is inaccurate when photons are within a critical depth z_c from the two boundaries of the slab, photons are tracked by the Monte Carlo method until they reach the center zone defined by $z_c \leq z \leq d - z_c$ (Figure 6.2).

The Monte Carlo step is based on the conventional Monte Carlo method described in Chapter 3. A photon packet with an initial weight of unity is launched perpendicularly onto the surface along the z axis (Figure 6.2). If the boundary is refractive-index-matched ($n_{\text{rel}} = 1$), all photon weight enters the scattering medium. Otherwise, only a portion enters after the Fresnel reflection. Then, a step size s is chosen statistically by

$$s = \frac{-\ln(\xi)}{\mu_a + \mu_s}, \qquad (6.14)$$

where ξ is a pseudorandom number evenly distributed between 0 and 1 ($0 < \xi \leq 1$). The photon packet loses some weight owing to absorption at the end of

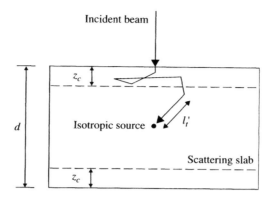

Figure 6.2. Illustration of the conversion from an incident photon packet to an isotropic point source in the Monte Carlo step of the hybrid model. The last step of l'_t in length converts the photon packet into an isotropic point source.

each step; the loss is equal to the photon weight at the beginning of the step multiplied by $1 - a$, where a denotes the albedo. The photon packet is then scattered in a new propagation direction that is statistically determined by the Henyey–Greenstein phase function with anisotropy g. When the photon weight is less than a threshold, the photon packet can be either terminated or continued as determined by Russian roulette. If reemitted into the ambient medium, the photon packet contributes to the diffuse reflectance $R_{MC}(r)$ (where the subscript denotes Monte Carlo) or the diffuse transmittance. The process is then repeated with multiple (N) photon packets.

If scattered in the center zone, the photon packet is conditionally converted to an isotropic point source. If one transport mean free path l'_t along the photon propagation direction fits in the center zone, the conversion is implemented (Figure 6.2); otherwise, the Monte Carlo step continues.

The conversion is based on the similarity relation, which converts the scattering medium from anisotropic to isotropic scattering while conserving the reduced scattering coefficient μ'_s. After taking the step of l'_t in length, the photon packet interacts with the isotropic scattering medium according to the transport albedo a'. With the weight reduced by a factor of $1 - a'$ due to absorption, the photon packet experiences isotropic scattering. The scattered photon packet then becomes an isotropic point source; its weight is recorded into a source function $S(r, z)$, which is guaranteed to be zero outside the center zone. Note that the step size for the conversion in a finite medium is slightly less than l'_t, but l'_t is used for simplicity.

In the diffusion step, the additional contribution to the diffuse reflectance that is due to the converted source is calculated by the diffusion theory. After the Monte Carlo step tracks all photon packets, S gives the total accumulated weight distribution. Next, S is converted to a relative source density function S_d, which represents the source strength per unit volume. Then, S_d is used to compute the additional diffuse reflectance R_{DT} based on the diffusion theory (DT)

$$R_{DT}(r) = \int_0^\infty \int_0^\infty \int_0^{2\pi} S_d(r', z')R(r, 0, 0; r', \phi', z')r' \, d\phi' \, dr' \, dz', \quad (6.15)$$

where R is given by Eq. (6.10). Because of the cylindrical symmetry, R_{DT} is independent of the azimuthal angle ϕ. The final diffuse reflectance R_d is given by

$$R_d(r) = R_{MC}(r) + R_{DT}(r). \quad (6.16)$$

The diffuse transmittance can be similarly computed.

6.5. NUMERICAL COMPUTATION

A grid system is set up in the cylindrical coordinates (Figure 6.1). The grid element sizes in the r and z directions are $\Delta r'$ and $\Delta z'$, respectively; the number of grid elements are N_r and N_z, respectively. The center coordinates of each grid element are given by

$$r'(i_r) = (i_r + 0.5)\Delta r', \quad (6.17)$$

$$z'(i_z) = (i_z + 0.5)\Delta z', \quad (6.18)$$

where $i_r = 0, 1, \ldots N_r - 1$ and $i_z = 0, 1, \ldots N_z - 1$. We can also use the optimized version of Eq. (6.17) as shown in Chapter 3. For brevity, the array elements for the physical quantities are referenced by either the location of the grid element or the indices of the grid element.

At the end of the Monte Carlo step, raw R_{MC} represents the total accumulated weight reflected into an annulus grid; it is converted to diffuse reflectance by

$$R_{MC}[i_r] \leftarrow \frac{R_{MC}[i_r]}{N \Delta a(i_r)}, \quad (6.19)$$

where Δa denotes the area of the annulus:

$$\Delta a(i_r) = 2\pi r'(i_r)\Delta r'. \quad (6.20)$$

Similarly, raw S is converted to S_d by

$$S_d[i_r, i_z] = \frac{S[i_r, i_z]}{N \Delta V(i_r)}, \quad (6.21)$$

where ΔV denotes the grid volume:

$$\Delta V(i_r) = \Delta a(i_r)\Delta z'. \quad (6.22)$$

The grid system for recording the source term S is also used to compute the integral over r' and z' in Eq. (6.15). From symmetry, the upper limit of the integral over ϕ' is lowered from 2π to π. Therefore, Eq. (6.15) is computed as follows:

$$R_{\text{DT}}(r) = 2 \sum_{i_r=0}^{N_r-2} \sum_{i_z=0}^{N_z-2} S_d[i_r, i_z] r'(i_r) \Delta r' \Delta z' \int_0^{\pi} R(r, 0, 0; r'(i_r), \phi', z'(i_z)) \, d\phi'.$$

(6.23)

The last grid elements in each direction are not used in the summation because they record weight deposited outside the grid system in the Monte Carlo step. The integration over ϕ' in Eq. (6.23) is implemented with Gaussian quadratures.

6.6. COMPUTATIONAL EXAMPLES

In this section, we compare the hybrid model with both the pure diffusion theory and the pure Monte Carlo method. Unless otherwise specified, 100,000 photon packets are tracked in both the Monte Carlo and the hybrid simulations.

The diffuse reflectance and the diffuse transmittance in response to an isotropic point source at $z' = l_t'$ in a scattering slab computed by the pure Monte Carlo method and the pure diffusion theory are shown in Figure 6.3. One to three point source pairs are used in the diffusion theory to satisfy the boundary conditions. The single pair is at $z_{0\pm}$ ($i = 0$); the double pairs are at $z_{0\pm}$ and $z_{1\pm}$; the triple pairs are at $z_{-1\pm}$, $z_{0\pm}$, and $z_{1\pm}$.

It is important to determine the number of point source pairs needed in the diffusion theory to accurately model diffuse reflectance and diffuse transmittance. As shown in Figure 6.3a, three pairs are required to achieve good accuracy in the diffuse reflectance. With fewer pairs, the accuracy is good until the radial distance is greater than approximately the slab thickness. As shown in Figure 6.3b, a single pair does not ensure accuracy to the diffuse transmittance because the boundary condition for the bottom surface is neglected altogether; two or three pairs, however, do provide accuracy. The number of pairs needed depends on the observation distance, the thickness of the slab, and the optical properties of the slab. In practice, more source pairs can be added until the new pair makes negligible contributions.

The diffuse reflectance and the diffuse transmittance in response to a pencil beam from both the pure Monte Carlo method and the pure diffusion theory are shown in Figure 6.4. The diffusion theory simulates an equivalent isotropic point source located at $z' = l_t'$ (see Chapter 5) using three source pairs ($i = -1, 0, 1$). The relative errors of the diffuse reflectance and the diffuse transmittance from the diffusion theory are shown in Figure 6.4c; they represent the differences between the results from the diffusion theory and the Monte Carlo method, divided point-by-point by the results from the Monte Carlo method. In this case, the diffuse reflectance from the diffusion theory is less than that from the accurate Monte

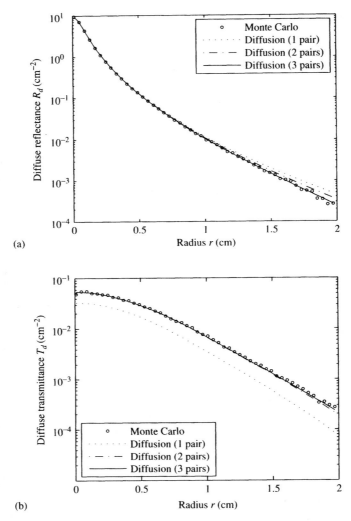

Figure 6.3. Comparison between the pure Monte Carlo method and the pure diffusion theory in terms of (a) the diffuse reflectance and (b) the diffuse transmittance in response to an isotropic source. The properties of the scattering slab are $n_{rel} = 1$, $\mu_a = 0.1$ cm^{-1}, $\mu_s = 100$ cm^{-1}, $g = 0.9$, and $d = 1$ cm.

Carlo method by as much as 75% near the source, but it becomes more accurate far from the source (Figures 6.4a and 6.4c). The diffuse transmittance, however, is accurate at all distances from the source (Figures 6.4b and 6.4c).

The diffuse reflectance data in response to an isotropic point source at various depths ($z' = 0.1l'_t, 0.3l'_t$, or $0.5l'_t$) from both the pure Monte Carlo method and the pure diffusion theory are shown in Figure 6.5. Three source pairs ($i = -1, 0, 1$)

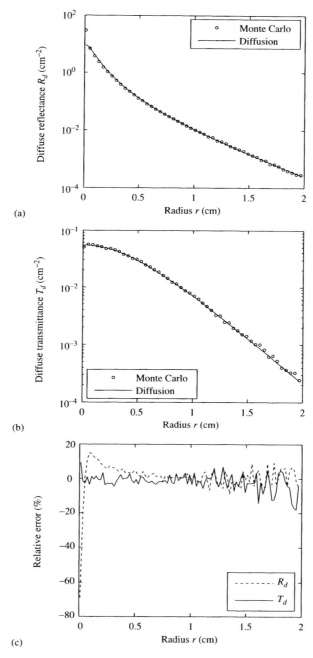

Figure 6.4. Comparison between the Monte Carlo method and the diffusion theory in terms of (a) the diffuse reflectance and (b) the diffuse transmittance in response to a pencil beam; (c) relative errors between the results. The properties of the scattering slab are described in Figure 6.3.

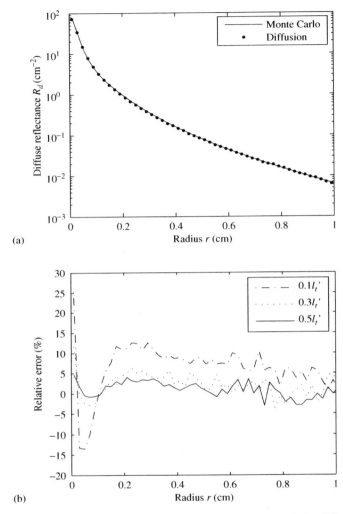

Figure 6.5. (a) Comparisons between the Monte Carlo method and the diffusion theory in terms of the diffuse reflectance when an isotropic point source is placed at $z' = 0.3l'_t$; (b) relative errors between the results when an isotropic point source is placed sequentially at $z' = 0.1l'_t, 0.3l'_t, 0.5l'_t$. The properties of the scattering slab are described in Figure 6.3.

are used in the diffusion theory. The relative errors between the diffusion theory and the Monte Carlo method in diffuse reflectance for these three source locations are shown in Figure 6.5b. Here, $0.5l'_t$ and $0.3l'_t$ give errors of 5% and 12%, respectively, whereas $0.1l'_t$ gives an error of up to 25%. Clearly, critical depth represents a tradeoff between the computational accuracy and efficiency of the hybrid model. Increasing the critical depth improves the computational accuracy at the expense of computational efficiency.

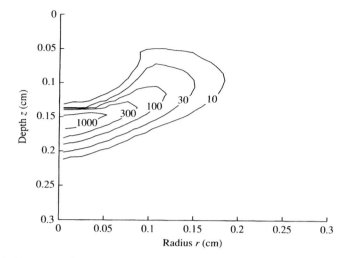

Figure 6.6. Contours of S_d in response to a pencil beam from the initial Monte Carlo step of the hybrid model. The critical depth z_c is set to 0.05 cm ($\approx 0.5 l_t'$). The contour values are in the unit of cm^{-3}. The properties of the scattering slab are described in Figure 6.3.

The contours of S_d from the Monte Carlo step of the hybrid model are shown in Figure 6.6. Since the critical depth is 0.05 cm ($\approx 0.5 l_t'$), S_d is densely populated near $z' = 1.5 l_t'$; it is also limited to within approximately $2 l_t'$ from the point of incidence in both the r and the z dimensions.

The diffuse reflectance data from both the pure Monte Carlo method and the hybrid model in response to a pencil beam, where $\Delta r' = \Delta z' = 0.01$ cm and $N_r = N_z = 30$, are shown in Figure 6.7a. Figure 6.7b plots the relative error in the diffuse reflectance from the hybrid model, which is within ±6% including both statistical and systematic differences. The statistical error can be further reduced by using either more photon packets at the expense of computation time or larger grid elements at the expense of resolution, whereas the systematic error can be further reduced by using a larger critical depth at the expense of computation time. In this example, if one million photon packets are tracked in each model, the hybrid model is 23 times faster than the Monte Carlo method. In other words, the hybrid model is significantly faster than the Monte Carlo method and almost as accurate.

The diffuse reflectance data from both the pure Monte Carlo method and the hybrid model in response to a pencil beam at various μ_a values, where $\Delta r' = 0.005$ cm, $\Delta z' = 0.003$ cm, and $N_r = N_z = 100$, are shown in Figure 6.8. When μ_a becomes comparable with μ_s' (e.g., $\mu_a = 10$ cm^{-1}, and $\mu_s' = 10$ cm^{-1}), the accuracy of the hybrid model is poor because the diffusion theory is valid only when $\mu_a \ll \mu_s'$. Therefore, the hybrid model is not expected to be accurate when $\mu_a \ll \mu_s'$ is not satisfied. If the critical depth increases, the accuracy of the hybrid model improves—even in the case of strong absorption—at the expense

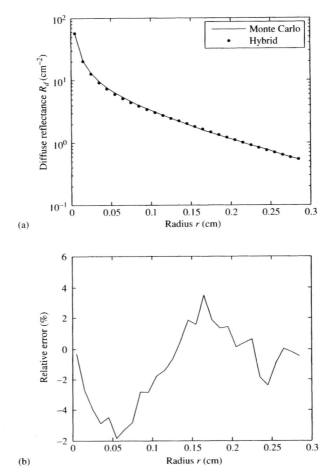

Figure 6.7. (a) Comparison between the pure Monte Carlo method and the hybrid model in terms of the diffuse reflectance in response to a pencil beam; (b) relative errors between the two approaches. The properties of the scattering slab are described in Figure 6.3.

of computation time, because the portion of the photon history tracked by the Monte Carlo step increases. The computer-dependent user times of the Monte Carlo method T_{MC} are 698, 583, and 136 s, respectively, for $\mu_a = 0.1$, 1, and 10 cm^{-1}. With the critical depth set to 0.05 cm, the user times of the hybrid model T_H are only 104, 99, and 76 s, respectively, for $\mu_a = 0.1$, 1, and 10 cm^{-1}. With the critical depth set to 0.1 cm, the user times of the hybrid model increase to 147, 142, and 114 s, respectively, for $\mu_a = 0.1$, 1, and 10 cm^{-1}.

User times for both the Monte Carlo method and the hybrid model under various conditions are listed in Table 6.1. While the other parameters are held constant, μ_a and d are varied. Here, $z_c = 0.05$ cm $\approx 0.5 l'_t$, $\Delta r' = \Delta z' = 0.01$ cm,

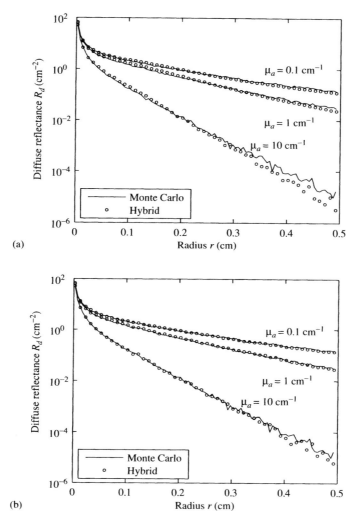

Figure 6.8. Comparisons between the Monte Carlo method and the hybrid model in terms of diffuse reflectance in response to a pencil beam when the critical depth is set to (a) 0.05 cm and (b) 0.1 cm. The absorption coefficient μ_a varies among 0.1, 1, and 10 cm^{-1}, while the other properties are held constant at $n_{rel} = 1.37$, $\mu_s = 100$ cm^{-1}, $g = 0.9$, and $d = 1$ cm.

and $N_r = N_z = 30$. The computation time of the hybrid model is insensitive to the optical properties unless μ_a becomes comparable with μ'_s as shown in the results associated with Figure 6.8. By contrast, the computation time of the Monte Carlo method is highly sensitive to the optical properties. A lower μ_a lengthens photon tracking because the chance of photon absorption per scattering event is

TABLE 6.1. User Times (Computer-Dependent) for Both Monte Carlo Method (T_{MC}) and Hybrid Model (T_H) and Their Ratio (T_{MC}/T_H) under Various Conditions.[a]

n_{rel}	d (cm)	μ_a (cm^{-1})	T_{MC} (s)	T_H (s)	T_{MC}/T_H
1.37	10	0.01	6684	23	291
1.37	10	0.1	2589	23	113
1.37	10	1	679	23	30
1.37	3	0.01	2095	23	91
1.37	3	0.1	1961	23	85
1.37	3	1	679	23	30
1.37	1	0.01	696	23	30
1.37	1	0.1	698	23	30
1.37	1	1	583	23	25
1	10	0.01	3992	19	210
1	10	0.1	1611	19	85
1	10	1	468	19	25
1	3	0.01	1253	19	66
1	3	0.1	1201	19	63
1	3	1	468	19	25
1	1	0.01	415	19	22
1	1	0.1	416	19	22
1	1	1	382	19	20

[a] The fixed optical properties include $\mu_s = 100$ cm^{-1} and $g = 0.9$.

reduced. Computation under the refractive-index-mismatched boundary condition takes longer because internal reflection at the boundary extends the lifetime of the photons in the scattering slab. In all cases, the hybrid model is faster than the Monte Carlo method by a factor of 20 to nearly 300, depending on the optical properties, the slab thickness, the number of photons tracked, the threshold for the Russian roulette, and the critical depth.

If the slab thickness reduces to several transport mean free paths, the diffusion theory becomes inaccurate. In this case, the pure Monte Carlo method should be used. Since the slab is relatively thin, T_{MC} is reasonably short. For example, $T_{MC} = 75$ s when the slab parameters are $n_{rel} = 1$, $\mu_a = 0.1$ cm^{-1}, $\mu_s = 100$ cm^{-1}, $g = 0.9$, and $d = 0.2$ cm; T_{MC} increases to 136 s when $n_{rel} = 1.37$.

PROBLEMS

6.1 Derive Eqs. (6.12) and (6.13).

6.2 Find the z coordinates of the first three source pairs (i.e., $i = 0, 1, -1$) for $n_{rel} = 1.37$, $\mu_a = 0.1$ cm^{-1}, $\mu_s = 100$ cm^{-1}, $g = 0.9$, $z' = 0.1$ cm, and $d = 1$ cm.

6.3 Calculate the weight of a photon packet that enters a scattering medium versus the angle of incidence if $n_{\rm rel} = 1.37$.

6.4 Explain how the computation time of the hybrid model depends on the optical properties.

6.5 Explain how the computation time of the Monte Carlo model depends on the scattering anisotropy in a scattering slab.

6.6 Implement the hybrid model. Update Table 6.1 with new computation times.

6.7 Implement the hybrid model. Adjust the threshold weight for Russian roulette and compare the computation times and accuracies.

6.8 Implement the hybrid model. Vary the critical depth in the hybrid model and compare the computation times and accuracies.

6.9 Extend the diffusion theory to the case of an infinitely narrow photon beam obliquely incident on a semiinfinite scattering medium.

6.10 Implement a hybrid model for an obliquely incident pencil beam on a semiinfinite scattering medium and compare with the diffuse reflectance computed from the diffusion theory.

READING

Wang LHV and Jacques SL (1993): Hybrid model of Monte Carlo simulation and diffusion theory for light reflectance by turbid media, *J. Opt. Soc. Am. A* **10**(8): 1746–1752. (All sections in this chapter.)

Wang LHV (1998): Rapid modeling of diffuse reflectance of light in turbid slabs, *J. Opt. Soc. Am. A* **15**(4): 936–944. (All sections in this chapter.)

FURTHER READING

Alexandrakis G, Busch DR, Faris GW, and Patterson MS (2001): Determination of the optical properties of two-layer turbid media by use of a frequency-domain hybrid Monte Carlo diffusion model, *Appl. Opt.* **40**(22): 3810–3821.

Alexandrakis G, Farrell TJ, and Patterson MS (2000): Monte Carlo diffusion hybrid model for photon migration in a two-layer turbid medium in the frequency domain, *App. Opt.* **39**(13): 2235–2244.

Carp SA, Prahl SA, and Venugopalan V (2004): Radiative transport in the delta-p-1 approximation: Accuracy of fluence rate and optical penetration depth predictions in turbid semi-infinite media, *J. Biomed. Opt.* **9**(3): 632–647.

Farrell TJ, Patterson MS, and Wilson B (1992): A diffusion-theory model of spatially resolved, steady-state diffuse reflectance for the noninvasive determination of tissue optical-properties invivo, *Med. Phys.* **19**(4): 879–888.

Flock ST, Wilson BC, and Patterson MS (1989): Monte Carlo modeling of light-propagation in highly scattering tissues. 2. Comparison with measurements in phantoms, *IEEE Trans. Biomed. Eng.* **36**(12): 1169–1173.

Gardner CM and Welch AJ (1994): Monte Carlo simulation of light transport in tissue—unscattered absorption events, *Appl. Opt.* **33**(13): 2743–2745.

Gardner CM, Jacques SL, and Welch AJ (1996): Light transport in tissue: Accurate expressions for one-dimensional fluence rate and escape function based upon Monte Carlo simulation, *Lasers Surg. Med.* **18**(2): 129–138.

Groenhuis RAJ, Ferwerda HA, and Tenbosch JJ (1983): Scattering and absorption of turbid materials determined from reflection measurements. 1. Theory, *Appl. Opt.* **22**(16): 2456–2462.

Groenhuis RAJ, Tenbosch JJ, and Ferwerda HA (1983): Scattering and absorption of turbid materials determined from reflection measurements. 2. Measuring method and calibration, *Appl. Opt.* **22**(16): 2463–2467.

Kim AD and Moscoso M (2005): Light transport in two-layer tissues, *J. Biomed. Opt.* **10**(3): 031015.

Schweiger M, Arridge SR, Hiraoka M, and Delpy DT (1995): The finite-element method for the propagation of light in scattering media—boundary and source conditions, *Med. Phys.* **22**(11): 1779–1792.

Spott T and Svaasand LO (2000): Collimated light sources in the diffusion approximation, *Appl. Opt.* **39**(34): 6453–6465.

Tarvainen T, Vauhkonen M, Kolehmainen V, and Kaipio JP (2005): Hybrid radiative-transfer-diffusion model for optical tomography, *Appl. Opt.* **44**(6): 876–886.

Tarvainen T, Vauhkonen M, Kolehmainen V, Arridge SR, and Kaipio JP (2005): Coupled radiative transfer equation and diffusion approximation model for photon migration in turbid medium with low-scattering and non-scattering regions, *Phys. Med. Biol.* **50**(20): 4913–4930.

Wang LHV and Jacques SL (1995): Use of a laser-beam with an oblique angle of incidence to measure the reduced scattering coefficient of a turbid medium, *Appl. Opt.* **34**(13): 2362–2366.

Wang LHV, Jacques SL, and Zheng LQ (1995): MCML—Monte Carlo modeling of light transport in multilayered tissues, *Comput. Meth. Prog. Biomed.* **47**(2): 131–146.

Wyman DR, Patterson MS, and Wilson BC (1989): Similarity relations for anisotropic scattering in Monte-Carlo simulations of deeply penetrating neutral particles, *J. Comput. Phys.* **81**(1): 137–150.

Wyman DR, Patterson MS, and Wilson BC (1989): Similarity relations for the interaction parameters in radiation transport, *Appl. Opt.* **28**(24): 5243–5249.

Yoon G, Prahl SA, and Welch AJ (1989): Accuracies of the diffusion-approximation and its similarity relations for laser irradiated biological media, *Appl. Opt.* **28**(12): 2250–2255.

CHAPTER 7
Sensing of Optical Properties and Spectroscopy

7.1. INTRODUCTION

Sensing the optical properties of biological tissue is important for diagnosis and therapy. After traversing biological tissue, reemitted light carries information about the optical properties of the scattering medium, which can be extracted using an inverse algorithm. Optical properties can be measured at multiple optical wavelengths for the production and investigation of spectra (spectroscopy).

7.2. COLLIMATED TRANSMISSION METHOD

The extinction coefficient μ_t, which is defined as the sum of the absorption coefficient μ_a and the scattering coefficient μ_s, can be measured by the collimated transmission method. In this method, a collimated lightbeam is incident perpendicularly on the surface of a sample. The sample can be a cuvette of liquid (e.g., an Intralipid® solution), a tissue-mimicking gel phantom (e.g., an agar gel containing polystyrene spheres), a solid phantom (e.g., a solidified resin containing TiO_2 particles), or a piece of biological tissue. The collimated (or ballistic) portion of the transmitted light is selected by apertures and then measured by a photodetector (Figure 7.1). First, a clear medium (e.g., water), the refractive index of which closely matches that of the sample to be tested, is measured to provide a reference ballistic-light signal I_0. Then, the sample is measured, which yields a transmitted light signal I_s. According to Beer's law, we have

$$I_s = I_0 \exp(-\mu_t d), \qquad (7.1)$$

where d denotes the sample thickness. Here, light absorption by the clear medium is neglected (see Problem 7.1). The ballistic transmittance T of the scattering

---

Biomedical Optics: Principles and Imaging, by Lihong V. Wang and Hsin-i Wu
Copyright © 2007 John Wiley & Sons, Inc.

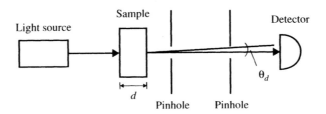

Figure 7.1. Schematic for the collimated transmission method.

medium is defined as

$$T = \frac{I_s}{I_0}. \quad (7.2)$$

Substituting Eq. (7.2) into Eq. (7.1), we obtain the extinction coefficient of the sample:

$$\mu_t = -\frac{1}{d} \ln T. \quad (7.3)$$

A key assumption in Eq. (7.1) is that the detected scattered light is much weaker than the detected ballistic light. Three factors affect this assumption: (1) the scattering optical depth of the sample $\mu_s d$, (2) the scattering phase function p, and (3) the acceptance angle (half-angle) of detection θ_d. The first factor determines the ratio of the number of scattered photons to the number of unscattered transmitted photons. The next two factors determine the collection fraction of the scattered photons to be detected χ.

7.2.1. Distribution of Scattering Count

We first consider an ideal scattering slab with the following optical properties: relative refractive index $n_{\text{rel}} = 1$ and scattering anisotropy $g = 1$. In this slab, specular reflection does not exist, and scattering does not deflect the photon. If the number of incident photons is N_{in}, the number of unscattered transmitted photons N_0 can be computed from Beer's law:

$$N_0 = N_{\text{in}} \exp(-\mu_t d). \quad (7.4)$$

The number of transmitted photons that have experienced i scattering events N_i is given by the following Poisson distribution (see Problem 7.2):

$$N_i = N_{\text{in}} \frac{(\mu_s d)^i \exp(-\mu_t d)}{i!}. \quad (7.5)$$

7.2.2. Collection Fraction

We next consider the collection fraction of singly scattered light in a real scattering medium. Distributed angularly, the singly scattered light is only partially received by the detector, which has a finite acceptance angle. The Henyey–Greenstein scattering phase function, which is the PDF of the cosine of the scattering polar angle θ, is assumed here (see Chapter 3)

$$p(\cos\theta) = \frac{1-g^2}{2(1+g^2-2g\cos\theta)^{3/2}}, \qquad (7.6)$$

where $\theta \in [0, \pi]$. Normalization requires that the integral of $p(\cos\theta)$ over $\cos\theta$ in the range of $[-1, 1]$ be unity.

For a slab with nonunity $n_{\rm rel}$, θ_d in air can be converted into an acceptance angle in the sample θ'_d by Snell's law:

$$\sin\theta_d = n_{\rm rel}\sin\theta'_d, \qquad (7.7)$$

which can be simplified to $\theta_d = n_{\rm rel}\theta'_d$ if $\theta_d \ll 1$. Integrating the phase function given by Eq. (7.6) over $\cos\theta$ in interval $[\cos\theta'_d, 1]$ yields the collection fraction χ of the singly scattered light

$$\chi = \frac{1+g}{2g}\left(1 - \frac{1-g}{\sqrt{1+g^2-2g\cos\theta'_d}}\right), \qquad (7.8)$$

which can be simplified to

$$\chi \approx \frac{\theta'^2_d}{2(1-g)^2}, \qquad (7.9)$$

if $\theta'_d \ll 1-g$ and $1-g \ll 1$ (i.e., $g \to 1$).

Example 7.1. Derive Eq. (7.9) from Eq. (7.8).

Because $\theta'_d \ll 1-g$ and $1-g \ll 1$, Eq. (7.8) can be approximated by repetitively using the Taylor expansion to the first order as follows:

$$\chi \approx \frac{1}{g}\left[1 - \frac{1-g}{\sqrt{(1-g)^2 + 2g(1-\cos\theta'_d)}}\right]$$

$$\approx \frac{1}{g}\left[1 - \frac{1-g}{\sqrt{(1-g)^2 + g\theta'^2_d}}\right]$$

$$\approx \frac{1}{g}\left[1 - \frac{1}{\sqrt{1 + [(g\theta_d'^2)/(1-g)^2]}}\right]$$

$$\approx \frac{1}{g}\left[1 - \frac{1}{1 + \frac{1}{2}[(g\theta_d'^2)/(1-g)^2]}\right]$$

$$\approx \frac{1}{g}\left[\frac{1}{2}\frac{g\theta_d'^2}{(1-g)^2}\right]$$

$$= \frac{\theta_d'^2}{2(1-g)^2}. \tag{7.10}$$

7.2.3. Error Expression

While the signal is from the unscattered light, the unwanted bias is from the scattered light. The unscattered light, assumed to be collimated, is completely collected by the detector. The scattered light contains both single- and multiple-scattered light. When $\mu_s d < 1$, multiple-scattered light is negligible; hence, only the single-scattered light is considered here. If $g \rightarrow 1$, the number of single-scattered photons can be estimated by N_1 from Eq. (7.5). Thus, the relative error due to the detected bias is approximately

$$\varepsilon_r = \frac{\chi N_1}{N_0}, \tag{7.11}$$

where the numerator and denominator represent the numbers of received single-scattered and unscattered photons, respectively. Although specular reflections on both slab surfaces are neglected explicitly, the unscattered and the collected single-scattered photons experience similar specular reflections. Therefore, the contributions of specular reflections to the numerator and the denominator in Eq. (7.11) partially offset each other.

Substituting the expressions for N_0 and N_1 from Eq. (7.5) into Eq. (7.11), we obtain

$$\varepsilon_r = \chi \mu_s d. \tag{7.12}$$

Although derived using the preceding approximate analytical expressions, Eq. (7.12) can be validated by the accurate Monte Carlo method (see Problem 7.3).

Substituting Eq. (7.9) into Eq. (7.12), we obtain

$$d = 2\varepsilon_r \frac{(1-g)^2}{\mu_s \theta_d'^2}, \tag{7.13}$$

which is valid if both $\theta'_d \ll 1-g$ and $1-g \ll 1$ hold. Therefore, the sample thickness must be limited on the basis of θ'_d as well as the optical properties so that the relative error is controlled within a given level. Although unknown initially, μ_s can be first measured and then checked for error using Eq. (7.12). If the error is unacceptable, one can modify the sample thickness—or the concentration of scatterers if the sample is liquid—and repeat the measurement.

Example 7.2. Apply Eq. (7.13) to a realistic case. Known parameters include $n_{\rm rel} = 1.37$, $g = 0.99$, and $\theta_d = 1$ mrad (milliradian). Assume $\mu_s \approx 100$ cm^{-1}.

If $\varepsilon_r \leq 1\%$ is desired, Eq. (7.13) leads to $d \leq 0.037$ cm.

7.3. SPECTROPHOTOMETRY

Spectrophotometry is based on the collimated transmission method. A spectrophotometer measures μ_t of a sample as a function of wavelength; if $\mu_s \ll \mu_a$, then $\mu_t \approx \mu_a$. The absorbance A, however, is typically reported; it is defined as

$$A = -\log_{10} T. \tag{7.14}$$

Substituting Eq. (7.3) into Eq. (7.14), we obtain

$$A = (\log_{10} e)\mu_t d = 0.4343 \mu_t d \tag{7.15}$$

or

$$\mu_t = (\ln 10)\frac{A}{d} = 2.303\frac{A}{d}. \tag{7.16}$$

The absorbance is also referred to as the *optical density* (OD), especially for a neutral-density filter, which has a nearly constant absorbance in a broad band. OD can be further related to dB (decibels) since the transmittance in dB is defined as $-10\log_{10} T$. For example, an OD of 1 means a 10-dB or 10-times attenuation, and an OD of 2 means a 20-dB or 100-times attenuation. OD is sometimes defined as the absorbance per unit length, however.

The unit of dB/cm (decibels per centimeter) is also used for various coefficients, such as the absorption and the extinction coefficients, although the unit of cm^{-1} (also expressed as nepers/cm in ultrasonics) is usually used in biomedical optics. The two units can be converted as follows:

$$1 \text{ cm}^{-1} = (10\log_{10} e) \text{ dB/cm} = 4.343 \text{ dB/cm}, \tag{7.17}$$

$$1 \text{ dB/cm} = (0.1\ln 10) \text{ cm}^{-1} = 0.2303 \text{ cm}^{-1}. \tag{7.18}$$

As a mnemonic aid for these two conversions, note that $e^{2.3} \approx 10$ and $10^{0.43} \approx e$.

A typical spectrophotometer contains one or more light sources that provide a broad spectrum. For example, a tungsten lamp can provide visible and infrared light, and a deuterium lamp can provide ultraviolet light. A diffraction grating angularly disperses the light emanating from the lamp into a spectrum. A narrow portion of the dispersed spectrum passes through a slit opening. The wavelength of the selected light can be tuned by rotating the grating with a knob. The grating in combination with the slit is also referred to as a *monochromator*. The "monochromatic" light is incident on the sample, and the transmitted light is detected by an optical detector such as a photodiode, which converts the optical signal into an electrical signal.

7.4. OBLIQUE-INCIDENCE REFLECTOMETRY

An oblique-incidence reflectometer can rapidly measure both the absorption coefficient μ_a and the reduced scattering coefficient μ'_s, where $\mu_a \ll \mu'_s$. As discussed in Chapter 5, a pencil beam normally incident on a semiinfinite scattering medium can be approximately represented by an isotropic point source (Figure 7.2a). The far diffuse reflectance—for which the observation points are beyond one transport mean free path l'_t from the point of incidence—in response to the pencil beam is well modeled using such an isotropic point source.

Similarly, an obliquely incident pencil beam can be approximated by an isotropic point source that is l'_t away from the point of incidence along the unscattered transmission path, as illustrated in Figure 7.2b. As a result, the isotropic point source is horizontally shifted from the point of incidence. Here, α_i and α_t are the angles of incidence and transmission, respectively; $n_{\rm rel}$ is the relative refractive index of the scattering medium. On the basis of Snell's law, we have

$$\sin \alpha_i = n_{\rm rel} \sin \alpha_t. \qquad (7.19)$$

From the geometry, we expect a horizontal shift of the far diffuse reflectance by

$$x_s = \frac{\sin \alpha_t}{\mu'_s + \mu_a}. \qquad (7.20)$$

A more accurate empirical expression is given below.

The schematic of an experimental oblique-incidence reflectometer is shown in Figure 7.3. A laser beam is incident on the object surface at an oblique angle. Diffusely reflected light is imaged by a CCD (charge-coupled device) camera. The CCD data are transferred to a computer and processed.

An experimentally measured diffuse reflectance distribution is shown in Figure 7.4. The midpoints of the left and right sides of curve M at all reflectance values are connected to form a centerline C. The shift x_m of the vertical portion of curve C represents the horizontal shift of the far diffuse reflectance; it agrees well with the theoretically predicted x_s from Eq. (7.20).

OBLIQUE-INCIDENCE REFLECTOMETRY   141

The diffuse reflectance versus x from the Monte Carlo method is shown in Figure 7.5. The shift of the vertical portion of curve C is $x_m = 0.174 \pm 0.009$ cm. From the optical properties used, Eq. (7.20) predicts $x_s = 0.167$ cm, which is in approximate agreement with the x_m predicted by the Monte Carlo method. Although not accurate, Eq. (7.20) is validated both experimentally and numerically.

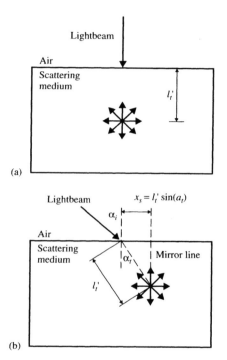

Figure 7.2. Lumped isotropic point sources for a pencil beam of (a) normal incidence ($\alpha_i = 0$) and (b) oblique incidence ($\alpha_i > 0$).

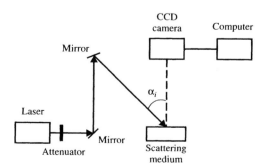

Figure 7.3. Schematic of a CCD-based oblique-incidence reflectometer.

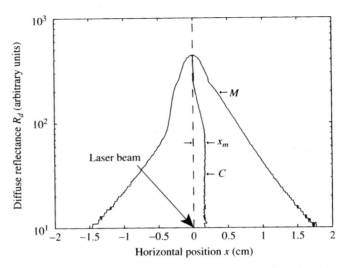

Figure 7.4. Experimentally measured diffuse reflectance as a function of x, where M represents the measured data and C represents the centerline.

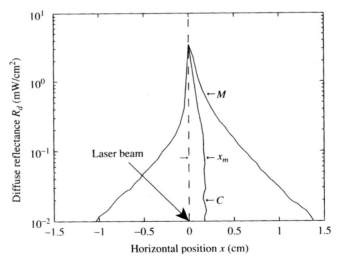

Figure 7.5. Curve M represents the Monte Carlo simulated diffuse reflectance of a 1-mW laser beam incident on a scattering medium with $\alpha_i = 45°$, and curve C represents the centerline of curve M. The optical properties of the scattering medium are $n_{rel} = 1.33$, $\mu_a = 0.25$ cm^{-1}, $\mu_s = 20$ cm^{-1}, and $g = 0.853$.

When compared with the accurate Monte Carlo simulated shifts for various μ_a values, the shift from Eq. (7.20) is highly accurate when $\mu_a \ll \mu_s'$, but it becomes progressively less accurate with increasing μ_a. The following empirical equation for the shift, however, significantly improves the accuracy:

$$x_s' = \frac{\sin \alpha_t}{\mu_s' + 0.35\mu_a}, \tag{7.21}$$

where $0.35\mu_a$ is used instead of μ_a. Equation (7.21) predicts $x_s' = 0.176$ cm, which is in better agreement with the value predicted by the Monte Carlo method. For simplicity, we define the diffusion coefficient for this section as

$$D = \frac{1}{3(\mu_s' + 0.35\mu_a)}. \tag{7.22}$$

Merging Eqs. (7.21) and (7.22), we obtain

$$x_s' = 3D \sin \alpha_t. \tag{7.23}$$

To measure the optical properties of the scattering medium, we first estimate the center x_s' of the far diffuse reflectance from the experimental data. From Eq. (7.23), we have

$$D = \frac{x_s'}{3 \sin \alpha_t}. \tag{7.24}$$

Since two independent optical properties are being measured, one more equation is needed.

The diffusion theory in Chapter 5 can be modified for the diffuse reflectance in response to an obliquely incident pencil beam:

$$R_d(x) = \frac{(1 - R_{\mathrm{sp}})a'}{4\pi}$$
$$\times \left[\frac{z_s'(1 + \mu_{\mathrm{eff}}\rho_1) \exp(-\mu_{\mathrm{eff}}\rho_1)}{\rho_1^3} + \frac{(z_s' + 2z_b)(1 + \mu_{\mathrm{eff}}\rho_2) \exp(-\mu_{\mathrm{eff}}\rho_2)}{\rho_2^3} \right]. \tag{7.25}$$

Here, R_{sp} denotes the specular reflectance; a' denotes the transport albedo; x denotes the distance between the observation point on the surface of the scattering medium and the point of incidence; ρ_1 and ρ_2 denote the distances from the two point sources (the original equivalent and the image sources) to the observation point, respectively; z_b denotes the distance between the extrapolated and the actual boundaries; μ_{eff} denotes the effective attenuation coefficient; and z_s' denotes the depth of the original point source:

$$z_s' = x_s' \cot(\alpha_t). \tag{7.26}$$

A nonlinear least-squares fit of the measured far diffuse reflectance to Eq. (7.25) yields μ_{eff}, which is defined as

$$\mu_{eff} = \sqrt{\mu_a/D}. \tag{7.27}$$

We have now quantified both D and μ_{eff} from the relative profile of the far diffuse reflectance, which can be obtained more easily than its absolute counterpart. From Eqs. (7.24) and (7.27), we obtain

$$\mu_a = D\mu_{eff}^2, \tag{7.28}$$

$$\mu_s' = \frac{1}{3D} - 0.35\mu_a. \tag{7.29}$$

7.5. WHITE-LIGHT SPECTROSCOPY

A spectroscopic oblique-incidence reflectometer (Figure 7.6) can measure absorption and reduced scattering spectra. White light from a lamp is coupled to a handheld probe made of 0.6-mm-diameter optical fibers. The source fiber is oriented at a 45° angle of incidence. Nine collection fibers, arranged in a linear array, collect the diffuse reflectance. Approximately 4.6 mW of white light is delivered

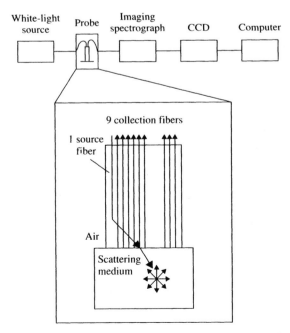

Figure 7.6. Schematic of a spectroscopic oblique-incidence reflectometer.

through the source fiber. The collection fibers are coupled to a connecting interface, the output of which is placed at the object plane of an imaging spectrograph. The imaging spectrograph spectrally disperses light from each detection fiber into a 1D spectrum and subsequently projects the 2D spatiospectral distribution onto a CCD camera. The CCD camera, which is controlled by a personal computer, records the 2D spatiospectral distribution. The CCD matrix has 512×512-pixels and measures 9.7×9.7 mm^2. With a 150 lines/mm grating, the CCD matrix is capable of accommodating a spectral range of 256 nm.

In the 2D spatiospectral distribution, the vertical and horizontal dimensions represent the spatial distribution of the diffuse reflectance at each wavelength and the spectral distribution of light from each collection fiber, respectively. The spatial distribution at each wavelength is used to fit for μ_a and μ_s', according to the theory described in the preceding section. As discussed in Chapter 1, the absorption spectrum can be used to assess the concentrations of oxy- and deoxyhemoglobin; on the basis of the Mie theory, the reduced scattering spectrum can be used to estimate the size distribution of the scatterers.

7.6. TIME-RESOLVED MEASUREMENT

Time-resolved diffuse reflectance can be used to measure the optical properties of biological tissue as well. With a short-pulsed collimated narrow laser beam normally incident on a semiinfinite scattering medium, a fast time-resolved detector—such as a streak camera or a single-photon counting system—measures the local diffuse reflectance $R_d(r, t)$ or the total diffuse reflectance $R_d(t)$. Here, r denotes the distance between the observation point and the point of incidence, and t denotes time. Under a simplified zero-boundary condition (the fluence rate on the real boundary is zero), the diffusion theory predicts

$$R_d(r, t) = \frac{z'}{(4\pi Dc)^{3/2} t^{5/2}} \exp\left(-\frac{r^2 + z'^2}{4Dct}\right) \exp(-\mu_a c t), \qquad (7.30)$$

$$R_d(t) = \int_0^\infty R_d(r, t) 2\pi r\, dr = \frac{z'}{\sqrt{4\pi Dc}\, t^{3/2}} \exp\left(-\frac{z'^2}{4Dct}\right) \exp(-\mu_a c t), \qquad (7.31)$$

where the source location $z' = l_t'$ and the diffusion coefficient $D = l_t'/3$. Whereas factors $t^{5/2}$ and $t^{3/2}$ dominate the early dynamics of the reflectance, $\exp(-\mu_a c t)$ dominates the later dynamics.

The absorption coefficient can be extracted by rewriting Eqs. (7.30) and (7.31) as follows:

$$-\frac{d \ln R_d(r, t)}{dt} = \mu_a c + \frac{5}{2t} - \frac{r^2 + z'^2}{4Dct^2}, \qquad (7.32)$$

$$-\frac{d \ln R_d(t)}{dt} = \mu_a c + \frac{3}{2t} - \frac{z'^2}{4Dct^2}. \qquad (7.33)$$

At large t values, each plot of $d \ln R_d$ versus t approaches a straight line with a slope equal to $-\mu_a c$. The second term in each of the equations above should be included for accuracy, while the third term can be neglected. For example, when $\mu_a = 0.1$ cm^{-1} and $\mu_s' = 10$ cm^{-1}, the second term is comparable with the first term for several nanoseconds, whereas the third term becomes negligible after only several hundred picoseconds. Therefore, the absorption coefficient can be estimated by either of the following expressions:

$$\mu_a \approx -\frac{1}{c}\left[\frac{d \ln R_d(r,t)}{dt} + \frac{5}{2t}\right], \quad (7.34)$$

$$\mu_a \approx -\frac{1}{c}\left[\frac{d \ln R_d(t)}{dt} + \frac{3}{2t}\right]. \quad (7.35)$$

7.7. FLUORESCENCE SPECTROSCOPY

Fluorescence spectroscopy provides a means for measuring the concentrations, quantum yields, and lifetimes of fluorescent molecules. Concentrations can provide morphologic information about biological tissue. Because quantum yields and lifetimes are related to the characteristics of biological molecules, they can provide biochemical information. These properties can reveal a variety of clinical problems such as epithelial neoplasia and atherosclerosis.

In a fluorescence spectroscopic system, light from a monochromatic excitation source is delivered through a flexible optical fiber bundle to the biological sample. The emitted fluorescent light from the sample is collected through another optical fiber bundle. The collected light is then separated into spectral components by a dispersing element. The dispersed fluorescence spectrum is finally detected by a detector array.

The system can be implemented with optical fibers or in free space. Whereas a handheld fiberoptic probe is typically used in contact with tissue, a free-space system is used in a noncontact fashion. A handheld probe may suffer from spectral dependence on the pressure applied by the probe on the in vivo tissue; a free-space system may suffer from intensity dependence on in vivo tissue motion. The contact approach is generally employed for small tissue areas, whereas the noncontact approach is more commonly used for relatively larger areas.

A fluorescence spectrum is related to both the excitation and the emission wavelengths. A fluorescence excitation spectrum can be produced by measuring the fluorescence intensity at a given emission wavelength for a range of excitation wavelengths. Conversely, a fluorescence emission spectrum can be produced by measuring the fluorescence intensity over a range of emission wavelengths at a given excitation wavelength. Ultimately, a fluorescence excitation–emission matrix (EEM) can be produced by measuring the fluorescence intensity over a range of emission wavelengths for a range of excitation wavelengths.

7.8. FLUORESCENCE MODELING

Although fluorescence is not discussed in Chapter 5, propagation of both excitation and fluorescent light in a scattering medium can be modeled by the diffusion theory. Assumed to be independent of fluorescence (actually, a Born approximation), the diffusion equation for the excitation light is given by

$$\frac{1}{c}\frac{\partial}{\partial t}\Phi_x(\vec{r}, t) + \mu_{ax}\Phi_x(\vec{r}, t) - \nabla \cdot D_x \nabla \Phi_x(\vec{r}, t) = S_x(\vec{r}, t), \quad (7.36)$$

where subscript x denotes the excitation wavelength. The other symbols are defined in Chapter 5.

The source term in Eq. (7.36) can be constructed from the first equivalent isotropic scattering events. From the similarity relation, an equivalent isotropic scattering medium is considered. Unscattered photons propagate along the ballistic path and constitute the primary beam, which has the following fluence rate distribution according to Beer's law:

$$\Phi_{px}(\vec{r}, t) = (1 - R_{sp})\Phi_{0x}(\vec{r}', t)\exp(-\mu'_{tx}l). \quad (7.37)$$

Here, the subscript p denotes the primary beam, R_{sp} denotes the specular reflectance from the surface of the scattering medium, Φ_{0x} denotes the incident fluence rate on the surface, and l denotes the ballistic path length into the scattering medium. The primary beam is converted into an isotropic source distribution, which serves as the source term for Eq. (7.36), by

$$S_x(\vec{r}, t) = \mu'_{sx}\Phi_{px}(\vec{r}, t), \quad (7.38)$$

where μ'_{sx} denotes the reduced scattering coefficient at the excitation wavelength.

Once an excitation photon is absorbed by a fluorophore, the probability that a fluorescence photon is emitted per unit time at time t ($t \geq 0$) can be modeled by

$$y(t) = \frac{Y}{\tau}\exp\left(-\frac{t}{\tau}\right), \quad (7.39)$$

were Y denotes the quantum yield for fluorescence emission and τ denotes the fluorescence lifetime.

Once fluorescent light is generated, its propagation can be modeled using another diffusion equation

$$\frac{1}{c}\frac{\partial}{\partial t}\Phi_m(\vec{r}, t) + \mu_{am}\Phi_m(\vec{r}, t) - \nabla \cdot D_m \nabla \Phi_m(\vec{r}, t) = S_m(\vec{r}, t), \quad (7.40)$$

where subscript m denotes the fluorescence emission wavelength. The source term is derived from the excitation light distribution using the following convolution

$$S_m(\vec{r}, t) = \int_0^t y(t - t')\mu_{afx}[\Phi_{px}(\vec{r}, t') + \Phi_x(\vec{r}, t')]dt', \quad (7.41)$$

where μ_{afx} denotes the absorption coefficient of the fluorophores at the excitation wavelength.

PROBLEMS

7.1 (a) Prove Eq. (7.1). (b) Explain why it is important to measure a clear medium first. (c) Assuming that the absorption coefficient of the clear medium is significant, modify Eq. (7.1).

7.2 Prove Eq. (7.5).

7.3 Write a Monte Carlo program to validate Eq. (7.12).

7.4 If the transmittance is given by absorbance A, express the transmittance in dB and then calculate the absorption coefficient in terms of A if the thickness of the sample is known.

7.5 In the collimated transmission method, assuming that measurements I_0, I_s, and d have independent uncertainties that are quantified by standard deviations σ_0, σ_s, and σ_d, respectively, derive the expected standard deviation in the predicted μ_t.

7.6 In a collimated transmission measurement, if the sample is optically thin ($d \ll 1/\mu_t$), the number of particles along the path can fluctuate significantly as a result of, for example, Brownian motion. Estimate the standard deviation of the number of received photons due to this fluctuation.

7.7 Write a Monte Carlo program to simulate the oblique-incidence diffuse reflectance from a semiinfinite medium. Duplicate Figure 7.5.

7.8 Write a Monte Carlo program to simulate the total time-resolved diffuse reflectance $R_d(t)$ from a semiinfinite scattering medium in response to a temporally ultrashort pencil beam. Compare it with the diffusion theory predicted values in response to an equivalent isotropic source located at (a) $1/(\mu_a + \mu_s')$, (b) $1/(0.35\mu_a + \mu_s')$, (c) $1/\mu_s'$ below the surface. Assume $n_{\text{rel}} = 1.38$, $\mu_a = 0.1 \text{ cm}^{-1}$, $\mu_s = 100 \text{ cm}^{-1}$, $g = 0.9$, and $\alpha_i = 45°$.

7.9 Derive the time-resolved diffuse reflectance equations in Section 7.6 assuming a zero boundary condition. Then derive them again using the extrapolated virtual boundary described in Chapter 5.

7.10 A fluorescent point object is placed at (x', y', z') below the surface of a semiinfinite scattering sample, where the z axis starts at the sample surface and points into the sample. A normally incident continuous-wave pencil beam at $(0, 0, 0)$ is used to excite the fluorophores. Use the diffusion theory to model the fluorescent reflectance measured on the sample surface at $(x, y, 0)$. Assume that the optical properties of the medium and the quantum yield of the fluorophores are known.

7.11 Explain why dB is sometimes defined by $10 \log_{10}$ instead of $20 \log_{10}$.

READING

Farrell TJ and Patterson MS (2003): Diffusion modeling of fluorescence in tissue, in *Handbook of Biomedical Fluorescence*, Mycek MA and Pogue, BW, eds., Marcel Dekker, New York, pp. 29–60. (See Section 7.8, above.)

Jacques SL, Wang LHV, and Hielscher AH (1995): Time-resolved photon propagation in tissues, in *Optical Thermal Response of Laser Irradiated Tissue*, Welch AJ and van Gemert MJC, eds., Plenum Press, New York, pp. 305–332. (See Section 7.6, above.)

Marquez G and Wang LHV (1997): White light oblique incidence reflectometer for measuring absorption and reduced scattering spectra of tissue-like turbid media, *Opt. Express* **1**: 454–460. (See Sections 7.4 and 7.5, above.)

Ramanujam N (2000): Fluorescence spectroscopy of neoplastic and non-neoplastic tissues, *Neoplasia* **2**(1–2): 89–117. (See Section 7.7, above.)

Richards-Kortum R and Sevick-Muraca E (1996): Quantitative optical spectroscopy for tissue diagnosis, *Annu. Rev. Phys. Chem.* **47**; 555–606. (See Section 7.7, above.)

Wang LHV and Jacques SL (1994): Error estimation of measuring total interaction coefficients of turbid media using collimated light transmission, *Phys. Med. Biol.* **39**: 2349–2354. (See Section 7.2, above.)

Wang LHV and Jacques SL (1995): Use of a laser beam with an oblique angle of incidence to measure the reduced scattering coefficient of a turbid medium, *Appl. Opt.* **34**: 2362–2366. (See Section 7.4, above.)

FURTHER READING

Baker SF, Walker JG, and Hopcraft KI (2001): Optimal extraction of optical coefficients from scattering media, *Opt. Commun.* **187**(1–3): 17–27.

Bevilacqua F and Depeursinge C (1999): Monte Carlo study of diffuse reflectance at source-detector separations close to one transport mean free path, *J. Opt. Soc. Am. A* **16**(12): 2935–2945.

Chang SK, Mirabal YN, Atkinson EN, Cox D, Malpica A, Follen M, and Richards-Kortum R (2005): Combined reflectance and fluorescence spectroscopy for in vivodetection of cervical pre-cancer, *J. Biomed. Opt.* **10**(2): 024031.

Collier T, Follen M, Malpica A, and Richards-Kortum R (2005): Sources of scattering in cervical tissue: Determination of the scattering coefficient by confocal microscopy, *Appl. Opt.* **44**(11): 2072–2081.

Dam JS, Pedersen CB, Dalgaard T, Fabricius PE, Aruna P, and Andersson-Engels S (2001): Fiber-optic probe for noninvasive real-time determination of tissue optical properties at multiple wavelengths, *Appl. Opt.* **40**(7): 1155–1164.

Garcia-Uribe A, Kehtarnavaz N, Marquez G, Prieto V, Duvic M, and Wang LHV (2004): Skin cancer detection by spectroscopic oblique-incidence reflectometry: Classification and physiological origins, *Appl. Opt.* **43**(13): 2643–2650.

Gobin L, Blanchot L, and Saint-Jalmes H (1999): Integrating the digitized backscattered image to measure absorption and reduced-scattering coefficients in vivo, *Appl. Opt.* **38**(19): 4217–4227.

Hull EL and Foster TH (2001): Steady-state reflectance spectroscopy in the p-3 approximation, *J. Opt. Soc. Am. A* **18**(3): 584–599.

Intes X, Le Jeune B, Pellen F, Guern Y, Cariou J, and Lotrian J (1999): Localization of the virtual point source used in the diffusion approximation to model a collimated beam source, *Waves Random Media* **9**(4): 489–499.

Jacques SL (1989): Time-resolved reflectance spectroscopy in turbid tissues, *IEEE Trans. Biomed. Eng.* **36**: 1155–1161.

Johns M, Giller CA, German DC, and Liu HL (2005): Determination of reduced scattering coefficient of biological tissue from a needle-like probe, *Opt. Express* **13**(13): 4828–4842.

Jones MR and Yamada Y (1998): Determination of the asymmetry parameter and scattering coefficient of turbid media from spatially resolved reflectance measurements, *Opt. Rev.* **5**(2): 72–76.

Kumar D and Singh M (2003): Characterization and imaging of compositional variation in tissues, *IEEE Trans. Biomed. Eng.* **50**(8): 1012–1019.

Liebert A, Wabnitz H, Grosenick D, Moller M, Macdonald R, and Rinneberg H (2003): Evaluation of optical properties of highly scattering media by moments of distributions of times of flight of photons, *Appl. Opt.* **42**(28): 5785–5792.

Lin SP, Wang LHV, Jacques SL, and Tittel FK (1997): Measurement of tissue optical properties by the use of oblique-incidence optical fiber reflectometry, *Appl. Opt.* **36**(1): 136–143.

Lin WC, Motamedi M, and Welch AJ (1996): Dynamics of tissue optics during laser heating of turbid media, *Appl. Opt.* **35**(19): 3413–3420.

Marquez G, Wang LHV, Lin SP, Schwartz JA, and Thomsen SL (1998): Anisotropy in the absorption and scattering spectra of chicken breast tissue, *Appl. Opt.* **37**(4): 798–804.

Mirabal YN, Chang SK, Atkinson EN, Malpica A, Follen M, and Richards-Kortum R (2002): Reflectance spectroscopy for in vivodetection of cervical precancer, *J. Biomed. Opt.* **7**(4): 587–594.

Mourant JR, Bigio IJ, Jack DA, Johnson TM, and Miller HD (1997): Measuring absorption coefficients in small volumes of highly scattering media: Source-detector separations for which path lengths do not depend on scattering properties, *Appl. Opt.* **36**(22): 5655–5661.

Mourant JR, Johnson TM, Los G, and Bigio LJ (1999): Non-invasive measurement of chemotherapy drug concentrations in tissue: Preliminary demonstrations of in vivomeasurements, *Phys. Med. Biol.* **44**(5): 1397–1417.

Nichols MG, Hull EL, and Foster TH (1997): Design and testing of a white-light, steady-state diffuse reflectance spectrometer for determination of optical properties of highly scattering systems, *Appl. Opt.* **36**(1): 93–104.

Nishidate I, Aizu Y, and Mishina H (2004): Estimation of melanin and hemoglobin in skin tissue using multiple regression analysis aided by Monte Carlo simulation, *J. Biomed. Opt.* **9**(4): 700–710.

Papaioannou T, Preyer NW, Fang QY, Brightwell A, Carnohan M, Cottone G, Ross R, Jones LR, and Marcu L (2004): Effects of fiber-optic probe design and probe-to-target distance on diffuse reflectance measurements of turbid media: An experimental and computational study at 337 nm, *Appl. Opt.* **43**(14): 2846–2860.

Patterson MS and Pogue BW (1994): Mathematical-model for time-resolved and frequency-domain fluorescence spectroscopy in biological tissue, *Appl. Opt.* **33**(10): 1963–1974.

Patterson MS, Chance B, and Wilson BC (1989): Time resolved reflectance and transmittance for the noninvasive measurement of tissue optical properties, *Appl. Opt.* **28**: 2331–2336.

Pham TH, Bevilacqua F, Spott T, Dam JS, Tromberg BJ, and Andersson-Engels S (2000): Quantifying the absorption and reduced scattering coefficients of tissuelike turbid media over a broad spectral range with noncontact Fourier-transform hyperspectral imaging, *Appl. Opt.* **39**(34): 6487–6497.

Rinzema K, Murrer LHP, and Star WM (1998): Direct experimental verification of light transport theory in an optical phantom, *J. Opt. Soc. Am. A—Optics Image Sci. Vision* **15**(8): 2078–2088.

Sefkow A, Bree M, and Mycek MA (2001): Method for measuring cellular optical absorption and scattering evaluated using dilute cell suspension phantoms, *Appl. Spectrosc.* **55**(11): 1495–1501.

Selden AC (2004): Photon transport parameters of diffusive media with highly anisotropic scattering, *Phys. Med. Biol.* **49**(13): 3017–3027.

Skala MC, Palmer GM, Zhu CF, Liu Q, Vrotsos KM, Marshek-Stone CL, Gendron-Fitzpatrick A, and Ramanujam N (2004): Investigation of fiber-optic probe designs for optical spectroscopic diagnosis of epithelial pre-cancers, *Lasers Surg. Med.* **34**(1): 25–38.

Swartling J, Dam JS, and Andersson-Engels S (2003): Comparison of spatially and temporally resolved diffuse-reflectance measurement systems for determination of biomedical optical properties, *Appl. Opt.* **42**(22): 4612–4620.

Takagi K, Haneishi H, Tsumura N, and Miyake Y (2000): Alternative oblique-incidence reflectometry for measuring tissue optical properties, *Opt. Rev.* **7**(2): 164–169.

Treweek SP and Barbenel JC (1996): Direct measurement of the optical properties of human breast skin, *Med. Biol. Eng. Comput.* **34**(4): 285–289.

Utzinger U and Richards-Kortum RR (2003): Fiber optic probes for biomedical optical spectroscopy, *J. Biomed. Opt.* **8**(1): 121–147.

Wan SK and Guo ZX (2006): Correlative studies in optical reflectance measurements of cerebral blood oxygenation, *J. Quantit. Spectrosc. Radiative Transfer* **98**(2): 189–201.

Wu T, Qu JNY, Cheung TH, Lo KWK, and Yu MY (2003): Preliminary study of detecting neoplastic growths in vivowith real time calibrated autofluorescence imaging, *Opt. Express* **11**(4): 291–298.

Yaroslavsky IV, Yaroslavsky AN, Goldbach T, and Schwarzmaier HJ (1996): Inverse hybrid technique for determining the optical properties of turbid media from integrating-sphere measurements, *Appl. Opt.* **35**(34): 6797–6809.

Zijp JR and ten Bosch JJ (1998): Optical properties of bovine muscle tissue in vitro; a comparison of methods, *Phys. Med. Biol.* **43**(10): 3065–3081.

Zonios G, Bykowski J, and Kollias N (2001): Skin melanin, hemoglobin, and light scattering properties can be quantitatively assessed in vivousing diffuse reflectance spectroscopy, *J. Invest. Dermatol.* **117**(6): 1452–1457.

Zuluaga AF, Utzinger U, Durkin A, Fuchs H, Gillenwater A, Jacob R, Kemp B, Fan J, and Richards-Kortum R (1999): Fluorescence excitation emission matrices of human tissue: A system for in vivomeasurement and method of data analysis, *Appl. Spectrosc.* **53**(3): 302–311.

CHAPTER 8

Ballistic Imaging and Microscopy

8.1. INTRODUCTION

Ideally, ballistic imaging is based on unscattered or singly backscattered ballistic photons. In reality, however, more-scattered quasiballistic photons are often measured as well to increase the signal strength. For brevity, subsequent use of the term *ballistic photons* in this chapter also refers to quasiballistic photons unless otherwise noted. Ballistic imaging provides high spatial resolution but suffers from limited imaging depth.

8.2. CHARACTERISTICS OF BALLISTIC LIGHT

The intensity of unscattered light I_T attenuates according to Beer's law:

$$I_T(z) = I_0 \exp(-\mu_t z). \tag{8.1}$$

Here, z denotes the ballistic path length in the scattering medium, μ_t denotes the extinction coefficient, and I_0 denotes the fluence rate of the incident light if specular reflection is negligible. The intensity of singly backscattered light is given by

$$I_R(z) = I_0 \exp(-2\mu_t z) R_b, \tag{8.2}$$

where R_b denotes the percentage of the backscattered light to be received by the detector, and the factor of 2 in the exponent is due to round-trip propagation. In both cases, the strictly ballistic signals decay exponentially with the path length.

The objective of ballistic imaging is to reject nonballistic photons and to retain ballistic photons on the basis of the following characteristic differences between them:

1. *Time of Flight*. Transmitted ballistic photons take shorter paths and arrive at the detector earlier than do nonballistic photons. Time-gated imaging and coherence-gated holographic imaging are based on this difference.

Biomedical Optics: Principles and Imaging, by Lihong V. Wang and Hsin-i Wu
Copyright © 2007 John Wiley & Sons, Inc.

2. *Collimation*. Transmitted ballistic light has better collimation (smaller divergence) than does nonballistic light. Spatial-frequency filtered imaging and optical heterodyne imaging are based on this difference.
3. *Polarization*. Ballistic light retains the incident polarization in a nonbirefringent scattering medium better than does nonballistic light. Polarization-difference imaging is based on this difference.
4. *Wavefront*. Ballistic light possesses a better-defined wavefront than does nonballistic light and hence can be better focused. Confocal microscopy and two-photon microscopy are based on this difference. Note that wavefront and collimation are related.

8.3. TIME-GATED IMAGING

Time-gated imaging, also referred to as *early-photon imaging*, takes advantage of the difference in arrival time between ballistic and nonballistic photons to select the early arriving component from the transmitted light. Figure 8.1 shows a block diagram of an experimental configuration for ultrafast time-gated imaging. A collimated ultrafast laser beam irradiates the scattering medium. The time gate is turned on for a short time to allow only the early-arriving photons to pass to the detector. The early-arriving photons carry information about optical attenuation along the optical axis. If the imaging system is raster scanned transversely, a 2D projection image (also termed a shadowgram) of the medium can be acquired.

Both the gating delay (the time lapse from the laser pulse to the rising edge of the time gate) and the open duration of the gate affect the image quality; a tradeoff exists between the spatial resolution and the signal strength. If the scattering medium is optically thin (thinner than the mean free path), gating at the arrival of the ballistic photons yields the best spatial resolution with good signal strength. If the scattering medium is optically thick (thicker than the mean free path), nonballistic light becomes significant. As the gating delay or the open duration increases, more nonballistic light contributes to the signal, yet the image becomes more blurred.

A time gate can be constructed using the Kerr effect. In the Kerr effect, birefringence is induced by an electric field applied to an isotropic transparent substance. Inducing a half-wave retardation in such a substance with the electric field of an auxiliary lightbeam can provide a high-speed shutter (Figure 8.2). The auxiliary beam is obtained by splitting the incoming laser light so that the

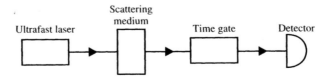

Figure 8.1. Experimental configuration for time-gated imaging.

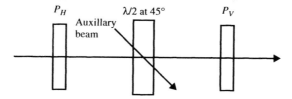

Figure 8.2. Schematic of a Kerr gate (P_H represents a horizontal polarizer; P_V, a vertical polarizer; $\lambda/2$, an activatable half-wave plate).

half-wave retardation can be synchronized with the signal beam. The activatable retarder is sandwiched between two cross-polarized linear polarizers; its fast axis is oriented diagonally ($\pm 45°$) between the orthogonal polarization axes of the two polarizers. When the retarder is not activated, no light can pass through the two polarizers; hence, the gate is closed. When the retarder is activated to provide half-wave retardation, the polarization orientation of the light ray is rotated $\pi/2$ and becomes aligned with the polarization axis of the second polarizer; hence, the gate is open. The shutter speed of such a gate can be as short as 100 fs.

A time gate can also be a single-photon counting system or a streak camera. The former is introduced in Chapter 11. A streak camera resolves the arrival time of ultrafast light by elegantly converting time into space (Figure 8.3). A photocathode plate converts incident photons into electrons by the photoelectric effect. The photoelectrons are accelerated toward a mesh and then deflected by a pair of fast sweep electrodes, the high voltage applied to which is synchronized to the incident light. Consequently, electrons arriving at different times bombard a microchannel plate (MCP) at different vertical locations. The MCP amplifies the current by generating secondary electrons. The amplified current then strikes the phosphor screen to produce photons. The vertical dimension of the phosphor screen thus provides the temporal resolution, which can reach about 200 fs. The horizontal dimension of the phosphor screen can provide either spatial or spectral resolution. For the former, a horizontal slit is added in front of the photocathode to form a narrow lightbeam. For the latter, a spectrometer is added in front of the photocathode to disperse the incident light horizontally into spectral components.

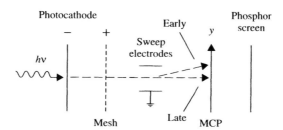

Figure 8.3. Basic components of a streak camera.

Example 8.1. Estimate the spatial resolution using some practical open durations of the time gate.

If the open duration of the gate is 100 fs, the spatial resolution is on the order of $100 \text{ fs} \times 3 \times 10^8 \text{ m/s} = 30 \text{ μm}$. If it is 5 ps, the resolution is on the order of $5 \text{ ps} \times 3 \times 10^8 \text{ m/s} = 1.5 \text{ mm}$.

8.4. SPATIOFREQUENCY-FILTERED IMAGING

Spatiofrequency-filtered imaging, also called *Fourier space-gated imaging*, takes advantage of the different spatiofrequency distributions between ballistic and nonballistic light to select the ballistic component from the transmitted light. Figure 8.4 is a schematic representation of two spatiofrequency-filtered imaging systems: (a) a narrowbeam scanning system and (b) a widebeam full-field system. A collimated laser beam—narrow in Figure 8.4a and broad in Figure 8.4b—irradiates the scattering medium. A lens focuses the ballistic component to a diffraction-limited point while dispersing the nonballistic component around the focus. A pinhole, placed at the focal plane of the lens, blocks most of the off-focus light and passes the ballistic component to the detector. In Figure 8.4a, optical attenuation through the medium is detected along one line at a time; transverse scanning yields a shadowgram. In Figure 8.4b, a 2D shadowgram is formed through a second lens with a single exposure as in X-ray projection imaging. In either case, the image signal carries information about the integrated attenuation along the optical path.

From the perspective of Fourier optics, the 2D spatial Fourier transform of a lightbeam over a cross section provides the spatiofrequency spectrum in the same way that the 1D temporal Fourier transform of a lightbeam at an observation point provides the temporal frequency spectrum. Different spatiofrequency

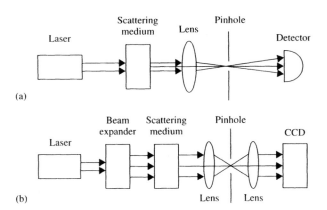

Figure 8.4. Schematic of spatiofrequency-filtered imaging; (a) a narrowbeam scanning system; (b) a widebeam full-field system.

components represent plane waves traveling in different directions, which can be focused to different points on the focal plane of a lens. Therefore, the lens functions as a spatial Fourier transformer, and the pinhole functions as a spatial filter. The major plane-wave component in the ballistic light propagates along the optical axis, whereas most plane-wave components in the nonballistic light propagate obliquely.

8.5. POLARIZATION-DIFFERENCE IMAGING

Polarization-difference imaging (PDI) takes advantage of the different polarization states between ballistic and nonballistic components to select the ballistic component from the transmitted light. A nonbirefringent scattering medium does not alter the polarization state of the ballistic light, whereas the medium randomizes the polarization state of the nonballistic light. In the PDI system shown in Figure 8.5, a polarizer linearly polarizes the source beam. The transmitted light passes through a linear polarization analyzer that is sequentially aligned in two orthogonal directions. Then, the light is detected by a photodetector.

When the polarization axis of the analyzer is parallel to the incident polarization, the intensity measurement is denoted by $I_{\parallel}(x, y)$, where (x, y) represent the transverse Cartesian coordinates. Likewise, when the polarization axis of the analyzer is perpendicular to the incident polarization, the intensity measurement is denoted by $I_{\perp}(x, y)$. We can approximately express the two intensities as

$$I_{\parallel}(x, y) = I_b(x, y) + \frac{1}{2} I_{nb}(x, y), \qquad (8.3)$$

$$I_{\perp}(x, y) = \frac{1}{2} I_{nb}(x, y). \qquad (8.4)$$

Here, $I_b(x, y)$ and $I_{nb}(x, y)$ denote the ballistic and nonballistic intensities, respectively. The nonballistic light is assumed to be completely unpolarized and

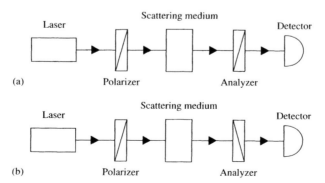

Figure 8.5. Schematic of a polarization-difference imaging system in two states. The polarization axis of the analyzer is (a) parallel and (b) perpendicular to the polarization axis of the incident polarizer.

hence passes through the analyzer with a 50% transmittance, regardless of the polarization orientation of the analyzer.

A PDI system generates an image by

$$I_{PD}(x, y) = I_{\parallel}(x, y) - I_{\perp}(x, y). \tag{8.5}$$

Substituting Eqs. (8.3) and (8.4) into (8.5), we obtain

$$I_{PD}(x, y) = I_b(x, y). \tag{8.6}$$

Therefore, the ballistic component is recovered.

PDI is simple and fast; it can also be implemented in reflection mode. However, a small number of scattering events lead to only partial randomization of polarization, which affects the efficacy of PDI. Furthermore, if birefringence is present, more complex quantities such as the Stokes vector (see Chapter 10) need to be measured.

8.6. COHERENCE-GATED HOLOGRAPHIC IMAGING

Coherence-gated holographic imaging takes advantage of the difference in arrival times between ballistic and nonballistic photons to select the early-arriving component from the transmitted light. To fully appreciate this technique, readers should review the principle of conventional holography (see Appendix 8A). Here, coherence-gated holographic imaging is based on digital holography, in which both recording and reconstruction are accomplished digitally (Figure 8.6). In the object arm, the lightbeam is filtered by a pinhole (spatial filter) and then expanded and collimated by a lens before irradiating the scattering medium. The transmitted object beam is first spatiofrequency-filtered and then collimated before it is recombined with the reference beam, which is oblique with a small angle θ (not shown in the figure). The path length of the reference beam is matched to that of the first-arriving ballistic light by adjusting a multimirror delay line. The interference pattern is imaged onto a CCD camera to form an image-plane hologram. Multiple CCD images are averaged to reduce speckle noise because speckles in successive holograms are assumed to be uncorrelated.

For monochromatic light, the reference and the object fields can be expressed with phasor representations as follows:

$$E_R(\omega, x) = E_0(\omega) \exp(ik_x x - i\omega(t - t_R)), \tag{8.7}$$

$$E_S(\omega, x, y) = E_0(\omega)\{a_1(x, y) \exp(-i\omega(t - t_1)) + a_2(x, y) \exp(-i\omega(t - t_2))\}. \tag{8.8}$$

Here, subscripts R and S denote the reference and the object (sample) beams, respectively; ω denotes the angular frequency; (x, y) denote the Cartesian coordinates on the detector surface; t denotes time; E_0 denotes the electric field

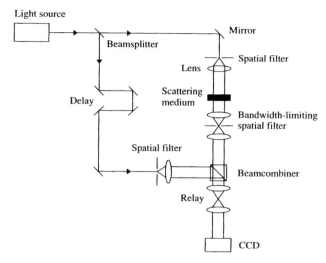

Figure 8.6. Schematic of a coherence-gated holographic imaging system. The reference beam is incident obliquely with a small angle θ (not shown).

amplitude; k_x denotes the x component of the wavevector, which appears because the reference beam is assumed to be tilted with respect to the x axis; t_R denotes the time delay in the reference beam; a_1 and t_1 denote the amplitude transmittance and the time delay of the ballistic light, respectively; and a_2 and t_2 denote the amplitude transmittance and the time delay of a representative group of nonballistic light, respectively. In addition to the t_2 component, more time-delay components can be added in a similar fashion. Since frequency tuning is needed, the frequency dependence of E_R and E_S is explicitly expressed here. The objective of this imaging technique is to retain the a_1 component and eliminate the a_2 component.

Note that the phasor expression, also referred to as the *complex expression*, is a convenient mathematical convention for representing oscillations. The actual oscillations are the real part of the phasor expression. In linear mathematical operations, real-part operators Re{ } on phasor expressions are implicit since real-part operators and linear operators are permutable. In nonlinear mathematical operations, however, one needs to exercise caution.

If $t_R = t_1$, the hologram can be expressed as

$$I(\omega, x, y) = |E_0(\omega)|^2 \{1 + |a_1|^2 + |a_2|^2\}$$
$$+ |E_0(\omega)|^2 \{a_1 \exp(-ik_x x) + a_1^* \exp(ik_x x)\}$$
$$+ |E_0(\omega)|^2 \{a_2 \exp(-ik_x x + i\omega(t_2 - t_1)) \quad (8.9)$$
$$+ a_2^* \exp(ik_x x - i\omega(t_2 - t_1))\}$$
$$+ |E_0(\omega)|^2 \{a_1^* a_2 \exp(i\omega(t_2 - t_1)) + a_1 a_2^* \exp(-i\omega(t_2 - t_1))\}.$$

The holographic image is reconstructed digitally as follows:

1. We take the spatial Fourier transformation of the hologram with respect to x:

$$\tilde{I}(\omega, k, y) = |E_0(\omega)|^2 \{1 + |a_1|^2 + |a_2|^2\} \delta(k)$$
$$+ |E_0(\omega)|^2 \{a_1 \delta(k - k_x) + a_1^* \delta(k + k_x)\}$$
$$+ |E_0(\omega)|^2 \{a_2 \delta(k - k_x) \exp(i\omega(t_2 - t_1))$$
$$+ a_2^* \delta(k + k_x) \exp(-i\omega(t_2 - t_1))\} \quad (8.10)$$
$$+ |E_0(\omega)|^2 \{a_1^* a_2 \exp(i\omega(t_2 - t_1))$$
$$+ a_1 a_2^* \exp(-i\omega(t_2 - t_1))\} \delta(k).$$

Here, the terms containing $\delta(k)$ represent the zero-frequency ("DC") component; the terms containing $\delta(k - k_x)$ represent the virtual image that has a spatial frequency of $+k_x$, whereas the terms containing $\delta(k + k_x)$ represent the real image that has a spatial frequency of $-k_x$.

2. We filter the signal to retain the $\delta(k - k_x)$ terms (first-order diffraction terms) and to reject the other components.
3. We take the inverse Fourier transformation of the filtered signal and then drop $\exp(-ik_x x)$, which results in

$$I'(\omega, x, y) = |E_0(\omega)|^2 \{a_1 + a_2 \exp(i\omega(t_2 - t_1))\}. \quad (8.11)$$

To separate the a_1 and a_2 terms in Eq. (8.11), we tune the laser frequency within a bandwidth $\Delta\omega$ while acquiring a hologram at each frequency. If $|E_0(\omega)|^2$ is slowly varying and $\Delta\omega \gg (t_2 - t_1)^{-1}$, integration of Eq. (8.11) over $\Delta\omega$ approximately averages out the a_2 term, which leads to

$$I''(x, y) = \int_{\Delta\omega} I'(\omega, x, y) \, d\omega \approx a_1(x, y) \int_{\Delta\omega} |E_0(\omega)|^2 \, d\omega. \quad (8.12)$$

Thus, I'' represents a ballistic image proportional to a_1. As will be seen in Chapter 9, this frequency-swept gating is equivalent to coherence gating based on a wideband light source.

8.7. OPTICAL HETERODYNE IMAGING

Optical heterodyne imaging takes advantage of the different spatial frequencies between ballistic and nonballistic light to select the collimated component from the transmitted light. *Optical heterodyne detection* means the superposition of two coherent optical beams that have slightly different temporal frequencies and the subsequent detection of the beat-frequency component of the superposed beam.

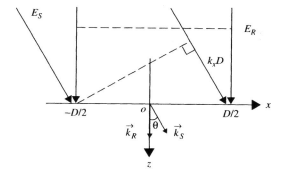

Figure 8.7. Illustration of the relationship between the reference and the sample beams with respect to the detector.

The antenna property of optical heterodyne detection is the basis for the ballistic imaging here. Figure 8.7 shows a schematic relationship between the reference and the sample beams with respect to the detector. The antenna theorem can be stated as follows. In the 1D case, $|\theta| \sim \lambda/D$, where θ is the acceptance angle between the two beams, λ is the optical wavelength, and D is the width of the aperture (limited by the size of the optical beam or the detector, whichever is smaller). In the 2D case, $\Omega \sim \lambda^2/A$, where Ω is the acceptance solid angle and A is the area of the aperture. The limitation on the acceptance angle is simply due to the cancellation of interference fringes on the detector surface when the two waves propagate in different directions.

Ballistic light, which propagates along the optical axis, has a nearly zero spatiofrequency bandwidth, while nonballistic light has a broad spectrum of spatial frequencies. Therefore, low spatiofrequency ballistic light can be retained and high spatiofrequency nonballistic light rejected according to the antenna theorem.

An experimental setup for heterodyne imaging is shown in Figure 8.8. The laser beam from the source is divided into two different paths by a beamsplitter. Before illuminating the sample, the sample beam is passed through an acoustooptic modulator to add a frequency shift of 80.0 MHz. The reference beam is passed through another acoustooptic modulator that introduces a slightly different frequency shift of 80.1 MHz. The two beams are then effectively superposed by a beam combiner before reaching a photomultiplier tube (PMT), which functions as a frequency mixer. The sum-frequency signal is filtered out automatically by the PMT, which is unable to respond to oscillations of optical frequencies. A beat-frequency signal of 0.1 MHz—the heterodyne frequency between the reference and the sample beams—from the PMT is amplified, digitized, and finally transferred to a computer. This signal represents the ballistic photons and can be used to form an image by transversely scanning the system across the sample.

In general, *heterodyne detection* refers to the use of a local oscillator (reference signal here) to mix a high-frequency signal (sample signal) with a more convenient intermediate-frequency signal (interference signal). The mixer (PMT)

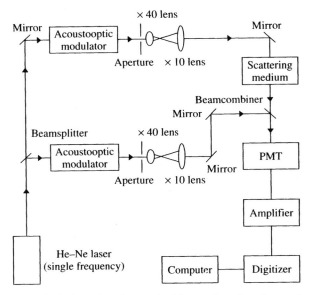

Figure 8.8. Schematic of an optical heterodyne imaging system.

generates both upper and lower sidebands, either of which may be filtered out if desired. In this case, the upper sideband is filtered out because the photodetector cannot respond to signals of optical frequency, and the lower sideband provides the interference signal.

Example 8.2. Show that the acceptance angle in the antenna theorem for a 1D case is approximately equal to λ/D.

In Figure 8.7, the electric fields of the sample and the reference waves can be expressed as

$$E_S = E_{S0} \exp(ik_x x - i\omega_S t + i\phi_{S0}), \tag{8.13}$$

$$E_R = E_{R0} \exp(-i\omega_R t + i\phi_{R0}), \tag{8.14}$$

respectively, where k_x is the projection of the wavevector $\vec{k}_S$ onto the x axis:

$$k_x = \frac{2\pi}{\lambda} \sin\theta. \tag{8.15}$$

If we consider only the 1D case, the light intensity on the detector surface ($z = 0$ plane) is

$$I(k_x, x) = |E_S + E_R|^2 = E_{S0}^2 + E_{R0}^2 + 2E_{S0}E_{R0}\cos(k_x x - \Delta\omega t + \Delta\phi_0), \tag{8.16}$$

where $\Delta\omega = \omega_S - \omega_R$ and $\Delta\phi_0 = \phi_{S0} - \phi_{R0}$. The first two terms represent the DC component while the last term represents the AC (interference) component. The AC photocurrent from the detector is proportional to the interference component integrated across the detector surface:

$$I_{AC}(k_x) \propto 2E_{S0}E_{R0} \int_{-D/2}^{D/2} \cos(k_x x - \Delta\omega t + \Delta\phi_0)\, dx$$

$$= \frac{2E_{S0}E_{R0}}{k_x}\left[\sin\left(k_x\frac{D}{2} - \Delta\omega t + \Delta\phi_0\right) - \sin\left(-k_x\frac{D}{2} - \Delta\omega t + \Delta\phi_0\right)\right]$$

$$= \frac{4E_{S0}E_{R0}}{k_x}\cos(-\Delta\omega t + \Delta\phi_0)\sin\left(k_x\frac{D}{2}\right) \tag{8.17}$$

$$= 2DE_{S0}E_{R0}\operatorname{sinc}\left(\frac{k_x D}{2\pi}\right)\cos(\Delta\omega t - \Delta\phi_0),$$

where $\operatorname{sinc}(x) = \sin(\pi x)/(\pi x)$, although sometimes the definition of $\operatorname{sinc}(x) = \sin(x)/x$ is used elsewhere. For effective detection, the argument of the sinc function must be less than unity, specifically

$$|k_x D| \leq 2\pi, \tag{8.18}$$

which leads to

$$|\sin\theta| \leq \frac{\lambda}{D}. \tag{8.19}$$

If $D \gg \lambda$, then $\sin\theta \approx \theta$. Thus

$$|\theta| \leq \frac{\lambda}{D}. \tag{8.20}$$

8.8. RADON TRANSFORMATION AND COMPUTED TOMOGRAPHY

In the aforementioned transmission-mode ballistic imaging, spatial resolution along the optical path can be achieved using the inverse Radon transformation, which is commonly used in X-ray CT for image reconstruction. Projections from multiple view angles are needed, however.

The Radon transform $p_\phi(x')$ of a function $f(x, y)$ is defined as the integral of the function along a line that is parallel to the y' axis at x' (Figure 8.9)

$$p_\phi(x') = \int_{-\infty}^{+\infty} f(x'\cos\phi - y'\sin\phi, x'\sin\phi + y'\cos\phi)\, dy', \tag{8.21}$$

where ϕ denotes the view angle—the angle between the x and x' axes. The Radon transform, also referred to as the *projection data* or the *sinogram*, is the input for image reconstruction.

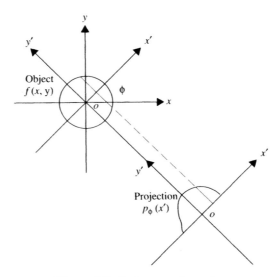

Figure 8.9. Radon transformation.

In ballistic imaging, the Radon transformation is related to the generalized Beer law:

$$I(z) = I(0)\exp(-\int_{-\infty}^{+\infty} \mu_t(z)\,dz), \tag{8.22}$$

where $I(z)$ denotes the light intensity, and z denotes the optical (ballistic path) axis. The following reformulation of Eq. (8.22) shows that the absorbance along the optical axis equals the Radon transform:

$$-\ln\frac{I(z)}{I(0)} = \int_{-\infty}^{+\infty} \mu_t(z)\,dz. \tag{8.23}$$

Various inverse algorithms can invert the Radon transformation for an image.

8.9. CONFOCAL MICROSCOPY

Confocal microscopy was invented in the 1950s but was not actively developed until the 1970s. A confocal microscope—in which both illumination and detection are focused on the same point in the object—provides optical sectioning for high-resolution 3D imaging of scattering samples. By contrast, a conventional microscope (Figure 8.10a)—in which the illumination is broadened by a condenser lens and the illuminated area of the object is mapped onto the image plane by an objective lens—provides no optical sectioning of planar features; it does, however, form a full-field image at once.

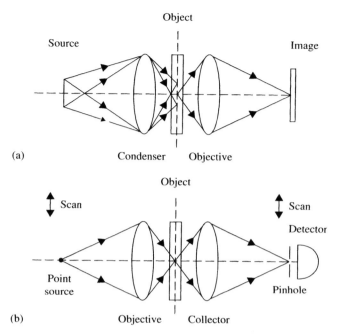

Figure 8.10. Schematic diagrams of (a) a conventional microscope and (b) a transmission confocal microscope.

In a transmission-mode confocal microscope (Figure 8.10b), a point source is imaged by a lens to a diffraction-limited spot to illuminate the 3D specimen. The illuminated spot is mapped by another lens to a pinhole to reject off-focus light. The filtered light is then detected by a photodetector. The detected signal is sensitive to the property of the sample at the illuminated point. Scanning point-by-point across the sample then forms an image. In other words, the focused light irradiates one tiny volume of the object at a time; the detector along with the pinhole collects light from the same region. The confocal illumination and detection effectively reject light from elsewhere. Consequently, both lenses play equally important roles in defining the spatial resolution.

A confocal microscope (Figure 8.11) can be implemented in reflection mode as well. As in the transmission mode, the illumination is focused by an objective lens to a spot. If elastically backscattered light is to be imaged, a beamsplitter is used to both partially reflect the source beam and partially transmit the reflected beam. If fluorescent light is to be imaged, a dichroic mirror can be used instead to both efficiently reflect the excitation light and efficiently transmit the fluorescent light. A reflection-mode confocal microscope can image hundreds of micrometers into scattering biological tissues.

The spatial resolution of a confocal microscope can be quantified by the PSF (the image of a point object). According to diffraction theory, the normalized

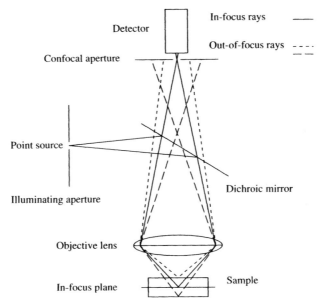

Figure 8.11. Schematic of a reflection-mode confocal microscope for fluorescence imaging.

optical coordinates u and v are defined by

$$u = \frac{8\pi \sin^2(\gamma/2)}{\lambda} z, \qquad (8.24)$$

$$v = \frac{2\pi \sin \gamma}{\lambda} r. \qquad (8.25)$$

Here, λ denotes the optical wavelength in the object, z denotes the distance from the focal point along the optical axis (defocus distance), γ denotes the angle that defines the numerical aperture NA along with the refractive index n (NA $= n \sin \gamma$), and r denotes the radial coordinate on the xy plane. If the system has circular symmetry, the field (or complex-amplitude) PSF of a lens in paraxial approximation is given by the following Hankel transform:

$$h(u,v) = 2 \int_0^1 \exp\left(\frac{i}{2} u \rho^2\right) J_0(\rho v) \rho \, d\rho. \qquad (8.26)$$

Here, ρ denotes the radial coordinate at the pupil normalized by the pupil radius and J_0 denotes the zeroth-order Bessel function of the first kind.

The PSF of a conventional microscope is

$$\text{PSF}(u,v) = |h(u,v)|^2, \qquad (8.27)$$

where the absolute squared operation is a consequence of conversion from complex amplitude to intensity. By contrast, the PSF of a confocal microscope is

$$\text{PSF}(u, v) = |h(u, v)|^4, \tag{8.28}$$

where the absolute-to-the-fourth-power operation is a result of double conversions from complex amplitude to intensity for both illumination and detection. Here, we have implicitly invoked the principle of reciprocity, which means that when the source and the observation points are exchanged, the observed field remains the same.

The field PSF on the optical axis can be calculated as follows:

$$h(u, 0) = 2 \int_0^1 \exp\left(\frac{i}{2} u \rho^2\right) \rho \, d\rho = \exp\left(\frac{i}{4} u\right) \text{sinc}\left(\frac{u}{4\pi}\right). \tag{8.29}$$

Therefore, the axial PSF of a conventional microscope is

$$\text{PSF}_z(u) = |h(u, 0)|^2 = \text{sinc}^2\left(\frac{u}{4\pi}\right), \tag{8.30}$$

and the axial PSF of a confocal microscope is

$$\text{PSF}_z(u) = |h(u, 0)|^4 = \text{sinc}^4\left(\frac{u}{4\pi}\right). \tag{8.31}$$

Likewise, the lateral field PSF on the in-focus plane can be calculated as follows:

$$h(0, v) = 2 \int_0^1 J_0(\rho v) \rho \, d\rho = \frac{2}{v}[\rho J_1(\rho v)]_0^1 = \frac{2}{v} J_1(v). \tag{8.32}$$

Therefore, the lateral PSF of a conventional microscope is

$$\text{PSF}_r(v) = |h(0, v)|^2 = \left|\frac{2 J_1(v)}{v}\right|^2, \tag{8.33}$$

and the lateral PSF of a confocal microscope is

$$\text{PSF}_r(v) = |h(0, v)|^4 = \left|\frac{2 J_1(v)}{v}\right|^4. \tag{8.34}$$

If the sample has an arbitrary complex-amplitude reflectivity distribution $o(u, v, \theta)$ (where θ is the polar angle), the image intensity distribution from a reflection-mode conventional microscope with incoherent illumination can be modeled by incoherent convolution $|h|^2 * |o|^2$. By contrast, the image intensity distribution from a reflection-mode confocal microscope can be modeled by coherent convolution $|h^2 * o|^2$. As expected, if the object reduces to a point, the convolution recovers the PSF.

If a fluorescent point object is imaged, the image intensity distribution from a conventional microscope remains $|h_m(u_m, v_m)|^2$, where subscript m indicates that the optical coordinates are defined at the fluorescence emission wavelength. However, the image intensity from a confocal microscope is modified to $|h_x(u_x, v_x) h_m(u_m, v_m)|^2$, where subscript x indicates that the optical coordinates are defined at the fluorescence excitation wavelength.

If the sample contains an arbitrary fluorophore density distribution $f(u, v, \theta)$, the image intensity distribution from a conventional microscope can still be modeled by incoherent convolution $|h_m|^2 * f$. The image intensity distribution

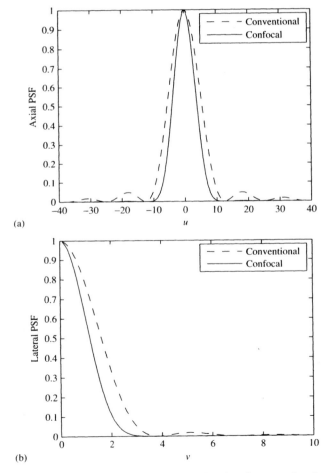

Figure 8.12. (a) Axial and (b) lateral PSFs in confocal and conventional microscopes.

from a confocal microscope can be modeled by incoherent convolution $|h_x h_m|^2 * f$ instead.

Example 8.3. Plot the axial and lateral PSFs for both conventional and confocal microscopes.

The following MATLAB code produces Figure 8.12:

```
u = linspace(-1,1)*3*4*pi;
subplot(2, 1, 1)
plot(u, (sinc(u/4/pi)).^2, 'k--', u, (sinc(u/4/pi)).^4, 'k-')
grid
xlabel('u')
ylabel('Axial PSF')
legend('Conventional', 'Confocal')

v = linspace(0,10);
subplot(2, 1, 2)
plot(v, (2*besselj(1,v)./v).^2, 'k--', v, (2*besselj(1,v)./v).^4, 'k-')
grid
xlabel('v')
ylabel('Lateral PSF')
legend('Conventional', 'Confocal')
```

8.10. TWO-PHOTON MICROSCOPY

Two-photon microscopy was initially developed in the early 1990s. A two-photon microscope (TPM) (Figure 8.13) achieves sectioning by nonlinear optical excitation. Unlike a confocal microscope, a TPM does not use a pinhole, although a pinhole can enhance the spatial resolution at the expense of signal strength.

To understand why a pinhole is unnecessary in a TPM, we first examine the difference between one-photon and two-photon excitations of fluorescence

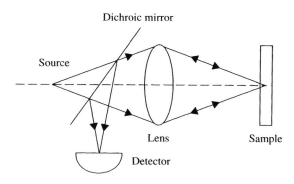

Figure 8.13. Schematic of a reflection-mode two-photon microscope.

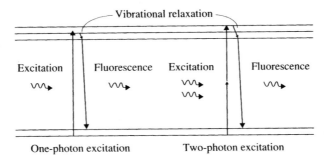

Figure 8.14. Jablonski energy diagrams for a one-photon excitation and two-photon excitation of fluorescence.

(Figure 8.14). In one-photon excitation, an electron is boosted to an excited state by absorbing a single photon. After a brief vibrational relaxation, it returns to a ground state while emitting a fluorescence photon. The probability of one-photon absorption is proportional to the light intensity. In two-photon excitation, an electron is pumped to an excited state by simultaneously absorbing two low-energy photons, followed by a similar process of fluorescence emission. Therefore, the probability of two-photon absorption is proportional to the square of the light intensity.

Compared with a confocal microscope, a TPM has the following characteristics in addition to the lack of a pinhole:

1. A more localized excitation volume defined by $|h_x(u_x, v_x)|^4$, rather than $|h_x(u_x, v_x)|^2$ as in the case of a fluorescence confocal microscope, leads to reduced photo-bleaching.
2. A longer excitation wavelength leads to increased penetration because both the absorption and the reduced scattering coefficients are decreased in the typical spectral region.
3. An ultra-short-pulsed laser is used.
4. Scattering contrast is not directly measured.

The diffraction-limited PSF is $|h_x(u_x, v_x)h_m(u_m, v_m)|^2$ for a fluorescence confocal microscope but becomes $|h_x(u_x, v_x)|^4$ for a TPM. In theory, when the two excitation wavelengths are the same and the single-photon emission wavelength is close to the single-photon excitation wavelength, the two PSFs are similar. However, when the two-photon excitation wavelength is twice as long as the single-photon excitation wavelength and the two-photon emission wavelength is the same as the single-photon emission wavelength, the PSF for a TPM is wider. If the sample contains an arbitrary fluorophore density distribution $f(u, v, \theta)$, the image intensity distribution from a two-photon microscope can be modeled by incoherent convolution $|h_x|^4 * f$.

Example 8.4. Estimate the number of excitation photons absorbed per fluorophore for a TPM.

Typically, the laser in a TPM has pulse duration $\tau_p \approx 100$ fs, pulse repetition rate $f_p \approx 80$ MHz, and average excitation power $P_0 \approx 50$ mW. Thus, the pulse energy is

$$E_p = \frac{P_0}{f_p} \approx 0.6 \text{ nJ}. \tag{8.35}$$

The number of excitation photons absorbed per fluorophore N_a can be estimated by

$$N_a = J_p^2 \frac{\sigma_{2p}}{\tau_p}. \tag{8.36}$$

Here, σ_{2p} denotes the two-photon absorption cross section ($\sim 10^{-58}$ m^4-s) and J_p denotes the photon flux per pulse (m^{-2}). We estimate J_p by

$$J_p = \frac{E_p}{h\nu A_f}, \tag{8.37}$$

where $h\nu$ is the photon energy and A_f is the area of the focused beam. We estimate A_f by

$$A_f = \pi \left(0.66 \frac{\lambda}{\text{NA}}\right)^2 = 1.44 \left(\frac{\lambda}{\text{NA}}\right)^2, \tag{8.38}$$

where λ is the excitation wavelength (~ 800 nm) and NA is the numerical aperture of the objective lens (~ 0.9). According to these parameters, $N_a \approx 0.005 \ll 1$. Only some of the absorbed photons are converted into fluorescence photons since the quantum yield ranges from $\sim 5\%$ to $\sim 90\%$. Nevertheless, fluorescence signals are detectable when the excitation volume contains enough fluorophores.

APPENDIX 8A. HOLOGRAPHY

The principle of conventional holography is described here. Holography records both the field amplitude E_0 and the phase ϕ of a lightbeam, whereas conventional photography records only the intensity. A hologram presents the effect of stereovision. When a hologram is recorded (Figure 8.15), a source beam is split into two parts: one illuminates the object, and the other serves as a coherent reference wave. When the object wave reaches the recording film, it interferes with the coherent reference wave. The intensity distribution of the combined beam forms an interferogram and is recorded on the film.

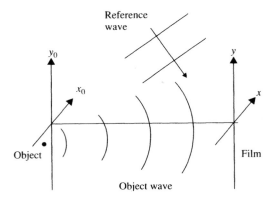

Figure 8.15. Recording of a hologram. The spherical wave from a point on the (x_0, y_0) plane is illustrated as the object wave.

To simplify our discussion, we assume a monochromatic light source so that the phase difference between the object and the reference waves is time-invariant. The object and the reference waves are denoted in phasor expressions by $E_S(x, y)$ and $E_R(x, y)$, respectively, where (x, y) is the Cartesian coordinates on the recording plane. The recorded intensity $I(x, y)$, referred to as a *hologram*, is then given by

$$I(x, y) = |E_S + E_R|^2 = (E_S + E_R)(E_S + E_R)^*$$
$$= |E_S|^2 + |E_R|^2 + E_S E_R^* + E_S^* E_R. \quad (8.39)$$

If E_R is zero, a hologram reduces to a conventional photograph.

Once the film is developed, the recorded hologram can be represented by the complex-amplitude transmittance of the film as follows:

$$t_f(x, y) = t_b + \beta I(x, y) = t_b' + \beta(|E_S|^2 + E_S E_R^* + E_S^* E_R), \quad (8.40)$$

where t_b denotes the film-dependent baseline (background transmittance with zero exposure) and β denotes the sensitivity that relates the transmittance to the recorded intensity. Since the reference beam contains no imaging information, the $|E_R|^2$ term is lumped with $t_b : t_b' = t_b + \beta |E_R|^2$.

Reconstruction of a hologram recovers the object wave—in either the original or the conjugated form—by illuminating the hologram. If the reconstruction beam is the same as the reference beam, we multiply E_R on both sides of Eq. (8.40) to obtain the field of the transmitted beam:

$$E_R t_f = E_R t_b' + \beta E_R |E_S|^2 + \beta E_S |E_R|^2 + \beta E_S^* E_R E_R. \quad (8.41)$$

The last two terms on the right-hand side are important because they contain information about both the amplitude and phase of either the object wave E_S or the complex-conjugated object wave E_S^*.

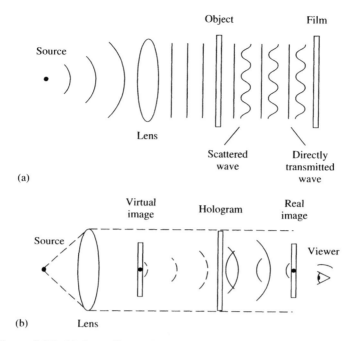

Figure 8.16. (a) Recording and (b) reconstruction in Gabor holography.

To illustrate the recording and reconstruction, we first describe the original Gabor holography (Figure 8.16a). For recording, a plane wave—converted from a spherical wave by a lens—is normally incident on an object. The transmitted light consists of two emerging waves—one is a directly transmitted plane wave that serves as the reference, and the other is a scattered object wave that carries imaging information about the object. The interference between these two waves results in a hologram on the film.

For reconstruction, a plane wave is incident on the hologram (Figure 8.16b). The field of the transmitted beam is given by Eq. (8.41). The first term on the right-hand side represents a homogeneous background since E_R for a plane wave is independent of x and y. The second term represents an intensity image of the object (a conventional photograph). The third term forms a virtual image because it replicates the original object wavefront and represents a divergent wave propagating from the hologram. Conversely, the fourth term forms a real image because it produces the complex conjugate of the original object wavefront and represents a converging wave propagating from the hologram. For a point object, the divergent and convergent waves for the virtual and real images are spherical as illustrated. The virtual image is so named because no photons actually reach the image location; by contrast, photons do reach the real image. Here, the real and virtual images appear together because their corresponding waves propagate in the same direction.

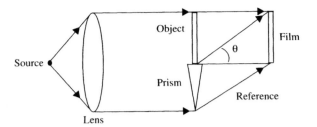

Figure 8.17. Schematic illustrating Leith–Upatnieks (offset-reference) holography.

To separate the real and the virtual images, Leith and Upatnieks used an obliquely incident reference wave for recording but a normally incident reference wave for reconstruction (Figure 8.17). For recording, a prism abutting the sample refracts the incident beam by an angle θ. We choose the vertical direction as the y axis and express the reference and the object fields as

$$E_R(y) = E_0(\omega) \exp(ik_y y - i\omega t), \tag{8.42}$$

$$E_S(x, y) = a(x, y) E_0(\omega) \exp(-i\omega t). \tag{8.43}$$

Here, E_0 denotes the amplitude of the source beam, ω denotes the angular frequency, t denotes time, $a(x, y)$ denotes the transverse distribution of the object wave, and k_y denotes the y component of the wavevector

$$k_y = \frac{2\pi \sin \theta}{\lambda}, \tag{8.44}$$

where λ is the optical wavelength. Because the reference wavefront is parallel to the x direction, E_R is independent of x. Of course, Gabor holography is recovered if $\theta = 0$.

The recorded intensity distribution on the film is

$$\begin{aligned} I(x, y) &= |E_S + E_R|^2 \\ &= E_0^2 [1 + |a(x, y)|^2 + a(x, y) \exp(-ik_y y) + a^*(x, y) \exp(ik_y y)]. \end{aligned} \tag{8.45}$$

The reconstruction with a plane wave of amplitude E_i normally incident on the hologram is illustrated in Figure 8.18. The complex amplitude of the transmitted field is given by

$$\begin{aligned} t_f E_i &= t'_b E_i + \beta E_0^2 E_i |a(x, y)|^2 + \beta E_0^2 E_i a(x, y) \exp(-ik_y y) \\ &\quad + \beta E_0^2 E_i a^*(x, y) \exp(ik_y y). \end{aligned} \tag{8.46}$$

As in Gabor holography, the last two terms on the right-hand side represent the virtual and the real images, respectively. The two complex-amplitude images,

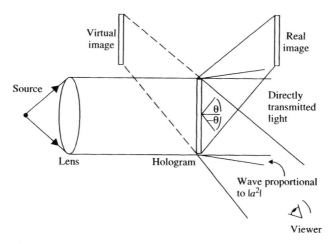

Figure 8.18. Reconstruction in Leith–Upatnieks holography.

however, have $\mp ik_y y$ in the exponents, indicating different spatial frequencies. As a result, the virtual and the real images can be viewed in the $\mp\theta$ directions, respectively; thus, one image may be viewed at a time, which is essentially a filtering process. In digital holography, this filtering is implemented computationally.

PROBLEMS

8.1 Use the Henyey–Greenstein phase function to compute the percentage of backscattered light R_b that can be received by the detector in a reflection-mode microscope. Assume that the diameter of the detector is 10 μm and the distance between the scatterer and the detector is $2l_t$, where $\mu_t = 100$ cm^{-1}. Set g to 0, 0.9, and 0.95 sequentially.

8.2 In transmission-mode ballistic imaging, assuming that the number of unscattered transmitted photons limits the maximum thickness of the biological tissue that can be imaged, derive the increase in this maximum thickness when the power of the incident source beam is doubled. Given an original maximum thickness of $30l_t$, compute the fractional improvement.

8.3 In time-gated transmission imaging, estimate the required temporal resolution of the time gate if a resolution better than 0.3 mm through a 3-mm-thick tissue sample is desired.

8.4 In spatiofrequency-filtered imaging, the smaller the pinhole, the better the rejection of scattered light, but the worse the transmission of ballistic light owing to light diffraction. Derive the transmittance of ballistic light through a pinhole of radius r_p. {*Hints:* (1) use $[2J_1(v)/v]^2$ for the diffracted intensity distribution and (2) use MATLAB to complete the Bessel function integration, e.g., `syms v; int(besselj(1,v)^2/v).`}

8.5 In polarization-difference imaging, assuming that the sample is equivalent to a half-wave retarder for the ballistic light due to tissue birefringence, modify the configuration for effective ballistic imaging.

8.6 If polarization-difference imaging is implemented in reflection mode, explain how to reject light scattered from deeper regions of the medium.

8.7 In Leith–Upatnieks holography, given a hologram recorded on a CCD camera with a pixel size of 5 μm, compute the maximum offset angle below which the Nyquist criterion (>2 pixels per cycle) is satisfied. Assume an optical wavelength of 0.5 μm.

8.8 In Leith–Upatnieks holography, assuming the reference beam to be tilted out of the xz plane, generalize the theory.

8.9 In coherence-gated holographic imaging, the a_2 term contributes much less to the signal than does the a_1 term when the tuned bandwidth of the laser is sufficiently wide. Explain how wide is considered sufficient.

8.10 Derive the 1D antenna theorem in heterodyne detection using real values of the electric fields rather than phasor expressions. (*Hint:* In this case, time averaging over the response time of the detector is explicit.)

8.11 In heterodyne imaging, assuming that the source wavelength is tuned over a range, show that even late-arriving light normally incident on the detector can be rejected as in the coherence-gated holographic imaging.

8.12 For both confocal and conventional microscopes, find the radius r at which the lateral PSF is zero. The center area within this radius is referred to as the *Airy disk*. [*Hint:* The first zero of $J_1(v)$ is at $v = 1.22\pi$.]

8.13 Define spatial resolution as the FWHM of the PSF. Set refractive index n to 1.0 and 1.5 sequentially. For both confocal and conventional microscopes, plot the axial resolutions versus NA in the range of 0.40–0.99. Repeat for the lateral resolutions versus NA in a separate figure. Plot the ratio of the conventional microscopic resolution to the confocal microscopic resolution versus NA in a third figure. (*Hint:* In MATLAB, type `help fzero`.)

8.14 Use the Monte Carlo method to simulate time-resolved transmitted light through a scattering slab. An infinitely short-pulsed pencil beam is normally incident on the slab from one side. Add a small pinhole on the other side in front of the detector. Set the thickness of the slab to 0.5, 1, 2, 4, ... times the mean free path sequentially.

8.15 In confocal and conventional microscopes, if a planar target perpendicular to the optical axis is imaged, what are the axial PSFs?

READING

Cho ZH, Jones JP, and Singh M (1993): *Foundations of Medical Imaging*, Wiley, New York. (See Section 8.8, above.)

Denk W, Strickler JH, and Webb WW (1990): 2-photon laser scanning fluorescence microscopy, *Science* **248**(4951): 73–76. (See Section 8.10, above.)

Dolne JJ, Yoo KM, Liu F, and Alfano RR (1994): IR Fourier space gate and absorption imaging through random media, *Lasers Life Sci.* **6**: 131–141. (See Section 8.4, above.)

Goodman JW (2004): *Introduction to Fourier Optics*, Roberts & Co., publishers, Englewood, CO., (See Appendix 8A, above.)

Gu M (1996): *Principles of Three Dimensional Imaging in Confocal Microscopes*, World Scientific, Singapore/River Edge, NJ. (See Section 8.9, above.)

Leith E, Chen C, Chen H, Chen Y, Dilworth D, Lopez J, Rudd J, Sun PC, Valdmanis J, and Vossler G (1992): Imaging through scattering media with holography, *J. Opt. Soc. Am. A* **9**(7): 1148–1153. (See Section 8.6, above.)

Rowe MP, Pugh EN, Tyo JS, and Engheta N (1995): Polarization-difference imaging—a biologically inspired technique for observation through scattering media, *Opt. Lett.* **20**(6): 608–610. (See Section 8.5, above.)

Toida M, Kondo M, Ichimura T, and Inaba H (1991): 2-dimensional coherent detection imaging in multiple-scattering media based on the directional resolution capability of the optical heterodyne method, *Appl. Phys. B—Photophys. Laser Chem.* **52**(6): 391–394. (See Section 8.7, above.)

Wang L, Ho PP, Liu C, Zhang G, and Alfano RR (1991): Ballistic 2-d imaging through scattering walls using an ultrafast optical kerr gate, *Science* **253**(5021): 769–771. (See Section 8.3, above.)

Wilson T, ed. (1990): *Confocal Microscopy*, Academic Press, New York. (See Section 8.9, above.)

FURTHER READING

Alfano RR, Liang X, Wang L, and Ho PP (1994): Time-resolved imaging of translucent droplets in highly scattering turbid media, *Science* **264**(5167): 1913–1915.

Bashkansky M and Reintjes J (1993): Imaging through a strong scattering medium with nonlinear-optical field cross-correlation techniques, *Opt. Lett.* **18**(24): 2132–2134.

Bohnke M and Masters BR (1999): Confocal microscopy of the cornea, *Progress Retinal Eye Res.* **18**(5): 553–628.

Cahalan MD, Parker I, Wei SH, and Miller MJ (2002): Two-photon tissue imaging: Seeing the immune system in a fresh light, *Nature Rev. Immunol.* **2**(11): 872–880.

Chen H, Shih M, Arons E, Leith E, Lopez J, Dilworth D, and Sun PC (1994): Electronic holographic imaging through living human tissue, *Appl. Opt.* **33**(17): 3630–3632.

Chen Y, Chen H, Dilworth D, Leith E, Lopez J, Shih M, Sun PC, and Vossler G (1993): Evaluation of holographic methods for imaging through biological tissue, *Appl. Opt.* **32**(23): 4330–4336.

Das BB, Yoo KM, and Alfano RR (1993): Ultrafast time-gated imaging in thick tissues—a step toward optical mammography, *Opt. Lett.* **18**(13): 1092–1094.

Demos SG and Alfano RR (1996): Temporal gating in highly scattering media by the degree of optical polarization, *Opt. Lett.* **21**(2): 161–163.

Demos SG and Alfano RR (1997): Optical polarization imaging, *Appl. Opt.* **36**(1): 150–155.

Demos SG, Savage H, Heerdt AS, Schantz S, and Alfano RR (1996): Time resolved degree of polarization for human breast tissue, *Opt. Commun.* **124**(5–6): 439–442.

Diaspro A (1999): Introduction to two-photon microscopy, *Microsc. Res. Tech.* **47**(3): 163–164.

Dunn AK, Wallace VP, Coleno M, Berns MW, and Tromberg BJ (2000): Influence of optical properties on two-photon fluorescence imaging in turbid samples, *Appl. Opt.* **39**(7): 1194–1201.

Emile O, Bretenaker F, and LeFloch A (1996): Rotating polarization imaging in turbid media, *Opt. Lett.* **21**(20): 1706–1708.

Gard DL (1999), Confocal microscopy and 3-D reconstruction of the cytoskeleton of Xenopus oocytes, *Microsc. Res. Tech.* **44**(6): 388–414.

Gauderon R, Lukins PB, and Sheppard CJR (1999): Effect of a confocal pinhole in two-photon microscopy, *Microsc. Res. Tech.* **47**(3): 210–214.

Guo YC, Ho PP, Savage H, Harris D, Sacks P, Schantz S, Liu F, Zhadin N, and Alfano RR (1997): Second-harmonic tomography of tissues, *Opt. Lett.* **22**(17): 1323–1325.

Hebden JC and Delpy DT (1994): Enhanced time-resolved imaging with a diffusion-model of photon transport, *Opt. Lett.* **19**(5): 311–313.

Hebden JC, Hall DJ, and Delpy DT (1995): The spatial-resolution performance of a time-resolved optical imaging-system using temporal extrapolation, *Med. Phys.* **22**(2): 201–208.

Hee MR, Izatt JA, Jacobson JM, Fujimoto JG, and Swanson EA (1993): Femtosecond transillumination optical coherence tomography, *Opt. Lett.* **18**(12): 950–952.

Hee MR, Izatt JA, Swanson EA, and Fujimoto JG (1993): Femtosecond transillumination tomography in thick tissues, *Opt. Lett.* **18**(13): 1107–1109.

Horinaka H, Hashimoto K, Wada K, Cho Y, and Osawa M (1995): Extraction of quasi-straightforward-propagating photons from diffused light transmitting through a scattering medium by polarization modulation, *Opt. Lett.* **20**(13): 1501–1503.

Kempe M, Genack AZ, Rudolph W, and Dorn P (1997): Ballistic and diffuse light detection in confocal and heterodyne imaging systems, *J. Opt. Soc. Am. A* **14**(1): 216–223.

Konig K (2000): Multiphoton microscopy in life sciences, *J. Microsc.—Oxford* **200**: 83–104.

Mahon R, Duncan MD, Tankersley LL, and Reintjes J (1993): Time-gated imaging through dense scatterers with a raman amplifier, *Appl. Opt.* **32**(36): 7425–7433.

Minsky M (1988): Memoir on inventing the confocal scanning microscope, *Scanning* **10**(4): 128–138.

Moon JA and Reintjes J (1994): Image-resolution by use of multiply scattered-light, *Opt. Lett.* **19**(8): 521–523.

Moon JA, Battle PR, Bashkansky M, Mahon R, Duncan MD, and Reintjes J (1996): Achievable spatial resolution of time-resolved transillumination imaging systems which utilize multiply scattered light, *Phys. Rev. E* **53**(1): 1142–1155.

Nakamura O (1999): Fundamental of two-photon microscopy, *Microsc. Res. Tech.* **47**(3): 165–171.

Nie SM and Zare RN (1997): Optical detection of single molecules, *Annu. Rev. Biophys. Biomolec. Struct.* **26**: 567–596.

Sappey AD (1994): Optical imaging through turbid media with a degenerate 4-wave-mixing correlation time gate, *Appl. Opt.* **33**(36): 8346–8354.

Schmidt A, Corey R, and Saulnier P (1995): Imaging through random-media by use of low-coherence optical heterodyning, *Opt. Lett.* **20**(4): 404–406.

Sheppard CJR and Shotton DM (1997): *Confocal Laser Scanning Microscopy*, Springer-Verlag, New York.

Shuman H, Murray JM, and Dilullo C (1989): Confocal microscopy—an overview, *Biotechniques* **7**(2): 154 FF.

So PTC, Dong CY, Masters BR, and Berland KM (2000): Two-photon excitation fluorescence microscopy, *Annu. Rev. Biomed. Eng.* **2**: 399–429.

Tyo JS, Rowe MP, Pugh EN, and Engheta N (1996): Target detection in optically scattering media by polarization-difference imaging, *Appl. Opt.* **35**(11): 1855–1870.

Wang L, Ho PP, and Alfano RR (1993): Time-resolved Fourier spectrum and imaging in highly scattering media, *Appl. Opt.* **32**(26): 5043–5048.

Wang LM, Ho PP, and Alfano RR (1993): Double-stage picosecond Kerr gate for ballistic time-gated optical imaging in turbid media, *Appl. Opt.* **32**(4): 535–540.

Wang QZ, Liang X, Wang L, Ho PP, and Alfano RR (1995): Fourier spatial filter acts as a temporal gate for light propagating through a turbid medium, *Opt. Lett.* **20**(13): 1498–1500.

Watson J, Georges P, Lepine T, Alonzi B, and Brun A (1995): Imaging in diffuse media with ultrafast degenerate optical parametric amplification, *Opt. Lett.* **20**(3): 231–233.

Wilson T and Sheppard C (1984): *Theory and Practice of Scanning Optical Microscopy*, Academic Press, London.

Yoo KM, Liu F, and Alfano RR (1991): Imaging through a scattering wall using absorption, *Opt. Lett.* **16**(14): 1068–1070.

Yoo KM, Xing QR, and Alfano RR (1991): Imaging objects hidden in highly scattering media using femtosecond 2nd-harmonic-generation cross-correlation time gating, *Opt. Lett.* **16**(13): 1019–1021.

Zaccanti G and Donelli P (1994): Attenuation of energy in time-gated transillumination imaging—numerical results, *Appl. Opt.* **33**(30): 7023–7030.

CHAPTER 9

Optical Coherence Tomography

9.1. INTRODUCTION

Optical coherence tomography (OCT), which was invented in the early 1990s, falls into the category of ballistic optical imaging. OCT is analogous to ultrasonography. The transverse resolution results from the confocal mechanism, albeit with a small numerical aperture. The axial resolution results from the arrival times of echoes. The detection in OCT, however, is based on interferometry since light speed is five orders of magnitude greater than sound speed. The maximum imaging depth in scattering biological tissue is 1–2 mm, and the spatial resolution ranges from 1 to 10 µm. As a result, the depth-to-resolution ratio is greater than 100, which qualifies OCT as a high-resolution imaging modality. The contrast originates primarily from backscattering (or backreflection) and polarization. Since the human eye provides an optically transparent window, noninvasive imaging of the retina is thus far the most competitive application of OCT.

9.2. MICHELSON INTERFEROMETRY

Michelson interferometry (Figure 9.1), the basis of OCT, is briefly introduced in this section. A monochromatic light source emits a source beam horizontally toward a beamsplitter that is inclined diagonally. The source beam is split into two halves. One half is reflected off the beamsplitter and then backreflected by a reference mirror. The other half is transmitted through the beamsplitter and then backreflected by an object surface (no internal backscattering exists here). These two backreflected beams are recombined by the beamsplitter and then received by a detector.

Biomedical Optics: Principles and Imaging, by Lihong V. Wang and Hsin-i Wu
Copyright © 2007 John Wiley & Sons, Inc.

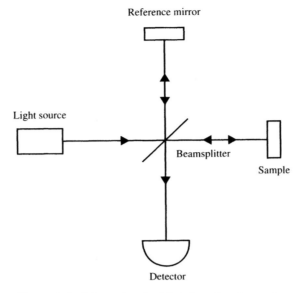

Figure 9.1. Schematic of a Michelson interferometer.

If polarization is neglected, the two backreflected electric fields can be represented by phasor expressions as follows:

$$E_R = E_{R0} \exp(i(2k_R l_R - \omega t)), \qquad (9.1)$$

$$E_S = E_{S0} \exp(i(2k_S l_S - \omega t)). \qquad (9.2)$$

Here, subscripts R and S denote the reference and the sample arms, respectively; E_{R0} and E_{S0} denote the electric field amplitudes of the two beams; k_R and k_S denote the propagation constants in the two beams; l_R and l_S denote the two arm lengths measured from the splitting point at the beamsplitter to the backreflection surfaces; ω denotes the optical angular frequency; and t denotes time. The factor 2 in front of k arises from the round-trip light propagation in each arm. The electric field E of the recombined beam is a superposition of the two monochromatic electric fields:

$$E = E_R + E_S.$$

The photocurrent $i(t)$ at the photodetector (a square-law detector) is given by

$$i(t) = \frac{\eta e}{h\nu} \frac{\langle |E_R + E_S|^2 \rangle}{2Z_0}. \qquad (9.3)$$

Here, η denotes the quantum efficiency of the detector (the ratio of the output number of electrons to the input number of photons), e denotes the electron charge, $h\nu$ denotes the photon energy, Z_0 denotes the intrinsic impedance of free space, and $\langle \rangle$ denotes averaging over the response time of the detection system (e.g., $10^{-12} - 10^{-9}$ s, or ps to ns). The response-time averaging is equivalent to lowpass filtering; thus, the photocurrent can still be a function of time t. For brevity, we neglect the constant factors and simply write

$$I(t) = \langle |E_R + E_S|^2 \rangle. \tag{9.4}$$

Here, $I(t)$ denotes the short-time-averaged light intensity; it is used instead of $i(t)$ from here on. For monochromatic light, we write

$$I(t) = |E_R + E_S|^2, \tag{9.5}$$

where another factor of $\frac{1}{2}$ is neglected.

Substituting Eqs. (9.1) and (9.2) into Eq. (9.5) yields

$$I(t) = E_{R0}^2 + E_{S0}^2 + 2E_{R0}E_{S0}\cos(2k_S l_S - 2k_R l_R). \tag{9.6}$$

The cosine term on the right-hand side results from the interference between the two lightbeams. We denote the phase difference between the two beams as $\Delta\phi$:

$$\Delta\phi = 2k_S l_S - 2k_R l_R. \tag{9.7}$$

With a varying $\Delta\phi$, this interference term becomes an alternating current (AC) that produces interference fringes; hence, the recorded I is also referred to as an *interferogram*.

If $k_R = k_S = k = 2\pi n/\lambda_0$, where n denotes the refractive index and λ_0 denotes the optical wavelength in vacuum, we have

$$\Delta\phi = 2k(l_S - l_R) = 2\pi \frac{2n\Delta l}{\lambda_0}, \tag{9.8}$$

where

$$\Delta l = l_S - l_R. \tag{9.9}$$

From here on, Δl is termed the *arm-length difference* (or *mismatch*) between the sample and the reference arms; $2\Delta l$ is termed the *(round-trip) path-length difference* (or *mismatch*) between the two beams; $2n\Delta l$ is termed the *(round-trip) optical path-length difference* (or *mismatch*) between the two beams. Therefore, the interference signal varies with Δl periodically. For monochromatic light, the fringes exhibit a sustained oscillation of constant amplitude.

9.3. COHERENCE LENGTH AND COHERENCE TIME

The coherence length l_c of light is defined as the spatial extent along the propagation direction over which the electric field is substantially correlated; it is related to the coherence time τ_c by $l_c = c\tau_c$, with c denoting the speed of light. In stationary states, where the statistical properties do not change with time, τ_c is defined as the FWHM of the autocorrelation function $G_1(\tau)$ of the electric field $E(t)$:

$$G_1(\tau) = \int_{-\infty}^{+\infty} E(t)E(t+\tau)\,dt. \tag{9.10}$$

Both the coherence length and the coherence time are inversely proportional to the frequency bandwidth for a given spectral shape according to the following Wiener–Khinchin theorem:

$$\int_{-\infty}^{+\infty} G_1(\tau)\exp(i\omega\tau)\,d\tau = |E(\omega)|^2, \tag{9.11}$$

where $E(\omega)$ is the Fourier transform of $E(t)$. Note that

$$|E(\omega)|^2 = S(\omega), \tag{9.12}$$

where $S(\omega)$ is the power spectral density distribution of the light. The Wiener–Khinchin theorem, a special case of the cross-correlation theorem, states that the autocorrelation function of the electric field and the power spectrum are a Fourier transform pair.

If $S(\omega)$ is Gaussian, we have

$$S(\omega) = \frac{1}{\sqrt{2\pi}\sigma_\omega} \exp\left(-\frac{(\omega-\omega_0)^2}{2\sigma_\omega^2}\right), \tag{9.13}$$

where ω_0 denotes the center angular frequency and σ_ω denotes the standard deviation of ω. Since the profile is of key interest, $S(\omega)$ is normalized to unit power:

$$\int_{-\infty}^{\infty} S(\omega)\,d\omega = 1. \tag{9.14}$$

It can be shown (see Problem 9.1) that the coherence length is given by

$$l_c = \frac{4\ln 2}{\pi}\frac{\lambda_0^2}{\Delta\lambda}, \tag{9.15}$$

where λ_0 denotes the center wavelength of the light source and $\Delta\lambda$ denotes the FWHM bandwidth in wavelength. The broader the bandwidth, the shorter the coherence length becomes. In an interferometer, the two beams are said to be coherent with each other when $2n\Delta l \leq l_c$. Note that Eq. (9.15) is for a Gaussian lineshape and the constant factor on the right-hand side varies with the spectral shape; $\lambda_0^2/\Delta\lambda$, however, is sometimes used to estimate l_c regardless of the spectral shape.

9.4. TIME-DOMAIN OCT

OCT is based on Michelson interferometry with a light source of short coherence length; it can be implemented either in free space or by using optical fibers (Figure 9.2). The optical fibers, however, must be single-mode because modal

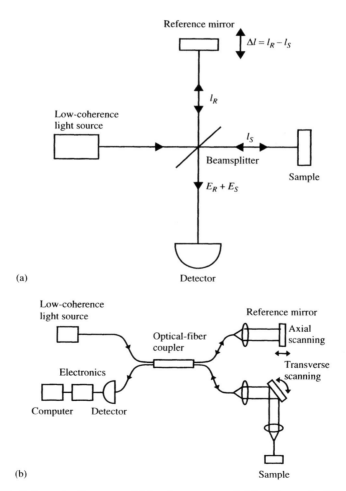

Figure 9.2. Schematic diagrams of (a) a free-space and (b) a fiberoptic OCT system.

dispersions in multimode fibers broaden the axial resolution. A superluminescent diode (SLD) is commonly used as the light source because of its high radiance and relatively low cost. While the reference beam is reflected from a mirror, the sample beam is backscattered from various depths in the biological tissue sample.

In time-domain OCT (TD-OCT), the reference arm length oscillates periodically, which causes a Doppler frequency shift in the reflected reference beam. The sample and the reference beams are recombined and subsequently detected by the photodetector; the detection falls into the category of heterodyne detection owing to the Doppler shift. The two beams coherently interfere only when their optical path-length difference is within the coherence length of the source, which is referred to as *coherence gating*, an effect that enables OCT to resolve the path-length distribution of the backscattered light. Therefore, the axial resolution is determined by the coherence length of the source. In the ballistic or quasiballistic regime, the path length distribution can be directly converted into a physical depth distribution; in the quasidiffusive or diffusive regime, however, this conversion breaks down.

Recording the interference fringes or the envelope as a function of l_R profiles the backscattering reflectance from the sample versus the depth. As a result, a 1D image—referred to as an *A-scan* or *A-line image*—is produced. Multiple A-scan images acquired by transverse scanning form a 2D B-scan image or a 3D volumetric image.

Before a more rigorous theory is described, a simple approach is presented to illuminate the basic principle of OCT. We let $E_S = E_C + E_I$, where E_C and E_I represent the components of the backscattered sample beam that are coherent and incoherent with the reference beam, respectively; thus, Eq. (9.4) becomes

$$I(t) = \langle |E_R + E_C + E_I|^2 \rangle. \tag{9.16}$$

Because E_I has a random phase difference relative to E_R, it does not contribute to the AC signal. Therefore, we have

$$I(t) = E_{R0}^2 + E_{C0}^2 + E_{I0}^2 + 2E_{R0}E_{C0}\cos\left(2\pi \frac{\Delta l_{CR}(t)}{\lambda_0/2}\right). \tag{9.17}$$

Here, E_{C0} and E_{I0} denote the amplitudes of E_C and E_I, respectively; Δl_{CR} denotes the arm-length difference between E_C and E_R; λ_0 denotes the center wavelength of the light source. In Eq. (9.17), the last term represents an AC signal I_{AC}, whose amplitude is proportional to E_{C0}. If the reference mirror is scanned axially, a depth-resolved distribution of E_{C0} can be acquired; it provides an A-scan image with an axial resolution that is limited by the coherence length.

A more rigorous theory is presented below. Any low-coherence light field $E(t)$ is a superposition of monochromatic waves of various frequencies by virtue of the inverse Fourier transformation:

$$E(t) = \frac{1}{2\pi}\int_{-\infty}^{+\infty} E(\omega)\exp(-i\omega t)\,d\omega. \tag{9.18}$$

Therefore, the electric fields of the reference and the sample beams can be expressed in the frequency domain as

$$E_R(\omega) = E_{R0}(\omega) \exp(i(2k_R(\omega)l_R - \omega t)), \tag{9.19}$$

$$E_S(\omega) = E_{S0}(\omega) \exp(i(2k_S(\omega)l_S - \omega t)). \tag{9.20}$$

For brevity, only a single backscatterer along each A-line is considered for E_S. The light intensity at angular frequency ω is

$$I(\omega) = |E_R(\omega) + E_S(\omega)|^2 = |E_R(\omega)|^2 + |E_S(\omega)|^2 + 2\text{Re}\left\{E_R(\omega)E_S^*(\omega)\right\}, \tag{9.21}$$

where the cross-term provides the interference signal at ω. Superposition of the interference signals at all angular frequencies yields the total interference signal I_{AC}:

$$I_{\text{AC}} = 2\text{Re}\left\{\int_{-\infty}^{\infty} E_R(\omega)E_S^*(\omega)\,d\omega\right\}. \tag{9.22}$$

Substituting Eqs. (9.19) and (9.20) into Eq. (9.22), we obtain

$$I_{\text{AC}} = 2\text{Re}\left\{\int_{-\infty}^{\infty} E_{R0}(\omega)E_{S0}^*(\omega)\exp(-i\Delta\phi(\omega))\,d\omega\right\}, \tag{9.23}$$

where

$$\Delta\phi(\omega) = 2k_S(\omega)l_S - 2k_R(\omega)l_R. \tag{9.24}$$

We have

$$S(\omega) \propto E_{R0}(\omega)E_{S0}^*(\omega), \tag{9.25}$$

where the proportionality constant is related to the amplitude reflectivities in the two arms. Thus, Eq. (9.23) can be rewritten as

$$I_{\text{AC}} \propto \text{Re}\left\{\int_{-\infty}^{\infty} S(\omega)\exp(-i\Delta\phi(\omega))\,d\omega\right\}. \tag{9.26}$$

If (1) the spectrum of the source light is bandlimited around a center frequency ω_0 and (2) the sample and the reference arms consist of a uniform nondispersive material, we can approximately express the propagation constant as a first-order Taylor series around ω_0

$$k_S(\omega) = k_R(\omega) = k(\omega) = k(\omega_0) + k'(\omega_0)(\omega - \omega_0), \tag{9.27}$$

where k' denotes the derivative of k with respect to ω. Thus, Eq. (9.24) becomes

$$\Delta\phi = k(\omega_0)(2\Delta l) + k'(\omega_0)(\omega - \omega_0)(2\Delta l), \tag{9.28}$$

which can be rewritten as

$$\Delta\phi = \omega_0 \Delta\tau_p + (\omega - \omega_0)\Delta\tau_g. \tag{9.29}$$

Here, $\Delta\tau_p$ denotes the round-trip phase delay between the two arms

$$\Delta\tau_p = \frac{k(\omega_0)}{\omega_0}(2\Delta l), \tag{9.30}$$

and $\Delta\tau_g$ denotes the round-trip group delay between the two arms:

$$\Delta\tau_g = k'(\omega_0)(2\Delta l). \tag{9.31}$$

From the definition of the phase velocity v_p

$$v_p = \frac{\omega_0}{k(\omega_0)}, \tag{9.32}$$

Eq. (9.30) can be rewritten as

$$\Delta\tau_p = \frac{2\Delta l}{v_p}, \tag{9.33}$$

From the definition of the group velocity v_g

$$v_g = \frac{1}{k'(\omega_0)}, \tag{9.34}$$

Eq. (9.31) can be rewritten as

$$\Delta\tau_g = \frac{2\Delta l}{v_g}. \tag{9.35}$$

Substituting Eq. (9.29) into Eq. (9.26) yields

$$I_{AC} \propto \text{Re}\left\{\exp(-i\omega_0\Delta\tau_p)\int_{-\infty}^{\infty} S(\omega)\exp(-i(\omega - \omega_0)\Delta\tau_g)\,d\omega\right\}. \tag{9.36}$$

If $S(\omega)$ is symmetric about ω_0, the integral in Eq. (9.36) is real; thus, Eq. (9.36) becomes

$$I_{AC} \propto \cos(\omega_0\Delta\tau_p)\int_{-\infty}^{\infty} S(\omega)\exp(-i(\omega - \omega_0)\Delta\tau_g)\,d\omega. \tag{9.37}$$

The cosine factor represents a carrier that oscillates with increasing $\Delta\tau_p$. The integral represents an envelope as a function of $\Delta\tau_g$; it determines the axial PSF of the interferometer. The envelope is equal to the inverse Fourier transform of $S(\omega)$, which is an outcome of the Wiener–Khinchin theorem. Note that taking the envelope of I_{AC} is a nonlinear operation.

If $S(\omega)$ is Gaussian, substituting Eq. (9.13) into Eq. (9.37) yields

$$I_{AC} \propto \exp\left(-\frac{(\Delta\tau_g)^2}{2\sigma_\tau^2}\right)\cos(\omega_0\Delta\tau_p). \tag{9.38}$$

The temporal Gaussian envelope has a standard deviation σ_τ:

$$\sigma_\tau = \frac{1}{\sigma_\omega}. \tag{9.39}$$

The source bandwidth is typically given by the FWHM in wavelength ($\Delta\lambda$). From $\omega = 2\pi c/\lambda$, we have approximately

$$\sigma_\omega = \frac{2\pi c}{\lambda_0^2}\sigma_\lambda, \tag{9.40}$$

where σ_λ denotes the standard deviation of λ. For any Gaussian distribution of ξ with standard deviation σ_ξ, its FWHM $\Delta\xi$ is given by

$$\Delta\xi = \left(2\sqrt{2\ln 2}\right)\sigma_\xi. \tag{9.41}$$

Thus, we have

$$\sigma_\lambda = \frac{\Delta\lambda}{2\sqrt{2\ln 2}}. \tag{9.42}$$

In free space, we have $k = \omega/c$ and $\Delta\tau_g = \Delta\tau_p = 2\Delta l/c$, where c denotes both the phase and the group velocities; thus, Eq. (9.38) becomes

$$I_{AC} \propto \exp\left(-\frac{(\Delta l)^2}{2\sigma_l^2}\right)\cos(2k_0\Delta l), \tag{9.43}$$

where k_0 is the propagation constant at the center wavelength λ_0 and the standard deviation σ_l is given by

$$\sigma_l = \frac{c\sigma_\tau}{2}. \tag{9.44}$$

The axial resolution of OCT in air Δz_R is commonly defined as the FWHM of the Gaussian envelope in Eq. (9.43). Using Eq. (9.41), we obtain

$$\Delta z_R = \left(2\sqrt{2\ln 2}\right)\sigma_l. \tag{9.45}$$

Sequentially substituting Eqs. (9.44), (9.39), (9.40), and (9.42) into Eq. (9.45), we obtain

$$\Delta z_R = \frac{2\ln 2}{\pi} \frac{\lambda_0^2}{\Delta\lambda}. \qquad (9.46)$$

Comparing Eqs. (9.15) and (9.46), we find

$$\Delta z_R = \frac{l_c}{2}. \qquad (9.47)$$

Therefore, the axial resolution in air equals half of the coherence length of the source owing to the round-trip propagation of the reference and the sample beams. If other factors are negligible, the axial resolution in biological tissue is the axial resolution in air divided by the index of refraction of the tissue.

The transverse resolution of OCT is commonly defined as the focal diameter of the incident sample beam, which is independent of the coherence length of the source. If the transverse distribution of the incident sample beam is Gaussian, the transverse resolution is given by

$$\Delta r_R = \frac{4\lambda_0}{\pi} \frac{f}{D}. \qquad (9.48)$$

Here, f denotes the focal length of the objective lens; D denotes the diameter of the beam on the lens or the diameter of the lens, whichever is smaller. From the following expression for the numerical aperture NA

$$\text{NA} \approx \frac{D}{2f}, \qquad (9.49)$$

Eq. (9.48) can be rewritten as

$$\Delta r_R = \frac{2\lambda_0}{\pi} \frac{1}{\text{NA}}. \qquad (9.50)$$

The depth range within which the lateral resolution is approximately maintained is defined by the depth of focus Δz_f

$$\Delta z_f = \frac{\pi \Delta r_R^2}{2\lambda_0}, \qquad (9.51)$$

which is twice the Rayleigh range of a Gaussian beam. This equation shows the tradeoff between the focal diameter and the focal zone of the sample beam—the smaller the focal diameter, the shorter the focal zone. Therefore, the use of a high-NA objective lens requires either transverse-priority scanning—as in en face (C-scan) imaging—or depth-priority scanning with dynamic focusing along the optical axis.

The preceding derivation for a single backscatterer can be extended to multiple backscatterers that are distributed along the optical axis. The time window for the Fourier transformation, however, is truncated according to the axial resolution.

A block diagram of a complete OCT system, where the reference mirror is scanned axially, is shown in Figure 9.3. The detection of OCT signals by demodulation involves the following steps: highpass filtering to remove the DC background, rectification to reverse the signs of the negative AC signals, lowpass filtering to recover the envelope of the interference fringes, and analog–digital conversion to record the data.

A schematic of a representative experimental embodiment for the demodulation of the interference signals is shown in Figure 9.4. The first highpass filter removes the DC background and passes the AC interference signal. The active full-wave rectifier takes the absolute value of the AC signal. The lowpass filter then recovers the envelope of the rectified signal.

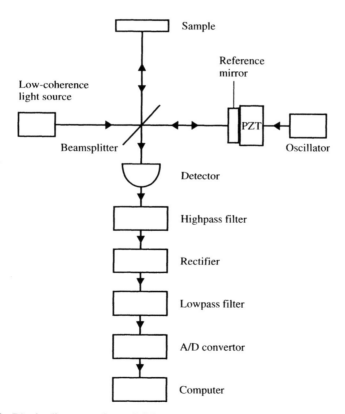

Figure 9.3. Block diagram of an OCT system, where PZT (lead zirconate titanate) represents a piezoelectric transducer that scans the reference mirror axially. A/D represents an analog-to-digital converter.

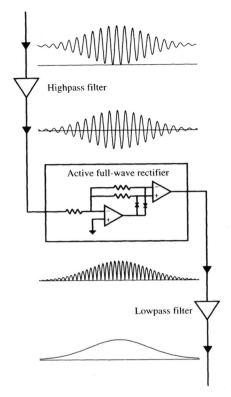

Figure 9.4. Schematic of a representative experimental embodiment for demodulating the interference signals.

Example 9.1. Derive Eq. (9.38) from Eq. (9.36).

From the identity

$$\int_{-\infty}^{\infty} \exp(-(ax^2 + bx + c))\,dx = \sqrt{\frac{\pi}{a}} \exp\left(\frac{b^2 - 4ac}{4a}\right), \qquad (9.52)$$

we derive

$$\int_{-\infty}^{\infty} S(\omega) \exp(-i(\omega - \omega_0)\Delta\tau_g)\,d\omega$$

$$= \frac{1}{\sqrt{2\pi}\sigma_\omega} \int_{-\infty}^{\infty} \exp\left(-\frac{(\omega - \omega_0)^2}{2\sigma_\omega^2} - i(\omega - \omega_0)\Delta\tau_g\right) d\omega \qquad (9.53)$$

$$= \exp\left(-\frac{1}{2}(\sigma_\omega \Delta\tau_g)^2\right).$$

From Eq. (9.36), we have

$$I_{AC} \propto \exp\left(-\frac{(\sigma_\omega \Delta \tau_g)^2}{2}\right) \mathrm{Re}\left\{\exp(-i\omega_0 \Delta \tau_p)\right\}$$
$$= \exp\left(-\frac{(\sigma_\omega \Delta \tau_g)^2}{2}\right) \cos(\omega_0 \Delta \tau_p), \quad (9.54)$$

which can be reformulated to Eq. (9.38) by using Eq. (9.39).

Example 9.2. Assume that $S(\omega)$ is Gaussian: (a) calculate Δz_R given $\lambda_0 = 830$ nm and $\Delta\lambda = 20$ nm; (b) plot Δz_R as a function of $\Delta\lambda$ at commonly used $\lambda_0 = 830$ nm and $\lambda_0 = 1300$ nm.

From Eq. (9.46), (a) $\Delta z_R = 15.2$ μm, and (b) the axial resolution as a function of $\Delta\lambda$ is plotted in Figure 9.5.

Example 9.3. Assume $S(\omega)$ to be Gaussian. Simulate the demodulation in TD-OCT in MATLAB. (a) Plot the interference signal versus $\Delta l/l_c$, where $\lambda_0 = 830$ nm and $\Delta\lambda = 60$ nm. Estimate the number of periods within the FWHM of the interference envelope. (b) Rectify the signal by taking the absolute value of the interference signal. (c) Plot the spectral amplitude of the rectified signal. (d) Filter the rectified signal to produce an envelope.

We reformulate Eq. (9.43) to

$$I_{AC} = \exp\left(-16\ln 2 \left(\frac{\Delta l}{l_c}\right)^2\right) \cos(2k_0 \Delta l). \quad (9.55)$$

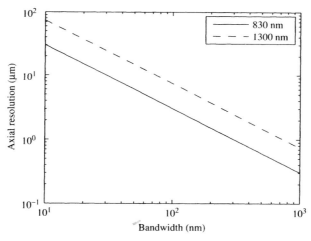

Figure 9.5. Axial resolution at two center wavelengths as a function of bandwidth.

From Eq. (9.15), we have

$$l_c = \frac{4\ln 2}{\pi}\frac{\lambda_0^2}{\Delta\lambda} = 10~\mu m.$$

When the interference signal is sampled in MATLAB, we must use a sufficiently high sampling rate to satisfy the Nyquist criterion (more than two data points per period).

Since the FWHM of the interference envelope is the axial resolution Δz_R and the period of the interference fringes is $\lambda_0/2$, the number of periods within the FWHM is

$$\frac{\Delta z_R}{\lambda_0/2} = \frac{l_c}{\lambda_0} \approx 12. \tag{9.56}$$

A representative MATLAB program is listed below:

```
% Use SI units throughout

lambda0 = 830E-9; % center wavelength
dlambda = 60E-9; % bandwidth (delta lambda)
c = 3E8; % speed of light

lc = 4*log(2)/pi*lambda0^2/dlambda % coherence length
Number_of_periods = 0.5*lc/(lambda0/2) % # of periods in FWHM

figure(1);

N = 2^12; % number of sampling points
dl = lc*linspace(-2,2, N); % array for Delta_l
k0 = 2*pi/lambda0; % propagation constant

subplot(4, 1, 1) % interferogram
Iac = exp(-16*log(2)*(dl/lc).^2) .* cos(2*k0 * dl);
plot(dl/lc, Iac, 'k')
title('(a) Interferogram')
xlabel('\Deltal/l_c')
ylabel('Signal')
axis([-0.6, 0.6, -1, 1])

subplot(4, 1, 2) % rectified interferogram
Irec = abs(Iac);
plot(dl/lc, Irec, 'k')
title('(b) Rectified interferogram')
xlabel('\Deltal/l_c')
ylabel('Signal')
axis([-0.6, 0.6, -1, 1])

subplot(4, 1, 3) % spectrum of the rectified interferogram
Frec1 = fft(Irec)/sqrt(N);
% order of frequencies: 0,1...(N/2-1),-N/2,-(N/2-1)...-1
```

```
Frec2 = fftshift(Frec1);
% shifted order of frequencies: -N/2,-(N/2-1)...-1, 0,1...(N/2-1)
dfreq = 1/(4*lc); % freq bin size = 1/sampling range
freq = dfreq*(-N/2:N/2-1); % frequency array
plot(freq*lambda0, abs(Frec2), 'k')
title('(c) Spectrum of the rectified interferogram')
xlabel('Frequency (1/\lambda_0)')
ylabel('Amplitude')
axis([-10, 10, 0, 5])

subplot(4, 1, 4) % envelope
freq_cut = 1/lambda0/2; % cut-off frequency for filtering
i_cut = round(freq_cut/dfreq); % convert freq_cut to an array index
Ffilt = Frec1; % initialize array
Ffilt(i_cut:N-i_cut+1) = 0; % filter
Ifilt = abs(ifft(Ffilt))*sqrt(N); % amplitude of inverse FFT

plot(dl/lc, Ifilt/max(Ifilt), 'k')
Iac_en = exp(-16*log(2)*(dl/lc).^2); % envelope
hold on;
plot(dl(1:N/32:N)/lc, Iac_en(1:N/32:N), 'ko')
hold off;
title('(d) Envelopes')
xlabel('\Deltal/l_c')
ylabel('Signals')
axis([-0.6, 0.6, -1, 1])
legend('Demodulated','Original')
```

The graphical output from the MATLAB program is shown in Figure 9.6.

9.5. FOURIER-DOMAIN RAPID-SCANNING OPTICAL DELAY LINE

In addition to the geometric means for varying the reference path length, a frequency-domain approach is also available based on the following inverse Fourier transformation:

$$E(t - \Delta\tau_g) = \frac{1}{2\pi} \int_{-\infty}^{+\infty} \left[E(\omega) \exp(i\omega\Delta\tau_g) \right] \exp(-i\omega t)\, d\omega. \quad (9.57)$$

This equation indicates that a linear phase ramp $\Delta\tau_g \omega$ in the frequency domain leads to a group delay $\Delta\tau_g$ in the time domain. In optics, a grating is a *temporal Fourier transformer* that can transform a time-domain signal into a temporofrequency-domain signal, whereas a lens is a *spatial Fourier transformer* that can transform a space-domain signal into a spatiofrequency-domain signal. Both optical elements can serve as *inverse Fourier transformers* as well.

According to this principle, a Fourier-domain rapid scanning optical delay line was developed using a grating–lens pair, a scanning planar mirror, and a static planar mirror (Figure 9.7). The grating Fourier transforms (disperses) the incident light into temporofrequency (chromatic) components propagating in various

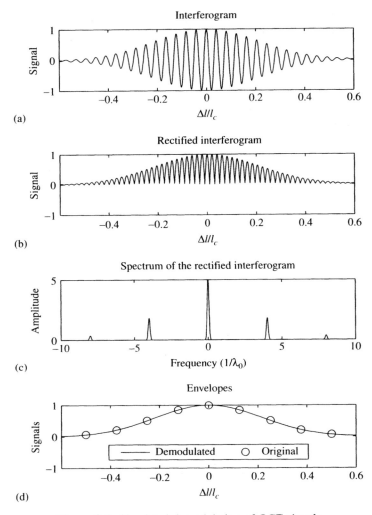

Figure 9.6. Simulated demodulation of OCT signals.

directions; each chromatic component has a spatial frequency. The component for the center wavelength λ_0 is aligned with the optical axis of the lens by adjusting the angle of incidence. The lens focuses each chromatic component into a point on the scanning mirror; hence, it Fourier-transforms the light into a spatiofrequency spectrum along the vertical direction, which represents the temporofrequency spectrum of the original light. The scanning mirror reflects the focused beam with a phase ramp across the temporofrequency spectrum because different temporofrequency components accumulate different round-trip path lengths. The lens and the grating then function in their inverse roles to produce a merged beam propagating onto the static mirror, which backreflects the merged beam. Then, the

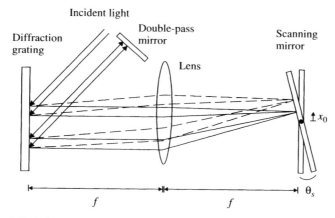

Figure 9.7. Schematic representation of the Fourier-domain optical delay line.

beam inversely propagates through the entire delay line, which can be analyzed according to the principle of reciprocity. The final beam propagates against the incident beam, but it has experienced a group delay.

The phase shift ϕ_s of each temporofrequency component can be expressed by

$$\phi_s(\lambda) = \frac{8\pi x_0 \theta_s}{\lambda} + \frac{8\pi f \theta_s (\lambda - \lambda_0)}{p_g \cos\theta_0 \lambda}. \tag{9.58}$$

Here, λ is the wavelength, x_0 is the center-wavelength displacement on the mirror surface relative to the pivot, θ_s is the tilt angle of the scanning mirror, f is the focal length of the lens, p_g is the pitch of the grating, and θ_0 is the first-order center-wavelength diffraction angle with respect to the normal vector of the grating. Since θ_0 is zero here, it is excluded in the following derivations. We rewrite Eq. (9.58) as a function of the optical angular frequency ω

$$\phi_s(\omega) = \frac{4x_0 \theta_s \omega}{c} - \frac{8\pi f \theta_s (\omega - \omega_0)}{p_g \omega_0}, \tag{9.59}$$

where ω_0 represents the center angular frequency.

The phase delay, defined as the delay at the center frequency, is given by

$$\Delta\tau_p = \frac{\phi_s(\omega_0)}{\omega_0} = \frac{4x_0 \theta_s}{c}, \tag{9.60}$$

which can be translated into the following free-space phase–path-length mismatch:

$$\Delta l_p = c\Delta\tau_p = 4x_0 \theta_s. \tag{9.61}$$

The group delay can be derived as follows:

$$\Delta\tau_g = \frac{\partial \phi_s(\omega)}{\partial \omega}\bigg|_{\omega=\omega_0} = \frac{4x_0\theta_s}{c} - \frac{4f\lambda_0\theta_s}{cp_g} = \Delta\tau_p - \frac{4f\lambda_0\theta_s}{cp_g}, \quad (9.62)$$

which can be translated into the following free-space group–path-length mismatch:

$$\Delta l_g = c\Delta\tau_g = 4x_0\theta_s - \frac{4f\lambda_0\theta_s}{p_g} = \Delta l_p - \frac{4f\lambda_0\theta_s}{p_g}. \quad (9.63)$$

The phase and the group delays are different, but both are proportional to θ_s. Thus, rotating the mirror provides a rapid delay line, which scans several millimeters at a repetition rate of several kilohertz.

9.6. FOURIER-DOMAIN OCT

For any time-domain method, a Fourier-domain equivalent usually exists. A Fourier-domain OCT (FD-OCT) system based on spectral interferometry is shown in Figure 9.8. FD-OCT avoids varying the reference optical path length altogether. The recombined beam, however, is dispersed by a spectrometer into spectral components. The corresponding spectral components interfere and form a spectral interferogram. The spectral interferogram is acquired by an optical detector array such as a 1D photodiode array. Taking the inverse Fourier transformation of the spectrum yields an A-line image in its entirety. As in TD-OCT,

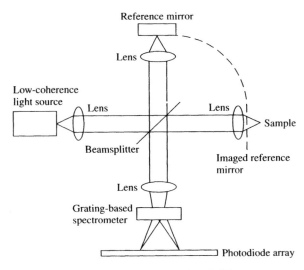

Figure 9.8. Schematic of an FD-OCT system.

transverse scanning across the sample provides 2D or 3D imaging. In FD-OCT, all backscatterers on the A-line are measured simultaneously; in TD-OCT, only some of the backscatterers are measured at any one time. Consequently, FD-OCT has a higher frame rate and a greater sensitivity.

Now, we extend the theory for TD-OCT to one for FD-OCT. Multiple backscatterers at various depths on the A-line are considered. Thus, the sample beam consists of multiple partial waves emanating from the backscatterers. The spectral components of the reference and the sample beams can be expressed as

$$E_R(\omega) = E_0(\omega) r_R \exp(i(2k_R(\omega)l_R - \omega t)), \quad (9.64)$$

$$E_S(\omega) = E_0(\omega) \int_{-\infty}^{+\infty} r'_S(l_S) \exp(i(2k_S(\omega)l_S - \omega t)) \, dl_S. \quad (9.65)$$

Here, $E_0(\omega)$ denotes the electric field incident on the reference mirror or the sample surface; r_R denotes the amplitude reflectivity of the reference mirror; $r'_S(l_S)$, the object function to be imaged, denotes the apparent amplitude reflectivity density (reflectivity per unit depth) of the backscatterers along the A-line in the sample. Since light incident on a scatterer may have been attenuated, $r'_S(l_S)$ represents the apparent—rather than the true—local amplitude reflectivity density. The amplitude reflectivity density r'_S of a discrete reflector is related to its amplitude reflectivity r_S through a delta function

$$r'_S(l_S) = r_S(l_{S0}) \delta(l_S - l_{S0}), \quad (9.66)$$

where l_{S0} denotes the location of the reflector. If $r'_S(l_S)$ is integrated around l_{S0}, then $r_S(l_{S0})$ is recovered.

If dispersion is neglected, we have

$$\frac{k_R}{n_R} = \frac{k_S}{n_S} = k = \frac{\omega}{c}, \quad (9.67)$$

where n_R denotes the refractive index of the medium in the reference arm and n_S denotes the average refractive index of the sample. For brevity, n_R is set to unity here. Apart from a constant scaling factor, the spectral interferogram is given by

$$I(k) = |E_R(kc) + E_S(kc)|^2. \quad (9.68)$$

Substituting Eqs. (9.64) and (9.65) into Eq. (9.68), we obtain

$$\begin{aligned} I(k) = {} & S(k) r_R^2 \\ & + 2S(k) r_R \int_{-\infty}^{+\infty} r'_S(l_S) \cos(2k(n_S l_S - l_R)) \, dl_S \\ & + S(k) \left| \int_{-\infty}^{+\infty} r'_S(l_S) \exp(i 2k(n_S l_S)) \, dl_S \right|^2, \end{aligned} \quad (9.69)$$

where the source power spectral density distribution $S(k) = |E_0(kc)|^2$. The first term on the right-hand side, referred to as the *reference-intensity term*, can be measured by blocking the sample arm—setting $r'_S(l_S)$ to zero. The second term, referred to as the *cross-interference term*, encodes $r'_S(l_S)$ in the integral with a cosine function—that has frequency $2(n_S l_S - l_R)$—of the wavenumber, which is $k/(2\pi) = 1/\lambda$. The third term, referred to as the *self-interference (sample-intensity) term*, originates from power spectrum $|E_S(kc)|^2$ and contains the interference among the partial waves from the various sample depths.

The cross-interference term can be decoded to extract $r'_S(l_S)$ by taking the inverse Fourier transformation. The deeper the origin of the backscattered partial wave, the higher the encoding frequency is. The shorter the reference arm, the higher the frequency as well. The Nyquist criterion requires that the frequencies of the spectral interferogram be less than half the spatial sampling frequency of the detector array. Thus, it is advantageous to minimize the frequency of the encoding cosine function. However, if the position of the "imaged reference mirror" is shifted into the sample to minimize the frequency, the two sides around this position will share encoding frequencies, which leads to ambiguity. Therefore, the "imaged reference mirror" must be placed outside the sample unless multiple interferograms with various phase differences between the two arms are measured to resolve the encoding ambiguity.

For brevity in notation, we set l_R to zero and shift the reference point for l_S to the "imaged reference mirror surface" in the sample arm. If the reference point is outside the sample, a new even function $\hat{r}'_S(l_S)$ can be crafted as follows so that $\hat{r}'_S(-l_S) = \hat{r}'_S(l_S)$:

$$\hat{r}'_S(l_S) = \begin{cases} r'_S(l_S) & \text{if } l_S \geq 0, \\ r'_S(-l_S) & \text{if } l_S < 0. \end{cases} \quad (9.70)$$

With this new function, Eq. (9.69) can be rewritten as

$$I(k) = S(k) \left\{ r_R^2 + r_R \int_{-\infty}^{+\infty} \hat{r}'_S(l_S) \exp(i 2k n_S l_S) \, dl_S \right. \\ \left. + \frac{1}{4} \left| \int_{-\infty}^{+\infty} \hat{r}'_S(l_S) \exp(i 2k n_S l_S) \, dl_S \right|^2 \right\}. \quad (9.71)$$

If we change the variable of the integrals by $l_S = l'_S/(2n_S)$, Eq. (9.71) can be rewritten as

$$I(k) = S(k) \left\{ r_R^2 + \frac{r_R}{2n_S} \mathcal{F}\left\{ \hat{r}'_S\left(\frac{l'_S}{2n_S}\right) \right\}(k) + \frac{1}{16 n_S^2} \left| \mathcal{F}\left\{ \hat{r}'_S\left(\frac{l'_S}{2n_S}\right) \right\}(k) \right|^2 \right\}, \quad (9.72)$$

where the following Fourier transformation is employed:

$$F(k) = \mathcal{F}\{f(l'_S)\}(k) = \int_{-\infty}^{+\infty} f(l'_S)\exp(ikl'_S)\,dl'_S. \tag{9.73}$$

Using the inverse Fourier transformation

$$f(l'_S) = \mathcal{F}^{-1}\{F(k)\}(l'_S) = \frac{1}{2\pi}\int_{-\infty}^{+\infty} F(k)\exp(-ikl'_S)\,dk, \tag{9.74}$$

we rewrite Eq. (9.72) as

$$\mathcal{F}^{-1}\{I(k)\}(l'_S) = \mathcal{F}^{-1}\{S(k)\}(l'_S)$$

$$* \left\{ \frac{r_R^2}{2n_S}\delta\left(\frac{l'_S}{2n_S}\right) + \frac{r_R}{2n_S}\hat{r}'_S\left(\frac{l'_S}{2n_S}\right) + \frac{1}{16n_S^2}\mathcal{C}\left\{\hat{r}'_S\left(\frac{l'_S}{2n_S}\right)\right\} \right\}. \tag{9.75}$$

Here, $*$ denotes convolution, and $\mathcal{C}\{\}$ denotes the autocorrelation-function operator:

$$\mathcal{C}\{f(l'_S)\} = \int_{-\infty}^{+\infty} f(l'_{S1})f(l'_{S1}+l'_S)\,dl'_{S1}. \tag{9.76}$$

In this derivation, the following property of the Dirac delta function is used:

$$\delta(l'_S) = \frac{1}{2n_S}\delta\left(\frac{l'_S}{2n_S}\right). \tag{9.77}$$

The following Wiener–Khinchin theorem is used as well:

$$\mathcal{C}\{f(l'_S)\} = \frac{1}{2\pi}\int_{-\infty}^{+\infty} |F(k)|^2 \exp(-ikl'_S)\,dk. \tag{9.78}$$

With a change of variable by $l'_S = 2n_S l_S$, Eq. (9.75) is converted to

$$\mathcal{F}^{-1}\{I(k)\}(2n_S l_S)$$

$$= \mathcal{F}^{-1}\{S(k)\}(2n_S l_S) * \left\{ \frac{r_R^2}{2n_S}\delta(l_S) + \frac{r_R}{2n_S}\hat{r}'_S(l_S) + \frac{1}{16n_S^2}\mathcal{C}\{\hat{r}'_S(l_S)\} \right\}. \tag{9.79}$$

The second term in the braces is the A-line image $\hat{r}'_S(l_S)$. The first and last terms, however, represent spurious images. The first term is nonzero only at $l_S = 0$, which is outside the sample; thus, it can be easily removed. Unfortunately, the last term can mingle with the second term; thus, it can be difficult to eliminate. In addition, the convolution with $\mathcal{F}^{-1}\{S(k)\}(2n_S l_S)$ blurs the image because $S(k)$ functions as a filter.

To recover the true image, one may take another interferogram with kl'_S shifted by π, which causes a sign change in Eq. (9.72):

$$I_2(k) = S(k)\left\{r_R^2 - \frac{r_R}{2n_S}\mathcal{F}\left\{\hat{r}'_S\left(\frac{l'_S}{2n_S}\right)\right\}(k) + \frac{1}{16n_S^2}\left|\mathcal{F}\left\{\hat{r}'_S\left(\frac{l'_S}{2n_S}\right)\right\}(k)\right|^2\right\}. \quad (9.80)$$

Taking the difference between Eqs. (9.72) and (9.80) yields

$$\Delta I(k) = S(k)\frac{r_R}{n_S}\mathcal{F}\left\{\hat{r}'_S\left(\frac{l'_S}{2n_S}\right)\right\}(k), \quad (9.81)$$

where $\Delta I(k) = I(k) - I_2(k)$. The A-line image can then be recovered by

$$\hat{r}'_S\left(\frac{l'_S}{2n_S}\right) = \frac{n_S}{r_R}\mathcal{F}^{-1}\left\{\frac{\Delta I(k)}{S(k)}\right\}(l'_S). \quad (9.82)$$

Changing the variable by $l'_S = 2n_S l_S$ leads to

$$\hat{r}'_S(l_S) = \frac{n_S}{r_R}\mathcal{F}^{-1}\left\{\frac{\Delta I(k)}{S(k)}\right\}(2n_S l_S). \quad (9.83)$$

This equation shows that the subtracted-and-deconvolved spectral interferogram in the braces recovers an ideal image; the deconvolution involves simply dividing $\Delta I(k)$ by $S(k)$. Although deconvolution can sharpen the image, one should exercise caution in the presence of noise.

An alternative to recovering the true image is to (1) measure the first term (reference-intensity term) in Eq. (9.69) by blocking the sample arm ($r'_S = 0$), (2) measure the third term (self-interference term) in Eq. (9.69) by blocking the reference arm ($r_R = 0$), and (3) subtract the measured first and third terms from the right-hand side of Eq. (9.69).

In practice, a spectrometer produces a spectrum with uniform wavelength spacing. This wavelength spacing is usually converted to uniform propagation-constant spacing by interpolation as required by the fast Fourier transformation (FFT) algorithm.

By using a single-element photodetector to measure the interference signal while a laser serially sweeps the wavelength, FD-OCT can also construct a spectral interferogram one wavelength at a time. A hardware "k clock" can be installed in the laser to achieve uniform propagation-constant spacing. Once a spectral interferogram is obtained, the theoretical analysis presented above is equally applicable.

Example 9.4. Simulate FD-OCT numerically using MATLAB. Assume a Gaussian source spectral density distribution $S(k)$.

A representative MATLAB program is shown below:

```
% Use SI units throughout

lambda0 = 830E-9; % center wavelength of source
dlambda = 20E-9; % FWHM wavelength bandwidth of source
ns=1.38; % refractive index of sample
ls1 = 100E-6; % location of backscatterer 1
ls2 = 150E-6; % location of backscatterer 2
rs1 = 0.5; % reflectivity of backscatterer 1
rs2 = 0.25; % reflectivity of backscatterer 2

k0=2*pi/lambda0; % center propagation constant
delta_k=2*pi*dlambda/lambda0^2; % FWHM bandwidth of k
sigma_k = delta_k/sqrt(2*log(2)); % standard deviation of k

N=2 10; % number of sampling points
nsigma = 5; % number of standard deviations to plot on each side of k0

subplot(4,1,1); % Generate the interferogram
k = k0 + sigma_k*linspace(-nsigma,nsigma, N); % array for k
S_k = exp(-(1/2)*(k-k0).^2/sigma_k^2); % Gaussian source PSD
E_s1 = rs1*exp(i*2*k*ns*ls1); % sample electric field from scatter 1
E_s2 = rs2*exp(i*2*k*ns*ls2); % sample electric field from scatter 2
I_k1 = S_k .* abs(1 + E_s1 + E_s2).^2; % interferogram (r_R = 1)
plot(k/k0,I_k1/max(I_k1), 'k');
title('Interferogram');
xlabel('Propagation constant k/k_0');
ylabel('Normalized intensity');
axis([0.9 1.1 0 1]);

subplot(4,1,2); % Inverse Fourier transform (IFT) of the interferogram
spec1=abs(fftshift(ifft(I_k1)))/sqrt(N);
dls_prime = 1/(2*nsigma*sigma_k/(2*pi)); % bin = 1/sampling range
ls_prime = dls_prime*(-N/2:N/2-1); % frequency array
plot(ls_prime/(2*ns),spec1/max(spec1), 'k'); % scale the frequency
title('IFT of the interferogram');
xlabel('Depth ls (m)');
ylabel('Relative reflectivity');
axis([-2*ls2 2*ls2 0 1]);

subplot(4,1,3); % IFT of the deconvolved interferogram
spec1_norm =abs(fftshift(ifft(I_k1./S_k)))/sqrt(N);
dls_prime = 1/(2*nsigma*sigma_k/(2*pi)); % bin size = 1/sampling range
ls_prime = dls_prime*(-N/2:N/2-1); % frequency array
plot(ls_prime/(2*ns),spec1_norm/max(spec1_norm), 'k');
title('IFT of the deconvolved interferogram');
xlabel('Depth ls (m)');
ylabel('Relative reflectivity');
axis([-2*ls2 2*ls2 0 1]);

subplot(4,1,4); % IFT of the deconvolved differential interferogram
I_k2 = S_k .* abs(-1 + E_s1 + E_s2). 2; % interferogram
delta_I_k = I_k1 - I_k2;
spec2=abs(fftshift(ifft(delta_I_k./S_k)))/sqrt(N);
plot(ls_prime/(2*ns),spec2/max(spec2), 'k');
title('IFT of the deconvolved differential interferogram');
```

```
xlabel('Depth ls (m)');
ylabel('Relative reflectivity');
axis([-2*ls2 2*ls2 0 1]);
```

The graphical output of this program is shown in Figure 9.9. The first panel shows the simulated spectral interferogram $I(k)$. The second panel shows the inverse Fourier transform of $I(k)$. Remember to change the independent variable to $2n_s l_s$ after taking the inverse Fourier transformation. The third panel shows the inverse Fourier transform of $I(k)/S(k)$. The fourth panel shows the inverse Fourier transform of $\Delta I(k)/S(k)$ as shown in Eq. (9.82). The locations and

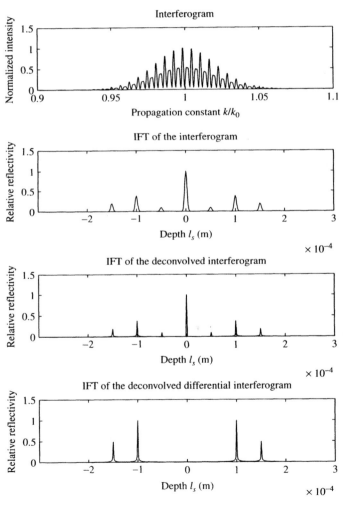

Figure 9.9. Simulated signal processing in FD-OCT with two backscatterers.

strengths of the backscatterers are recovered in all the inverse Fourier transforms. In the second panel, however, a DC component and a spurious backscatter at 50 μm appear; the latter is due to interference between the partial waves from the two backscatterers. The third panel shows that deconvolution sharpens the peaks. The fourth panel shows that the inverse Fourier transformation of the subtracted-and-deconvolved spectral interferogram yields a much cleaner 1D image.

If the second backscatterer is obliterated by setting r_{S2} to zero, we obtain the results shown in Figure 9.10 instead. The DC component still appears in the inverse Fourier transform of $I(k)$. However, no self-interference signal of the sample beam appears because only a single backscatterer exists.

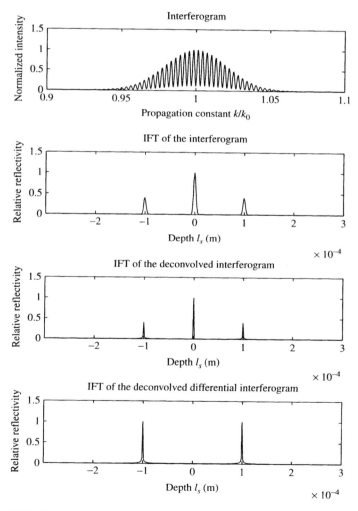

Figure 9.10. Simulated signal processing in FD-OCT with a single backscatterer.

9.7. DOPPLER OCT

Like ultrasonography, OCT can image blood flow on the basis of the Doppler effect. Doppler OCT has measured blood flow at a flow rate on the order of 10 pL/s in 10-µm-diameter vessels positioned up to 1 mm beneath the tissue surface. The basic principle of Doppler OCT is illustrated by the following analysis in a nondispersive medium. A moving reference mirror varies the arm-length difference by

$$\Delta l(t) = \Delta l_0 - v_R t, \qquad (9.84)$$

where Δl_0 denotes the arm-length difference at time $t = 0$ and v_R denotes the velocity of the reference mirror. The phase difference between the two arms at the center wavelength λ_0 in Eq. (9.8) is modified to

$$\Delta\phi(t) = 2k_0(\Delta l_0 - v_R t) = \frac{4\pi}{\lambda_0}(\Delta l_0 - v_R t). \qquad (9.85)$$

This time-varying phase difference leads to a Doppler shift f_R given by

$$f_R(\lambda_0) = \frac{1}{2\pi}\left|\frac{d\Delta\phi(t)}{dt}\right| = \frac{2v_R}{\lambda_0}. \qquad (9.86)$$

The Doppler shift f_R is the beat frequency, also referred to as the *carrier frequency*, of the interference fringes. The other spectral components experience similar Doppler shifts given by

$$f_R(\lambda) = \frac{2v_R}{\lambda}, \qquad (9.87)$$

which can be used to compute the bandwidth of the interference signal as

$$\Delta f_R = \frac{2v_R}{\lambda_0^2}\Delta\lambda. \qquad (9.88)$$

In the presence of flow, an additional Doppler shift leads to the following carrier frequency:

$$f_{RS}(\lambda_0) = \frac{1}{2\pi}\left|\frac{d\Delta\phi(t)}{dt}\right| = \left|\frac{2}{\lambda_0}(v_S \cos\theta_S - v_R)\right| = |f_S(\lambda_0) - f_R(\lambda_0)|. \qquad (9.89)$$

Here, v_S denotes the velocity of the scatterers, θ_S denotes the angle between the flow direction and the light-incidence direction, and f_S denotes the Doppler shift due to the backscatterers:

$$f_S(\lambda_0) = \frac{2v_S \cos\theta_S}{\lambda_0}. \qquad (9.90)$$

If $f_S < f_R$, we have

$$f_{RS}(\lambda_0) = f_R(\lambda_0) - f_S(\lambda_0) \tag{9.91}$$

or

$$f_S(\lambda_0) = f_R(\lambda_0) - f_{RS}(\lambda_0). \tag{9.92}$$

Once the original (predemodulated) interference fringes are acquired, taking the Fourier transformation of the fringes yields f_{RS}, which further yields f_S through Eq. (9.92), since f_R is known. If θ_S is known, v_S can be computed from Eq. (9.90). Because the Fourier transformation is usually performed in sliding short-time windows, a tradeoff between axial and velocity resolutions exists in the flow estimation.

9.8. GROUP VELOCITY DISPERSION

Group velocity dispersion (GVD), which is neglected in previous sections, deteriorates the axial resolution of OCT. GVD causes polychromatic light to experience nonlinearly frequency-dependent phase delays. As a result, GVD broadens ultrashort laser light. More relevantly, any GVD mismatch between the reference and the sample arms of an OCT system broadens the axial PSF.

To analyze the effect of GVD on OCT signals, we first expand the propagation constant into a Taylor series to the second order around the center angular frequency ω_0:

$$k(\omega) = k(\omega_0) + k'(\omega_0)(\omega - \omega_0) + \frac{1}{2}k''(\omega_0)(\omega - \omega_0)^2. \tag{9.93}$$

From the definitions of the phase and the group velocities [Eqs. (9.32) and (9.34)], Eq. (9.93) is reformulated to

$$k(\omega) = \frac{1}{v_p(\omega_0)}\omega_0 + \frac{1}{v_g(\omega_0)}(\omega - \omega_0) + \frac{1}{2}k''(\omega_0)(\omega - \omega_0)^2. \tag{9.94}$$

The frequency-dependent phase mismatch between the reference and the sample beams is

$$\Delta\phi(\omega) = 2k_S(\omega)l_S - 2k_R(\omega)l_R. \tag{9.95}$$

If the propagation constants in the reference and the sample paths are equal except for a GVD mismatch in arm length l_d, substitution of Eq. (9.94) into Eq. (9.95) yields

$$\Delta\phi(\omega) = \frac{1}{v_p(\omega_0)}\omega_0(2\Delta l) + \frac{1}{v_g(\omega_0)}(\omega - \omega_0)(2\Delta l)$$
$$+ \frac{1}{2}\Delta k''(\omega_0)(\omega - \omega_0)^2(2l_d), \tag{9.96}$$

where $\Delta l = l_S - l_R$ and

$$\Delta k''(\omega_0) = k_S''(\omega_0) - k_R''(\omega_0). \tag{9.97}$$

Substituting Eqs. (9.33) and (9.35) into Eq. (9.96), we obtain

$$\Delta\phi(\omega) = \omega_0 \Delta\tau_p + (\omega - \omega_0)\Delta\tau_g + \frac{1}{2}\Delta k''(\omega_0)(\omega - \omega_0)^2(2l_d). \tag{9.98}$$

Substitution of Eq. (9.98) into Eq. (9.26) leads to

$$I_{AC} \propto \text{Re}\left\{\exp(-i\omega_0\Delta\tau_p)\int_{-\infty}^{\infty} S(\omega)\right.$$
$$\left. \times \exp\left(-i\left[(\omega-\omega_0)\Delta\tau_g + \frac{\Delta k''}{2}(\omega-\omega_0)^2(2l_d)\right]\right) d\omega\right\}. \tag{9.99}$$

If $S(\omega)$ is Gaussian [Eq. (9.13)], Eq. (9.99) describes an interference signal modulated by a complex Gaussian envelope:

$$I_{AC} \propto \text{Re}\left\{\frac{\sigma_\tau}{\Gamma(2l_d)}\exp\left(-\frac{1}{2}\frac{\Delta\tau_g^2}{\Gamma^2(2l_d)}\right)\exp(-i\omega_0\Delta\tau_p)\right\}. \tag{9.100}$$

Here, $\Gamma(2l_d)$ represents the standard deviation of the axial PSF:

$$\Gamma^2(2l_d) = \sigma_\tau^2 + i\tau_d^2, \tag{9.101}$$

where the GVD time constant is defined by

$$\tau_d = \sqrt{\Delta k''(\omega_0)(2l_d)}. \tag{9.102}$$

From Eq. (9.101), we have

$$\frac{1}{\Gamma^2(2l_d)} = \frac{\sigma_\tau^2}{\sigma_\tau^4 + \tau_d^4} - i\frac{\tau_d^2}{\sigma_\tau^4 + \tau_d^4}. \tag{9.103}$$

Substituting Eq. (9.103) into Eq. (9.100), we discover that the real and imaginary components on the right-hand side of Eq. (9.103) cause broadening and chirping, respectively, in the interference signal. The original standard deviation σ_τ is broadened to

$$\tilde{\sigma}_\tau = \sigma_\tau\sqrt{1 + \left(\frac{\tau_d}{\sigma_\tau}\right)^4}. \tag{9.104}$$

Thus, the envelope broadens with increasing τ_d and by a factor of $\sqrt{2}$ when $\tau_d = \sigma_\tau$. The interference signal chirps with increasing Δl, as can be seen by differentiating the total phase ϕ_{AC} in the two exponents in Eq. (9.100):

$$\frac{d\phi_{AC}}{d\Delta l} = 2k(\omega_0) - \frac{4\tau_d^2}{\sigma_\tau^4 + \tau_d^4} k'^2(\omega_0)\Delta l. \tag{9.105}$$

Here, l_d is assumed to be independent of Δl. If Δl is uniformly scanned, substituting Eq. (9.84) into Eq. (9.105) leads to

$$\frac{d\phi_{AC}}{dt} = 2k(\omega_0)v_R - \frac{4\tau_d^2}{\sigma_\tau^4 + \tau_d^4} k'^2(\omega_0)v_R(\Delta l_0 - v_R t). \tag{9.106}$$

Thus, the angular frequency of the interference signal varies with time, which is chirping.

Because the amplitude of $1/\Gamma(2l_d)$ decreases the peak magnitude of the interference envelope, the *system sensitivity*—defined as the ratio of the incident light power to the weakest measurable sample light power, usually in dB—degrades. The degradation in the photocurrent amplitude is given by the following multiplicative factor:

$$\frac{\sigma_\tau}{|\Gamma(2l_d)|} = \frac{1}{[1 + (\tau_d/\sigma_\tau)^4]^{1/4}} = \sqrt{\frac{\sigma_\tau}{\tilde{\sigma}_\tau}}, \tag{9.107}$$

which indicates an inverse proportionality to $\sqrt{\tilde{\sigma}_\tau}$.

In practice, the effect of GVD can be reduced by minimizing l_d. For example, the optical fiber lengths in the reference and the sample arms should be matched as closely as possible. In retinal imaging, the clear path in the eye can be matched with one in an optically similar medium—such as water—in the reference arm.

Example 9.5. Estimate the GVD mismatch length l_d of a fused-silica fiber relative to air beyond which envelope broadening becomes significant. An SLD is used as the light source, where center wavelength $\lambda_0 = 800$ nm and bandwidth $\Delta\lambda = 20$ nm.

From Eqs. (9.39), (9.40), and (9.42), we obtain

$$\sigma_\tau = \frac{\sqrt{2\ln 2}}{2\pi c} \frac{\lambda_0^2}{\Delta\lambda} = 20 \text{ fs}. \tag{9.108}$$

In a fused-silica fiber, $k'' = 350$ fs^2/cm at 800-nm wavelength. Therefore, $\Delta k'' = 350$ fs^2/cm relative to air. When $\tau_d = \sigma_\tau$, envelope broadening is considered significant. From Eq. (9.102), we obtain

$$l_d = \frac{\tau_d^2}{2\Delta k''} = 0.57 \text{ cm}. \tag{9.109}$$

Example 9.6. Derive Eq. (9.105).

The total phase in Eq. (9.100) is

$$\phi_{AC} = \omega_0 \Delta \tau_p - \frac{1}{2} \frac{\tau_d^2}{\sigma_\tau^4 + \tau_d^4} (\Delta \tau_g)^2. \quad (9.110)$$

Substituting Eqs. (9.33) and (9.35) into Eq. (9.110), we obtain

$$\phi_{AC}(\Delta l) = 2k(\omega_0)\Delta l - \frac{1}{2} \frac{\tau_d^2}{\sigma_\tau^4 + \tau_d^4} [k'(\omega_0) 2\Delta l]^2, \quad (9.111)$$

which leads to Eq. (9.105) by differentiation with respect to Δl.

9.9. MONTE CARLO MODELING OF OCT

Although singly backscattered photons are more desirable in OCT, multiple-scattered photons can contribute to OCT signals as well. Here, both single- and multiple-scattered contributions are simulated using the Monte Carlo method. Since only ensemble-averaged quantities are modeled, certain features of OCT—such as speckles—are excluded.

OCT signals are divided into two classes as shown in Figure 9.11. Both classes are based on sample light that is coherent with the reference light. Class I originates from backscattering in a target layer whose central sample arm length z_c and thickness Δz are determined as follows:

$$n_S z_c = n_R l_R, \quad (9.112)$$

$$\Delta z = \frac{l_c}{2n_S}. \quad (9.113)$$

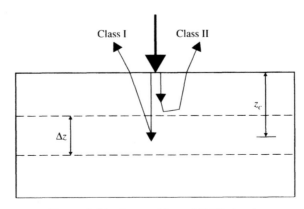

Figure 9.11. Composition of OCT signals.

Here, n_S and n_R denote the refractive indices in the sample and the reference arms, respectively. Class II originates from multiple scattering above the target layer. Class I provides useful imaging information about the target layer, whereas class II does not. Since both classes mingle in the interference signal, class II deteriorates both the contrast and the resolution of OCT.

We assume

$$I_{AC} \propto \sqrt{I_S}, \tag{9.114}$$

where

$$I_S = I_1 + I_2. \tag{9.115}$$

Here, I_1 and I_2 denote the ensemble-averaged intensity of class I light and class II light, respectively.

Angle-biased sampling, a variance reduction technique, is employed to accelerate the computation of backscattering. Standard sampling of the scattering angle is inefficient for backscattering because scattering in biological tissue is highly forward-peaked. The angle-biased sampling technique samples an artificially biased scattering phase function in lieu of the true function and then compensates for the bias with a photon-weight correction given by

$$w^* = \frac{P(\theta, \phi)}{P^*(\theta, \phi)} w. \tag{9.116}$$

Here, θ $(0 \leq \theta \leq \pi)$ and ϕ $(0 \leq \phi < 2\pi)$ denote the photon deflection polar and azimuthal angles, respectively; P and P^* denote the true and biased phase functions, respectively; w and w^* denote the photon weights associated with P and P^*, respectively.

The Henyey–Greenstein phase function $p(\cos \theta)$ (see Chapter 3) is adopted here for $P(\theta, \phi)$:

$$p(\cos \theta) = \frac{1 - g^2}{2(1 + g^2 - 2g \cos \theta)^{3/2}}, \tag{9.117}$$

where g denotes the scattering anisotropy. Further, $p(-\cos \theta)$ is used for $P^*(\theta, \phi)$. Thus, once $\cos \theta$ is sampled with $p(\cos \theta)$, $-\cos \theta$ is actually used to propagate the photon packet. From Eqs. (9.116) and (9.117), the photon-weight correction is given by

$$w^* = \left(\frac{1 + g^2 + 2g \cos \theta}{1 + g^2 - 2g \cos \theta} \right)^{3/2} w. \tag{9.118}$$

In the simulation, a photon packet is launched from a pencil beam and then tracked by the conventional Monte Carlo method (see Chapter 3). If the photon packet reaches a scattering site in the target layer, it is first labeled and then scattered with the aforementioned angle-biased sampling. The photon packet is terminated whenever $n_S l_S > n_R l_R + (l_c/2)$. When reaching the detector, photon packets with $n_S l_S < n_R l_R - (l_c/2)$ are discarded; then, the labeled ones are recorded into class I and the unlabeled ones into class II.

Parameters used in this simulation include $l_c = 15$ µm, $n_R = n_S = 1.5$, absorption coefficient $\mu_a = 1.5$ cm^{-1}, scattering coefficient $\mu_s = 60$ cm^{-1}, and $g = 0.9$. The detector has a radius of 10 µm and an acceptance angle of 5°. Figure 9.12 shows that class II is smaller than class I at small probing depths but becomes greater than class I at probing depths beyond ~500 µm. Figure 9.13 shows that the average number of scattering events in class II is greater than in class I and increases faster with probing depth. Since scattering randomizes polarization, class II can be rejected using cross-polarization detection by as much as 50%. If the class I light is completely polarized while the class II light is completely unpolarized, the intersection can be extended from ~500 to ~700 µm as indicated in Figure 9.12.

Furthermore, Figure 9.12 shows that class I signal decays at a rate related to the extinction coefficient μ_t. The detected signal intensity I_S depends on three factors sequentially: (1) the number of photons reaching the target layer, (2) the proportion that is subsequently backscattered, and (3) the proportion that ultimately reaches the detector. Singly backscattered light from different depths has the same angular distribution because the scatterers are assumed to have the same phase function. However, singly backscattered light from a greater depth is broadened over a larger area on the tissue surface and hence is not captured as much by the detector. Consequently, the singly backscattered portion in class I has a decay rate greater than μ_t. When the multiple-scattered portion is included, class I ends up with a decay rate that is only slightly different from μ_t.

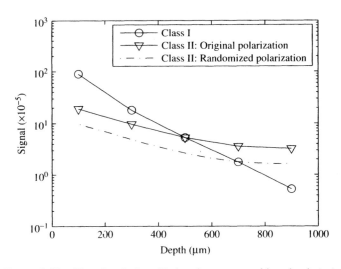

Figure 9.12. Class I and class II signals versus probing depth (z_c).

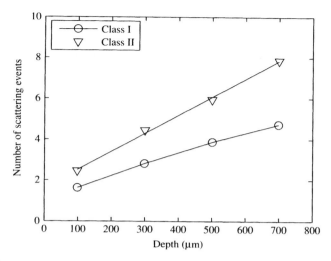

Figure 9.13. Average numbers of scattering events in class I and class II signals versus probing depth (z_c).

As shown in Figure 9.13, class I also contains contributions from multiple-scattered light; the average number of scattering events increases linearly with the probing depth and reaches ~2 at 200 μm. In principle, only singly backscattered photons can provide exact localized imaging information because they experience no interactions outside the target layer. Conversely, multiple-scattered photons cannot directly furnish localized imaging information because they experience interactions at multiple sites.

PROBLEMS

9.1 Show Eq. (9.15) based on the definition of coherence length. Further, prove $l_c = 8 \ln 2 (c/\Delta\omega)$.

9.2 If direct electronic detection were used to provide a 10-μm axial resolution, what would be the required temporal resolution?

9.3 Derive Eq. (9.6). Plot the interference signal versus the arm-length mismatch to show that the AC interference signal rides on a DC background. Under what condition does the contrast (AC amplitude/DC amplitude) reach the maximum? What is the maximum value?

9.4 Prove Eq. (9.41).

9.5 Verify Eq. (9.14).

9.6 (a) Given that if the scatterers are static and the reference mirror is translated at a speed of 40 mm/s toward the incident beam, calculate the beat frequency in OCT. The center wavelength of the light source is 830 nm. **(b)** Assuming that if the scatterers flow toward the incident beam at a speed of 0.5 mm/s with an angle of 75° with respect to the backscattered light, calculate the new beat frequency.

9.7 Implement the simulated TD-OCT shown in Example 9.3 using your own code **(a)** for $\Delta\lambda = 100$, 50, and 25 nm sequentially and **(b)** for double backscatterers that are separated by $4l_c$, $2l_c$, l_c, and $l_c/2$ sequentially. Add a DC background to the interference signal and assume a 5% modulation depth (AC/DC).

9.8 Show analytically that the demodulation demonstrated in Example 9.3 recovers the envelope apart from a constant scaling factor.

9.9 Prove Eq. (9.58) and the subsequent equations in Section 9.5.

9.10 Show that the Fourier-domain rapid scanning optical delay line leads to the following center frequency and bandwidth in the OCT interference signals:

$$f_0 = \frac{4x_0}{\lambda_0}\frac{d\theta_s(t)}{dt} \quad \text{and} \quad \Delta f = \frac{2\Delta\lambda}{\lambda_0^2}\left(2x_0 - \frac{2f\lambda_0}{p_g}\right)\frac{d\theta_s(t)}{dt}.$$

9.11 Extend Example 9.3 to Doppler OCT. Assume that the reference mirror is translated toward the incident beam at a speed of 40 mm/s and the scatterers flow toward the incident beam at a speed of 0.5 mm/s with an angle of 75° with respect to the backscattered light. In addition, take the Fourier transformation of the interference fringes before rectification and recover the flow velocity of the scatterers when the flow direction and the reference mirror velocity are given.

9.12 Extend the theory in Section 9.4 to one for Doppler OCT.

9.13 Derive Eq. (9.100).

9.14 Derive Eq. (9.103).

9.15 Use MATLAB to demonstrate numerically the broadening and chirping from GVD. Assume a set of realistic parameters.

9.16 Use MATLAB to demonstrate the concepts of phase and group velocities using **(a)** two copropagating plane waves that have the same magnitude but a small frequency difference and **(b)** three copropagating plane waves that have the same magnitude but a small frequency difference. Generalize the expressions for the phase and the group velocities.

9.17 Implement a Monte Carlo simulation of OCT to duplicate Figures 9.12 and 9.13.

9.18 Replace the $\propto$ sign with an $=$ sign in Eq. (9.25) by adding appropriate parameters.

9.19 Relate the interference signal I_{AC} in OCT to the autocorrelation function G_1 of the electric field.

9.20 Derive the transverse resolution in OCT on the basis of the confocal mechanism.

READING

Ai J and Wang LHV (2005): Synchronous self-elimination of autocorrelation interference in Fourier-domain optical coherence tomography, *Opt. Lett.* **30**(21): 2939–2941. (See Section 9.6, above.)

Fujimoto JG (2002): Optical coherence tomography: Introduction, in *Handbook of Optical Coherence Tomography*, Bouma BE and Tearney GJ, eds., Dekker, New York. Marcel (See Section 9.4, above.)

Hee MR (2002): Optical coherence tomography: Theory, in *Handbook of Optical Coherence Tomography*, Bouma BE and Tearney GJ, eds., Marcel Dekker, New York. (See Sections 9.4 and 9.8, above.)

Huang D, Swanson EA, Lin CP, Schuman JS, Stinson WG, Chang W, Hee MR, Flotte T, Gregory K, Puliafito CA, and Fujimoto JG (1991): Optical coherence tomography, *Science* **254**(5035): 1178–1181. (See Sections 9.1 and 9.4, above.)

Lindner MW, Andretzky P, Kiesewetter F, and Hausler G (2002): Spectral radar: Optical coherence tomography in the Fourier domain, in *Handbook of Optical Coherence Tomography*, Bouma BE and Tearney GJ, eds., Marcel Dekker, New York. (See Section 9.6, above.)

Milner TE, Yazdanfar S, Rollins AM, Izatt JA, Lindmo T, Chen ZP, Nelson JS, and Wang XJ (2002): Doppler optical coherence tomography, in *Handbook of Optical Coherence Tomography*, Bouma BE and Tearney GJ, eds., Marcel Dekker, New York. (See Section 9.7, above.)

Rollins AM and Izatt JA (2002): Reference optical delay scanning, in *Handbook of Optical Coherence Tomography*, Bouma BE and Tearney GJ, eds., Marcel Dekker, New York. (See Section 9.5, above.)

Yao G and Wang LHV (1999): Monte Carlo simulation of optical coherence tomography in homogeneous turbid media, *Phys. Med. Biol.* **44**(9): 2307–2320. (See Section 9.9, above.)

FURTHER READING

Bouma BE and Tearney GJ (1999): Power-efficient nonreciprocal interferometer and linear-scanning fiber-optic catheter for optical coherence tomography, *Opt. Lett.* **24**(8): 531–533.

Bouma BE and Tearney GJ (2002): *Handbook of Optical Coherence Tomography*, Marcel Dekker, New York.

Cense B and Nassif NA (2004): Ultrahigh-resolution high-speed retinal imaging using spectral-domain optical coherence tomography, *Opt. Express* **12**(11): 2435–2447.

Chen ZP, Milner TE, Srinivas S, Wang XJ, Malekafzali A, van Gemert MJC, and Nelson JS (1997): Noninvasive imaging of in vivo blood flow velocity using optical Doppler tomography, *Opt. Lett.* **22**(14): 1119–1121.

Chiang HP, Chang WS, and Wang JP (1993): Imaging through random scattering media by using cw-broad-band interferometry, *Opt. Lett.* **18**(7): 546–548.

Chinn SR, Swanson EA, and Fujimoto JG (1997): Optical coherence tomography using a frequency-tunable optical source, *Opt. Lett.* **22**(5): 340–342.

Choma MA, Sarunic MV, Yang CH, and Izatt JA (2003): Sensitivity advantage of swept source and Fourier domain optical coherence tomography, *Opt. Express* **11**(18): 2183–2189.

de Boer JF, Cense B, Park BH, Pierce MC, Tearney GJ, and Bouma BE (2003): Improved signal-to-noise ratio in spectral-domain compared with time-domain optical coherence tomography, *Opt. Lett.* **28**(21): 2067–2069.

Drexler W, Morgner U, Kartner FX, Pitris C, Boppart SA, Li XD, Ippen EP, and Fujimoto JG (1999): In vivo ultrahigh-resolution optical coherence tomography, *Opt. Lett.* **24**(17): 1221–1223.

Drexler W, Morgner U, Ghanta RK, Kartner FX, Schuman JS, and Fujimoto JG (2001): Ultrahigh-resolution ophthalmic optical coherence tomography, *Nature Med.* **7**(4): 502–507.

Dubois A, Vabre L, Boccara AC, and Beaurepaire E (2002): High-resolution full-field optical coherence tomography with a Linnik microscope, *Appl. Opt.* **41**(4): 805–812.

Fercher AF, Hitzenberger CK, Sticker M, Moreno-Barriuso E, Leitgeb R, Drexler W, and Sattmann H (2000): A thermal light source technique for optical coherence tomography, *Opt. Commun.* **185**(1–3): 57–64.

Fujimoto JG: Optical coherence tomography for ultrahigh resolution in vivo imaging, *Nature Biotechnol.* **21**(11): 1361–1367.

Hartl I, Li XD, Chudoba C, Ghanta RK, Ko TH, Fujimoto JG, Ranka JK, and Windeler RS (2001): Ultrahigh-resolution optical coherence tomography using continuum generation in an air-silica microstructure optical fiber, *Opt. Lett.* **26**(9): 608–610.

Hee MR, Izatt JA, Swanson EA, and Fujimoto JG (1993): Femtosecond transillumination tomography in thick tissues, *Opt. Lett.* **18**(13): 1107–1109.

Hitzenberger CK, Baumgartner A, and Fercher AF (1998): Dispersion induced multiple signal peak splitting in partial coherence interferometry, *Opt. Commun.* **154**(4): 179–185.

Hitzenberger CK and Fercher AF (1999): Differential phase contrast in optical coherence tomography, *Opt. Lett.* **24**(9): 622–624.

Izatt JA, Hee MR, Owen GM, Swanson EA, and Fujimoto JG (1994): Optical coherence microscopy in scattering media, *Opt. Lett.* **19**(8): 590–592.

Izatt JA, Kulkami MD, Yazdanfar S, Barton JK, and Welch AJ (1997): In vivo bidirectional color Doppler flow imaging of picoliter blood volumes using optical coherence tomography, *Opt. Lett.* **22**(18): 1439–1441.

Knuttel A and Boehlau-Godau M (2000): Spatially confined and temporally resolved refractive index and scattering evaluation in human skin performed with optical coherence tomography, *J. Biomed. Opt.* **5**(1): 83–92.

Kowalevicz AM, Ko T, Hartl I, Fujimoto JG, Pollnau M, and Salathe RP (2002): Ultrahigh resolution optical coherence tomography using a superluminescent light source, *Opt. Express* **10**(7): 349–353.

Kulkarni MD, Thomas, CW and Izatt JA (1997): Image enhancement in optical coherence tomography using deconvolution, *Electron. Lett.* **33**(16): 1365–1367.

Kulkarni MD, van Leeuwen TG, Yazdanfar S, and Izatt JA (1998): Velocity-estimation accuracy and frame-rate limitations in color Doppler optical coherence tomography, *Opt. Lett.* **23**(13): 1057–1059.

Lee TM, Oldenburg AL, Sitafalwalla S, Marks DL, Luo W, Toublan FJJ, Suslick KS, and Boppart SA (2003): Engineered microsphere contrast agents for optical coherence tomography, *Opt. Lett.* **28**(17): 1546–1548.

Leitgeb R, Hitzenberger CK, and Fercher AF (2003): Performance of Fourier domain vs. time domain optical coherence tomography, *Opt. Express* **11**(8): 889–894.

Leitgeb R, Wojtkowski M, Kowalczyk A, Hitzenberger CK, Sticker M, and Fercher AF (2000): Spectral measurement of absorption by spectroscopic frequency-domain optical coherence tomography, *Opt. Lett.* **25**(11): 820–822.

Li XD, Chudoba C, Ko T, Pitris C, and Fujimoto JG (2000): Imaging needle for optical coherence tomography, *Opt. Lett.* **25**(20): 1520–1522.

Liu HH, Cheng PH, and Wang JP (1993): Spatially coherent white-light interferometer based on a point fluorescent source, *Opt. Lett.* **18**(9): 678–680.

Morgner U, Drexler W, Kartner FX, Li XD, Pitris C, Ippen EP, and Fujimoto JG (2000): Spectroscopic optical coherence tomography, *Opt. Lett.* **25**(2): 111–113.

Pan YT, Birngruber R, and Engelhardt R (1997): Contrast limits of coherence-gated imaging in scattering media, *Appl. Opt.* **36**(13): 2979–2983.

Pan YT, Xie HK, and Fedder GK (2001): Endoscopic optical coherence tomography based on a microelectromechanical mirror, *Opt. Lett.* **26**(24): 1966–1968.

Podoleanu AG, Rogers JA, and Jackson DA (1999): OCT en-face images from the retina with adjustable depth resolution in real time, *IEEE J. Select. Topics Quantum Electron.* **5**(4): 1176–1184.

Povazay B, Bizheva K, Unterhuber A, Hermann B, Sattmann H, Fercher AF, Drexler W, Apolonski A, Wadsworth WJ, Knight JC, Russell PSJ, Vetterlein M, and Scherzer E (2002): Submicrometer axial resolution optical coherence tomography, *Opt. Lett.* **27**(20): 1800–1802.

Rollins AM, Kulkarni MD, Yazdanfar S, Ung-arunyawee R, and Izatt JA (1998): In vivo video rate optical coherence tomography, *Opt. Express* **3**(6): 219–229.

Rollins AM and Izatt JA (1999): Optimal interferometer designs for optical coherence tomography, *Opt. Lett.* **24**(21): 1484–1486.

Schmitt JM, Knuttel A, Yadlowsky M, and Eckhaus MA (1994): Optical-coherence tomography of a dense tissue—statistics of attenuation, and backscattering, *Phys. Med. Biol.* **39**(10): 1705–1720.

Schmitt JM and Knuttel A (1997): Model of optical coherence tomography of heterogeneous tissue, *J. Opt. Soc. Am. A* **14**(6): 1231–1242.

Schmitt JM (1999): Optical coherence tomography (OCT): A review, *IEEE J. Select. Topics Quantum Electron.* **5**(4): 1205–1215.

Schmitt JM, Xiang SH, and Yung KM (1999): Speckle in optical coherence tomography, *J. Biomed. Opt.* **4**(1): 95–105.

Swanson EA, Huang D, Hee MR, Fujimoto JG, Lin CP, and Puliafito CA (1992): High-speed optical coherence domain reflectometry, *Opt. Lett.* **17**(2): 151–153.

Tearney GJ, Brezinski ME, Southern JF, Bouma BE, Hee MR, and Fujimoto JG (1995): Determination of the refractive-index of highly scattering human tissue by optical coherence tomography, *Opt. Lett.* **20**(21): 2258–2260.

Tearney GJ, Boppart SA, Bouma BE, Brezinski ME, Weissman NJ, Southern JF, and Fujimoto JG (1996): Scanning single-mode fiber optic catheter-endoscope for optical coherence tomography, *Opt. Lett.* **21**(7): 543–545.

Tearney GJ, Bouma BE, and Fujimoto JG (1997): High-speed phase- and group-delay scanning with a grating-based phase control delay line, *Opt. Lett.* **22**(23): 1811–1813.

Tripathi R, Nassif N, Nelson JS, Park BH, and de Boer JF (2002): Spectral shaping for non-Gaussian source spectra in optical coherence tomography, *Opt. Lett.* **27**(6): 406–408.

Vabre L, Dubois A, and Boccara AC (2002): Thermal-light full-field optical coherence tomography, *Opt. Lett.* **27**(7): 530–532.

Wang RKK, Xu XQ, Tuchin VV, and Elder JB (2001): Concurrent enhancement of imaging depth and contrast for optical coherence tomography by hyperosmotic agents, *J. Opt. Soc. Am. B* **18**(7): 948–953.

Wang YM, Zhao YH, Nelson JS, Chen ZP, and Windeler RS (2003): Ultrahigh-resolution optical coherence tomography by broadband continuum generation from a photonic crystal fiber, *Opt. Lett.* **28**(3): 182–184.

Wojtkowski M, Kowalczyk A, Leitgeb R, and Fercher AF (2002): Full range complex spectral optical coherence tomography technique in eye imaging, *Opt. Lett.* **27**(16): 1415–1417.

Wojtkowski M, Bajraszewski T, Targowski P, and Kowalczyk A (2003): Real-time in vivo imaging by high-speed spectral optical coherence tomography, *Opt. Lett.* **28**(19): 1745–1747.

Youngquist RC, Carr S, and Davies DEN (1987): Optical Coherence-domain reflectometry—a new optical evaluation technique, *Opt. Lett.* **12**(3): 158–160.

Yun SH, Tearney GJ, de Boer JF, Iftimia N, and Bouma BE (2003): High-speed optical frequency-domain imaging, *Opt. Express* **11**(22): 2953–2963.

CHAPTER 10
Mueller Optical Coherence Tomography

10.1. INTRODUCTION

Mueller optical coherence tomography (Mueller OCT) was conceived to image the polarization properties of biological tissue on the basis of polarization-sensitive detection. Although OCT is analogous to ultrasonography in general, polarization exists in transverse optical waves but not in longitudinal ultrasonic waves. As a result, Mueller OCT has no counterpart in ultrasonography.

10.2. MUELLER CALCULUS VERSUS JONES CALCULUS

The term *Mueller calculus* refers to the use of the Stokes vector and the Mueller matrix in polarimetry, whereas *Jones calculus* refers to the use of the Jones vector and matrix. The Stokes vector can quantify the polarization state of any light, whereas the Jones vector can quantify the polarization state of completely polarized light only. The effect of a medium on the polarization state of light can be represented by a Mueller or a Jones matrix; the former operates on a Stokes vector, and the latter operates on a Jones vector. If the medium does not degrade the degree of polarization, the Mueller and the Jones matrices are equivalent, and both are applicable; otherwise, only the Mueller matrix is applicable. Because coherent detection in polarization-sensitive OCT always reports a unity degree of polarization, both matrices apply to OCT. The Mueller matrix, however, clearly separates the two major contrast mechanisms in OCT—backscattering (or backreflection) and polarization—and hence is preferable for presenting final images. In this book, *Mueller OCT* refers to polarization-sensitive OCT based on either the Mueller matrix or the Jones matrix.

10.3. POLARIZATION STATE

Polarization of light refers to the orientation of the electric-field vector $\vec{E}(z, t)$ on the transverse xy plane, where the z axis is aligned with the wave propagation

---

Biomedical Optics: Principles and Imaging, by Lihong V. Wang and Hsin-i Wu
Copyright © 2007 John Wiley & Sons, Inc.

direction and t is time. The Cartesian coordinates (x, y, z) are right-handed, and the x axis is typically horizontal. An electric field vector of monochromatic light can be decomposed into two orthogonal components along the x and y axes:

$$\vec{E}(z, t) = \vec{E}_x(z, t) + \vec{E}_y(z, t). \tag{10.1}$$

The two components can be expressed as

$$\vec{E}_x(z, t) = \hat{e}_x E_{x0} \cos(kz - \omega t + \phi_x), \tag{10.2}$$

$$\vec{E}_y(z, t) = \hat{e}_y E_{y0} \cos(kz - \omega t + \phi_y). \tag{10.3}$$

Here, $\hat{e}_x$ and $\hat{e}_y$ denote unit vectors along the x and y axes, respectively; E_{x0} and ϕ_x denote the amplitude and the phase, respectively, of the horizontal component; E_{y0} and ϕ_y denote the amplitude and the phase, respectively, of the vertical component; ω denotes the angular frequency; and t denotes time. The relative phase between the two components is

$$\Delta\phi = \phi_y - \phi_x. \tag{10.4}$$

We rewrite Eq. (10.3) as

$$\vec{E}_y(z, t) = \hat{e}_y E_{y0} \cos(kz + \phi_x - (\omega t - \Delta\phi)). \tag{10.5}$$

For $-\pi < \Delta\phi < 0$, $\vec{E}_y(z, t)$ has a phase lead over $\vec{E}_x(z, t)$ at a given z; for $0 < \Delta\phi < \pi$, $\vec{E}_y(z, t)$ has a phase lag behind $\vec{E}_x(z, t)$.

In general, the tip of the electric field vector at a given z rotates with time along an ellipse as shown in Figure 10.1a; hence, the light is said to be elliptically polarized. The major axis of the ellipse makes an orientation angle θ_o with respect to the x axis

$$\theta_o = \frac{1}{2} \arctan_2 \frac{2 E_{x0} E_{y0} \cos(\Delta\phi)}{E_{x0}^2 - E_{y0}^2}, \tag{10.6}$$

where $0 \leq \theta_o < \pi$. Here, $\arctan_2$ denotes four-quadrant inverse tangent, yielding an angle dependent on the quadrant of $(E_{x0}^2 - E_{y0}^2, \cos\Delta\phi)$.

An ellipticity angle θ_e is defined to quantify the shape and the handedness of the ellipse:

$$\theta_e = \mp \arctan \frac{b}{a}. \tag{10.7}$$

Here, a and b denote the semimajor and semiminor axes, respectively, of the ellipse; the negative and positive signs represent left- and right-handed polarizations, respectively (to be discussed below). Since $a \geq b$, $-\pi/4 \leq \theta_e \leq \pi/4$. It can be shown that

$$\theta_e = -\frac{1}{2} \arcsin \frac{2 E_{x0} E_{y0} \sin(\Delta\phi)}{E_{x0}^2 + E_{y0}^2}. \tag{10.8}$$

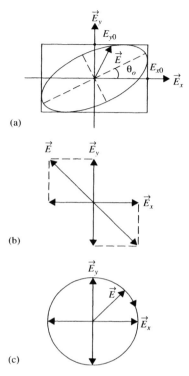

Figure 10.1. (a) Elliptical polarization, where the endpoint of the electric field vector traces an ellipse; (b) $-\pi/4$ linear polarization; (c) right circular polarization.

An auxiliary angle is introduced as

$$\theta_d = \arctan \frac{E_{y0}}{E_{x0}}, \tag{10.9}$$

where $0 \leq \theta_d \leq \pi/2$. From Eq. (10.9), Eqs. (10.6) and (10.8) can be reformulated to

$$\theta_o = \frac{1}{2} \arctan(\tan(2\theta_d) \cos(\Delta\phi)), \tag{10.10}$$

$$\theta_e = -\frac{1}{2} \arcsin(\sin(2\theta_d) \sin(\Delta\phi)). \tag{10.11}$$

The ellipse can reduce to a line. For instance, if $E_{y0} = 0$, the ellipse reduces to a horizontal line; hence, the light is said to be horizontally linearly polarized or simply horizontally polarized. Likewise, if $E_{x0} = 0$, the light is said to be vertically linearly polarized. If $E_{x0} = E_{y0}$ and $\Delta\phi = 0$, the light is $+\pi/4$ linearly polarized. If $E_{x0} = E_{y0}$ and $\Delta\phi = \pi$, the light is $-\pi/4$ linearly polarized as shown in Figure 10.1b.

If $E_{x0} = E_{y0}$ and $\Delta\phi = \pm\pi/2$, the ellipse reduces to a circle; hence, the light is said to be circularly polarized. When $\Delta\phi = -\pi/2$, Eqs. (10.2) and (10.3) become

$$\vec{E}_x(z, t) = \hat{e}_x E_{x0} \cos(kz + \phi_x - \omega t), \quad (10.12)$$

$$\vec{E}_y(z, t) = \hat{e}_y E_{y0} \sin(kz + \phi_x - \omega t). \quad (10.13)$$

Angle $-\omega t$ enlarges clockwise with time according to the definition of planar angle. Therefore, the electric field vector at a given z rotates clockwise when an observer looks back at the source (Figure 10.1c). In this case, the light is said to be right circularly polarized because a snapshot of the electric vectors along the z axis resembles a right-handed screw. On the contrary, when $\Delta\phi = \pi/2$, Eqs. (10.2) and (10.3) become

$$\vec{E}_x(z, t) = \hat{e}_x E_{x0} \cos(\omega t - kz_0 - \phi_x), \quad (10.14)$$

$$\vec{E}_y(z, t) = \hat{e}_y E_{y0} \sin(\omega t - kz_0 - \phi_x). \quad (10.15)$$

Angle $+\omega t$ widens counterclockwise with time; hence, the light is said to be left circularly polarized. For any elliptically polarized beam, when $-\pi < \Delta\phi < 0$, the polarization is said to be right-handed; when $0 < \Delta\phi < \pi$, the polarization is said to be left-handed.

In an alternative convention, $\cos(\omega t - kz + \phi_x)$ and $\cos(\omega t - kz + \phi_y)$ represent the x and y components of the electric field, respectively; in this case, the interpretation of $\Delta\phi$ must be reversed. Although either convention can be adopted as long as consistency is maintained, the coexistence of the two conventions has been a source of confusion in the literature. Here, we use the convention in Eqs. (10.2) and (10.3).

10.4. STOKES VECTOR

The polarization state of any light can be quantified by the Stokes vector, which can be constructed from six intensities measured with different polarization analyzers in front of the detector. Light intensities measured with a horizontal linear analyzer, a vertical linear analyzer, a $+\pi/4$ linear analyzer, a $-\pi/4$ linear analyzer, a right circular analyzer, and a left circular analyzer are denoted by I_H, I_V, $I_{+\pi/4}$, $I_{-\pi/4}$, I_R, and I_L, respectively. The Stokes vector $\mathbf{S}$ is defined by

$$\mathbf{S} = \begin{pmatrix} S_0 \\ S_1 \\ S_2 \\ S_3 \end{pmatrix} = \begin{pmatrix} I_H + I_V \\ I_H - I_V \\ I_{+\pi/4} - I_{-\pi/4} \\ I_R - I_L \end{pmatrix}. \quad (10.16)$$

All four Stokes parameters are real numbers; parameter S_0 represents the original intensity of the light, and each of the other three parameters (S_1, S_2, and S_3)

represents the difference between the two intensities that are measured using analyzers with orthogonal polarization states.

The four Stokes parameters are constrained by

$$S_0^2 \geq S_1^2 + S_2^2 + S_3^2; \tag{10.17}$$

the equality sign applies for completely polarized light and the inequality sign, for partially polarized light; for completely unpolarized light, $S_1 = S_2 = S_3 = 0$. A normalized form of the Stokes vector is

$$\mathbf{S} = \left(1, \frac{S_1}{S_0}, \frac{S_2}{S_0}, \frac{S_3}{S_0}\right)^T, \tag{10.18}$$

where the superscript T stands for transposition.

From the Stokes vector, the degree of polarization (DOP), the degree of linear polarization (DOLP), and the degree of circular polarization (DOCP) can be defined as follows:

$$\text{DOP} = \frac{\sqrt{S_1^2 + S_2^2 + S_3^2}}{S_0},$$

$$\text{DOLP} = \frac{\sqrt{S_1^2 + S_2^2}}{S_0}, \tag{10.19}$$

$$\text{DOCP} = \frac{|S_3|}{S_0}.$$

If the DOP of light remains unity after interaction along a path with a medium, the medium is said to be nondepolarizing. Otherwise, the medium is said to be depolarizing.

Because any light can be decomposed into two orthogonally polarized waves, the law of energy conservation requires

$$I_H + I_V = I_{+\pi/4} + I_{-\pi/4} = I_R + I_L = S_0. \tag{10.20}$$

As a result, we can express $\mathbf{S}$ with four independent measurements, such as $I_H, I_V, I_{+\pi/4}$, and I_R:

$$\mathbf{S} = \begin{pmatrix} I_H + I_V \\ I_H - I_V \\ 2I_{+\pi/4} - (I_H + I_V) \\ 2I_R - (I_H + I_V) \end{pmatrix}. \tag{10.21}$$

For monochromatic light (completely polarized), we have

$$\mathbf{S} = \begin{pmatrix} E_{x0}^2 + E_{y0}^2 \\ E_{x0}^2 - E_{y0}^2 \\ 2E_{x0}E_{y0}\cos(\Delta\phi) \\ -2E_{x0}E_{y0}\sin(\Delta\phi) \end{pmatrix}, \quad (10.22)$$

which can be related to the polarization ellipse by

$$\mathbf{S} = S_0 \begin{pmatrix} 1 \\ \cos(2\theta_e)\cos(2\theta_o) \\ \cos(2\theta_e)\sin(2\theta_o) \\ \sin(2\theta_e) \end{pmatrix}. \quad (10.23)$$

If a Stokes vector is represented geometrically by vector (S_1, S_2, S_3) in Cartesian coordinates, the endpoints of the geometric vectors for all possible polarization states of constant S_0 construct a sphere of radius S_0, which is referred to as a *Poincaré sphere*. The surface of the sphere represents completely polarized states (DOP = 1), whereas the inside represents partially polarized states (DOP < 1). If DOP = 1, the polar and the azimuthal angles of a vector equal $\pi/2 - 2\theta_e$ and $2\theta_o$, respectively [Eqs. (10.23)]. On the spherical surface, the equator represents linear polarizations; the top and the bottom hemispheres represent right- and left-handed polarizations, respectively; the north and the south poles represent right and left circular polarizations, respectively.

Example 10.1. Derive the normalized Stokes vectors for right and left circularly polarized monochromatic light.

For right and left circularly polarized monochromatic light, we have $E_{x0} = E_{y0}$ and $\Delta\phi = \mp\pi/2$. From Eqs. (10.22), the Stokes vectors are

$$\mathbf{S}_R = \begin{pmatrix} 1 \\ 0 \\ 0 \\ 1 \end{pmatrix}, \quad \mathbf{S}_L = \begin{pmatrix} 1 \\ 0 \\ 0 \\ -1 \end{pmatrix}, \quad (10.24)$$

where the subscripts R and L denote the right and left circular polarizations, respectively.

10.5. MUELLER MATRIX

The Mueller matrix $\mathbf{M}$ can represent the effect of a given medium on a Stokes vector. For an incident Stokes vector $\mathbf{S}_{in}$, the output Stokes vector $\mathbf{S}_{out}$ is given by

$$\mathbf{S}_{out} = \mathbf{M}\mathbf{S}_{in}, \quad (10.25)$$

where

$$\mathbf{M} = \begin{pmatrix} M_{00} & M_{01} & M_{02} & M_{03} \\ M_{10} & M_{11} & M_{12} & M_{13} \\ M_{20} & M_{21} & M_{22} & M_{23} \\ M_{30} & M_{31} & M_{32} & M_{33} \end{pmatrix}. \quad (10.26)$$

In general, a Mueller matrix has 16 independent elements.

The Mueller matrix is determined only by the intrinsic properties of the medium and the optical path. Conversely, the Mueller matrix can fully characterize the optical polarization properties of the medium along a given path. Since element M_{00} explicitly represents only the intensity-based property of the medium, the Mueller matrix can clearly separate backscattering contrast from polarization contrast in Mueller OCT.

10.6. MUELLER MATRICES FOR A ROTATOR, A POLARIZER, AND A RETARDER

The Mueller matrix for a rotator, which rotates the incident electric field by an angle θ, can be expressed as

$$\mathbf{M}_r(\theta) = \begin{pmatrix} 1 & 0 & 0 & 0 \\ 0 & \cos(2\theta) & -\sin(2\theta) & 0 \\ 0 & \sin(2\theta) & \cos(2\theta) & 0 \\ 0 & 0 & 0 & 1 \end{pmatrix}. \quad (10.27)$$

A polarizer, also referred to as a *diattenuator*, has polarization-dependent attenuation—also termed *dichroism*. If the two orthogonal eigenpolarization states—states that are unaffected by the polarizing element apart from a constant factor—are linear or circular, the diattenuator is referred to as a *linear* or *circular diattenuator*.

If the two eigenpolarization axes of a linear polarizer are defined as the x and y axes, the electric field transmittances along the x and y axes can be represented by

$$p_x = T_p \cos\theta_p, \quad (10.28)$$

$$p_y = T_p \sin\theta_p, \quad (10.29)$$

where T_p denotes the total electric field transmittance. Taking the ratio of these two equations yields $\tan\theta_p = p_y/p_x$. The Mueller matrix for the linear polarizer is given by

$$\mathbf{M}_p = \frac{T_p^2}{2} \begin{pmatrix} 1 & \cos(2\theta_p) & 0 & 0 \\ \cos(2\theta_p) & 1 & 0 & 0 \\ 0 & 0 & \sin(2\theta_p) & 0 \\ 0 & 0 & 0 & \sin(2\theta_p) \end{pmatrix}. \quad (10.30)$$

If $\theta_p = 0$ (i.e., $p_y = 0$), $\mathbf{M}_p$ represents a horizontal polarizer, which can convert any—even random—polarization into horizontal polarization. Likewise, if $\theta_p = \pi/2$ (i.e., $p_x = 0$), $\mathbf{M}_p$ represents a vertical polarizer.

A phase retarder (also referred to as a *wave plate*, a *phase shifter*, or a *compensator*) has polarization-dependent phase delays. If the two orthogonal eigenpolarization states are linear or circular, the retarder is called a *linear* or *circular retarder*.

If the two axes of a linear retarder are defined as the x and y axes, the x and y components of an incident optical beam are phase-shifted differently:

$$\phi'_x = \phi_x - \frac{\phi}{2},$$
$$\phi'_y = \phi_y + \frac{\phi}{2}, \tag{10.31}$$

where the primed phases are for the retarded optical beam and ϕ represents the phase shift between the two orthogonal components. If ϕ is positive, the x and y axes are referred to as the *fast* and *slow axes*, respectively. The Mueller matrix for the linear retarder is given by

$$\mathbf{M}_\phi = \begin{pmatrix} 1 & 0 & 0 & 0 \\ 0 & 1 & 0 & 0 \\ 0 & 0 & \cos\phi & \sin\phi \\ 0 & 0 & -\sin\phi & \cos\phi \end{pmatrix}. \tag{10.32}$$

For a quarter-wave retarder, $\phi = \pm\pi/2$. For a half-wave retarder, $\phi = \pm\pi$.

The Mueller matrix of a polarizing element whose axes are rotated by θ in the xy plane is given by

$$\mathbf{M}(\theta) = \mathbf{M}_r(\theta)\mathbf{M}(0)\mathbf{M}_r(-\theta), \tag{10.33}$$

where $\mathbf{M}(0)$ and $\mathbf{M}(\theta)$ represent the Mueller matrices of the polarizing element before and after the rotation.

Example 10.2. Show that a quarter-wave retarder converts a $\pm\pi/4$ linearly polarized beam into a circularly polarized beam and vice versa.

According to Eq. (10.32) with $\phi = -\pi/2$, a Mueller matrix operation on the incident Stokes vector leads to

$$\begin{pmatrix} 1 & 0 & 0 & 0 \\ 0 & 1 & 0 & 0 \\ 0 & 0 & 0 & -1 \\ 0 & 0 & 1 & 0 \end{pmatrix} \begin{pmatrix} 1 \\ 0 \\ \pm 1 \\ 0 \end{pmatrix} = \begin{pmatrix} 1 \\ 0 \\ 0 \\ \pm 1 \end{pmatrix}. \tag{10.34}$$

The Stokes vectors on the right-hand side represent right and left circular polarizations, respectively. Conversely, we have

$$\begin{pmatrix} 1 & 0 & 0 & 0 \\ 0 & 1 & 0 & 0 \\ 0 & 0 & 0 & -1 \\ 0 & 0 & 1 & 0 \end{pmatrix} \begin{pmatrix} 1 \\ 0 \\ 0 \\ \pm 1 \end{pmatrix} = \begin{pmatrix} 1 \\ 0 \\ \mp 1 \\ 0 \end{pmatrix}, \quad (10.35)$$

which means that a circular polarization is converted into a linear polarization.

This reciprocal conversion between linear and circular polarizations is guaranteed by the principle of reciprocity of light. For example, the $+\pi/4$ linear polarization is converted into the right circular polarization as shown in Eq. (10.34). When light propagation is time reversed, the right circular polarization becomes a left circular polarization, which is then converted back into the $+\pi/4$ linear polarization as shown in Eq. (10.35).

10.7. MEASUREMENT OF MUELLER MATRIX

A Mueller matrix can be measured with various combinations of source polarizers and detection analyzers. One possible measurement scheme is described here. Four incident polarization states—horizontal polarization (H), vertical polarization (V), $+\pi/4$ linear polarization ($+\pi/4$), and right circular polarization (R)—are used sequentially for the incident optical beam. Their normalized Stokes vectors are

$$\mathbf{S}_H^i = \begin{pmatrix} 1 \\ 1 \\ 0 \\ 0 \end{pmatrix}, \quad \mathbf{S}_V^i = \begin{pmatrix} 1 \\ -1 \\ 0 \\ 0 \end{pmatrix}, \quad \mathbf{S}_{+\pi/4}^i = \begin{pmatrix} 1 \\ 0 \\ 1 \\ 0 \end{pmatrix}, \quad \mathbf{S}_R^i = \begin{pmatrix} 1 \\ 0 \\ 0 \\ 1 \end{pmatrix}, \quad (10.36)$$

where superscript i stands for the incident beams. From Eq. (10.25), the four corresponding output Stokes vectors can be determined as follows:

$$\begin{aligned} \mathbf{S}_H^o &= \mathbf{M}\mathbf{S}_H^i = \mathbf{M}_0 + \mathbf{M}_1, \\ \mathbf{S}_V^o &= \mathbf{M}\mathbf{S}_V^i = \mathbf{M}_0 - \mathbf{M}_1, \\ \mathbf{S}_{+\pi/4}^o &= \mathbf{M}\mathbf{S}_{+\pi/4}^i = \mathbf{M}_0 + \mathbf{M}_2, \\ \mathbf{S}_R^o &= \mathbf{M}\mathbf{S}_R^i = \mathbf{M}_0 + \mathbf{M}_3. \end{aligned} \quad (10.37)$$

Here, superscript o stands for the output beams; $\mathbf{M}_0, \mathbf{M}_1, \mathbf{M}_2$, and $\mathbf{M}_3$ denote the four column vectors of matrix $\mathbf{M}$:

$$\mathbf{M} = (\mathbf{M}_0 \quad \mathbf{M}_1 \quad \mathbf{M}_2 \quad \mathbf{M}_3). \quad (10.38)$$

The four column vectors (each has four elements) can be obtained from Eqs. (10.37):

$$\mathbf{M}_0 = \frac{1}{2}(\mathbf{S}_H^o + \mathbf{S}_V^o),$$
$$\mathbf{M}_1 = \frac{1}{2}(\mathbf{S}_H^o - \mathbf{S}_V^o),$$
$$\mathbf{M}_2 = \frac{1}{2}[2\mathbf{S}_{+\pi/4}^o - (\mathbf{S}_H^o + \mathbf{S}_V^o)],$$
$$\mathbf{M}_3 = \frac{1}{2}[2\mathbf{S}_R^o - (\mathbf{S}_H^o + \mathbf{S}_V^o)].$$

(10.39)

At least four independent Stokes vectors must be measured to fully determine a general Mueller matrix, and each Stokes vector requires at least four independent intensity measurements using different analyzers. Therefore, at least 16 independent intensity measurements must be acquired to completely determine a Mueller matrix. If a Mueller matrix has less than 16 independent elements, fewer intensity measurements are required.

Example 10.3. Given that

$$\mathbf{S}_H^o = \begin{pmatrix} 1 \\ 0 \\ 0 \\ 1 \end{pmatrix}, \quad \mathbf{S}_V^o = \begin{pmatrix} 1 \\ 0 \\ 0 \\ -1 \end{pmatrix}, \quad \mathbf{S}_{+\pi/4}^o = \begin{pmatrix} 1 \\ 0 \\ 1 \\ 0 \end{pmatrix}, \quad \mathbf{S}_R^o = \begin{pmatrix} 1 \\ -1 \\ 0 \\ 0 \end{pmatrix},$$

(10.40)

construct the Mueller matrix **M**.

We have

$$\mathbf{M}_0 = \frac{1}{2}(\mathbf{S}_H^o + \mathbf{S}_V^o) = (1 \quad 0 \quad 0 \quad 0)^T,$$
$$\mathbf{M}_1 = \frac{1}{2}(\mathbf{S}_H^o - \mathbf{S}_V^o) = (0 \quad 1 \quad 0 \quad 0)^T,$$
$$\mathbf{M}_2 = \frac{1}{2}[2\mathbf{S}_{+\pi/4}^o - (\mathbf{S}_H^o + \mathbf{S}_V^o)] = (0 \quad 0 \quad 0 \quad -1)^T,$$
$$\mathbf{M}_3 = \frac{1}{2}[2\mathbf{S}_R^o - (\mathbf{S}_H^o + \mathbf{S}_V^o)] = (0 \quad 0 \quad 1 \quad 0)^T.$$

(10.41)

Therefore, we construct

$$\mathbf{M} = \begin{pmatrix} 1 & 0 & 0 & 0 \\ 0 & 1 & 0 & 0 \\ 0 & 0 & 0 & 1 \\ 0 & 0 & -1 & 0 \end{pmatrix},$$

(10.42)

which is a quarter-wave plate with $\phi = \pi/2$.

10.8. JONES VECTOR

The polarization state of fully polarized light can be quantified by the Jones vector **E**. A Jones vector is a two-element column vector, representing the horizontal (*x*) and vertical (*y*) components of the electric field in phasor expression. From Eqs. (10.2) and (10.3), the Jones vector for monochromatic light is given by

$$\mathbf{E} = \begin{pmatrix} E_{x0} \exp(i\phi_x) \\ E_{y0} \exp(i\phi_y) \end{pmatrix} \xrightarrow{\text{Normalized}} \frac{1}{\sqrt{E_{x0}^2 + E_{y0}^2}} \begin{pmatrix} E_{x0} \\ E_{y0} \exp(i\Delta\phi) \end{pmatrix}. \quad (10.43)$$

The normalized form is simpler; however, the absolute values of the amplitudes and the phases are forfeited. Unlike the full Jones vector, neither the normalized Jones vector nor the Stokes vector can be used to treat interference between coherent lightbeams.

From Eq. (10.43), the horizontal linear polarization state is expressed as

$$\mathbf{E}_H = \begin{pmatrix} E_{x0} \exp(i\phi_x) \\ 0 \end{pmatrix} \xrightarrow{\text{Normalized}} \begin{pmatrix} 1 \\ 0 \end{pmatrix}. \quad (10.44)$$

Similarly, the vertical linear polarization state is expressed as

$$\mathbf{E}_V = \begin{pmatrix} 0 \\ E_{y0} \exp(i\phi_y) \end{pmatrix} \xrightarrow{\text{Normalized}} \begin{pmatrix} 0 \\ 1 \end{pmatrix}. \quad (10.45)$$

The $+\pi/4$ linear polarization state, in which $E_{y0} = E_{x0}$ and $\phi_y = \phi_x$, is expressed as

$$\mathbf{E}_{+\pi/4} = \begin{pmatrix} E_{x0} \exp(i\phi_x) \\ E_{x0} \exp(i\phi_x) \end{pmatrix} \xrightarrow{\text{Normalized}} \frac{1}{\sqrt{2}} \begin{pmatrix} 1 \\ 1 \end{pmatrix}. \quad (10.46)$$

The $-\pi/4$ linear polarization state is expressed as

$$\mathbf{E}_{-\pi/4} = \begin{pmatrix} E_{x0} \exp(i\phi_x) \\ -E_{x0} \exp(i\phi_x) \end{pmatrix} \xrightarrow{\text{Normalized}} \frac{1}{\sqrt{2}} \begin{pmatrix} 1 \\ -1 \end{pmatrix}. \quad (10.47)$$

The right circular polarization state is expressed as

$$\mathbf{E}_R = \begin{pmatrix} E_{x0} \exp(i\phi_x) \\ E_{x0} \exp(i\phi_x - i\pi/2) \end{pmatrix} \xrightarrow{\text{Normalized}} \frac{1}{\sqrt{2}} \begin{pmatrix} 1 \\ -i \end{pmatrix}. \quad (10.48)$$

The left circular polarization state is expressed as

$$\mathbf{E}_L = \begin{pmatrix} E_{x0} \exp(i\phi_x) \\ E_{x0} \exp(i\phi_x + i\pi/2) \end{pmatrix} \xrightarrow{\text{Normalized}} \frac{1}{\sqrt{2}} \begin{pmatrix} 1 \\ i \end{pmatrix}. \quad (10.49)$$

If the convention of using $\cos(\omega t - kz + \phi_x)$ and $\cos(\omega t - kz + \phi_y)$ for the x and y components of the electric field is adopted, the signs of the second elements in Eqs. (10.48) and (10.49) must be reversed.

A normalized Jones vector satisfies the following inner-product identity:

$$\mathbf{E}^{T*} \cdot \mathbf{E} = 1. \quad (10.50)$$

The normalized Jones vectors of two orthogonal polarization states satisfy the following orthonormal identity:

$$\mathbf{E}_1^{T*} \cdot \mathbf{E}_2 = 0. \quad (10.51)$$

For example

$$\mathbf{E}_H^{T*} \cdot \mathbf{E}_V = \mathbf{E}_{+\pi/4}^{T*} \cdot \mathbf{E}_{-\pi/4} = \mathbf{E}_R^{T*} \cdot \mathbf{E}_L = 0. \quad (10.52)$$

10.9. JONES MATRIX

A Jones matrix $\mathbf{J}$ can convert an input Jones vector $\mathbf{E}_{in}$ into an output Jones vector $\mathbf{E}_{out}$:

$$\mathbf{E}_{out} = \mathbf{J}\mathbf{E}_{in}, \quad (10.53)$$

or

$$\begin{pmatrix} E_H^o \\ E_V^o \end{pmatrix} = \begin{pmatrix} J_{11} & J_{12} \\ J_{21} & J_{22} \end{pmatrix} \begin{pmatrix} E_H^i \\ E_V^i \end{pmatrix}. \quad (10.54)$$

Because both $\mathbf{E}_{in}$ and $\mathbf{E}_{out}$ can represent fully polarized light only, $\mathbf{J}$ is applicable to nondepolarizing media only.

10.10. JONES MATRICES FOR A ROTATOR, A POLARIZER, AND A RETARDER

The Jones matrix for a rotator, which rotates the incident electric field by an angle θ, is given by

$$\mathbf{J}_r(\theta) = \begin{pmatrix} \cos\theta & -\sin\theta \\ \sin\theta & \cos\theta \end{pmatrix}. \quad (10.55)$$

The Jones matrix for a linear polarizer that is aligned with the x axis is

$$\mathbf{J}_p(0) = \begin{pmatrix} p_x & 0 \\ 0 & p_y \end{pmatrix}. \quad (10.56)$$

For a linear polarizer oriented at an angle θ with respect to the x axis, we apply the rotational transformation matrix in Eq. (10.55) to derive its Jones matrix

$$\mathbf{J}_p(\theta) = \mathbf{J}_r(\theta)\mathbf{J}_p(0)\mathbf{J}_r(-\theta), \tag{10.57}$$

which leads to

$$\mathbf{J}_p(\theta) = \begin{pmatrix} p_x \cos^2\theta + p_y \sin^2\theta & (p_x - p_y)\sin\theta\cos\theta \\ (p_x - p_y)\sin\theta\cos\theta & p_x \sin^2\theta + p_y \cos^2\theta \end{pmatrix}. \tag{10.58}$$

The Jones matrix for a linear retarder whose fast axis is aligned with the x axis can be expressed as

$$\mathbf{J}_\phi(0) = \begin{pmatrix} \exp(-i\phi/2) & 0 \\ 0 & \exp(i\phi/2) \end{pmatrix}. \tag{10.59}$$

For a linear retarder oriented at an angle θ with respect to the x axis, the Jones matrix becomes

$$\mathbf{J}_\phi(\theta) = \mathbf{J}_r(\theta)\mathbf{J}_\phi(0)\mathbf{J}_r(-\theta). \tag{10.60}$$

10.11. EIGENVECTORS AND EIGENVALUES OF JONES MATRIX

For a linear polarizer, we notice that

$$\mathbf{J}_p(0)\mathbf{E}_H = p_x \mathbf{E}_H, \tag{10.61}$$

$$\mathbf{J}_p(0)\mathbf{E}_V = p_y \mathbf{E}_V. \tag{10.62}$$

Both equations are eigenequations. $\mathbf{E}_H$ and $\mathbf{E}_V$ are the eigenvectors of $\mathbf{J}_p(0)$ and are also referred to as *eigenpolarizations*; p_x and p_y are the associated eigenvalues.

If the two eigenvectors are orthogonal, the polarizing element is considered polarization homogeneous. Otherwise, the polarizing element is considered polarization inhomogeneous. Common polarization-homogeneous optical elements include linear polarizers, linear retarders, and circular retarders. A common polarization-inhomogeneous optical element is a circular polarizer that is constructed with a linear polarizer inclined at $\pi/4$ with respect to the x axis followed by a $\lambda/4$ retarder inclined horizontally.

For a polarization-homogeneous element, the Jones matrix can be constructed from its eigenpolarizations and eigenvalues. We denote the first normalized eigenvector as

$$\mathbf{E}_1 = \begin{pmatrix} E_{1x} \\ E_{1y} \end{pmatrix}. \tag{10.63}$$

On the basis of the orthonormality, the second orthogonal eigenvector is given by

$$\mathbf{E}_2 = \begin{pmatrix} -E_{1y}^* \\ E_{1x}^* \end{pmatrix}. \tag{10.64}$$

The corresponding eigenvalues are λ_1 and λ_2:

$$\mathbf{J}\mathbf{E}_1 = \lambda_1 \mathbf{E}_1, \tag{10.65}$$

$$\mathbf{J}\mathbf{E}_2 = \lambda_2 \mathbf{E}_2. \tag{10.66}$$

A new matrix, termed the *modal matrix*, is constructed from the eigenvectors

$$\mathbf{K} = (\mathbf{E}_1, \quad \mathbf{E}_2) = \begin{pmatrix} E_{1x} & -E_{1y}^* \\ E_{1y} & E_{1x}^* \end{pmatrix}, \tag{10.67}$$

which can be easily inverted by

$$\mathbf{K}^{-1} = \mathbf{K}^{T*} = \begin{pmatrix} E_{1x}^* & E_{1y}^* \\ -E_{1y} & E_{1x} \end{pmatrix}. \tag{10.68}$$

Another new matrix, termed the *diagonal eigenvalue matrix*, is constructed from the eigenvalues:

$$\mathbf{\Lambda} = \begin{pmatrix} \lambda_1 & 0 \\ 0 & \lambda_2 \end{pmatrix}. \tag{10.69}$$

We rewrite Eqs. (10.65) and (10.66) as

$$\mathbf{J}\mathbf{K} = \mathbf{K}\mathbf{\Lambda}. \tag{10.70}$$

Therefore, we have

$$\mathbf{J} = \mathbf{K}\mathbf{\Lambda}\mathbf{K}^{-1}. \tag{10.71}$$

Example 10.4. Show that a circular polarizer that is constructed with a linear polarizer inclined at $\pi/4$ with respect to the x axis followed by a $\lambda/4$ retarder inclined horizontally is polarization-inhomogeneous.

From Eqs. (10.57) and (10.59), the Jones matrix for the circular polarizer is given by

$$\mathbf{J}_{pc} = \mathbf{J}_\phi(0)\mathbf{J}_p(\pi/4), \tag{10.72}$$

where $\phi = \pi/2$, $p_x = 1$, and $p_y = 0$. Matrix operation leads to

$$\mathbf{J}_{pc} = \frac{\exp(-i\pi/4)}{2}\begin{pmatrix} 1 & 1 \\ i & i \end{pmatrix}. \tag{10.73}$$

We can show that

$$\mathbf{J}_{pc}\mathbf{E}_L = \frac{1}{\sqrt{2}}\mathbf{E}_L, \tag{10.74}$$

$$\mathbf{J}_{pc}\mathbf{E}_{-\pi/4} = 0\mathbf{E}_{-\pi/4}. \tag{10.75}$$

The two eigenvectors, $\mathbf{E}_L$ and $\mathbf{E}_{-\pi/4}$, are nonorthogonal because

$$\mathbf{E}_L^{T*} \cdot \mathbf{E}_{-\pi/4} \neq 0. \tag{10.76}$$

Therefore, $\mathbf{J}_{pc}$ is polarization inhomogeneous. We can further show that the linear polarizer and the $\lambda/4$ retarder are not permutable.

Example 10.5. Construct Jones matrices for homogeneous polarizers $\mathbf{J}_H$, $\mathbf{J}_V$, $\mathbf{J}_{+\pi/4}$, $\mathbf{J}_{-\pi/4}$, $\mathbf{J}_R$, and $\mathbf{J}_L$.

Since the electric field of light can be decomposed into two orthogonal components, only one of which can be transmitted through an ideal homogeneous polarizer, we use orthogonal eigenvectors to construct the Jones matrices.

For $\mathbf{J}_H$ and $\mathbf{J}_V$, we may consider the orthogonal pair:

$$\begin{pmatrix} 1 \\ 0 \end{pmatrix} \quad \text{and} \quad \begin{pmatrix} 0 \\ 1 \end{pmatrix}.$$

From

$$\begin{pmatrix} J_{11} & J_{12} \\ J_{21} & J_{22} \end{pmatrix}\begin{pmatrix} 1 \\ 0 \end{pmatrix} = \begin{pmatrix} 1 \\ 0 \end{pmatrix}, \tag{10.77}$$

$$\begin{pmatrix} J_{11} & J_{12} \\ J_{21} & J_{22} \end{pmatrix}\begin{pmatrix} 0 \\ 1 \end{pmatrix} = \begin{pmatrix} 0 \\ 0 \end{pmatrix}, \tag{10.78}$$

we obtain

$$\mathbf{J}_H = \begin{pmatrix} 1 & 0 \\ 0 & 0 \end{pmatrix}. \tag{10.79}$$

Likewise, from

$$\begin{pmatrix} J_{11} & J_{12} \\ J_{21} & J_{22} \end{pmatrix}\begin{pmatrix} 1 \\ 0 \end{pmatrix} = \begin{pmatrix} 0 \\ 0 \end{pmatrix}, \tag{10.80}$$

$$\begin{pmatrix} J_{11} & J_{12} \\ J_{21} & J_{22} \end{pmatrix}\begin{pmatrix} 0 \\ 1 \end{pmatrix} = \begin{pmatrix} 0 \\ 1 \end{pmatrix}, \tag{10.81}$$

we obtain

$$\mathbf{J}_V = \begin{pmatrix} 0 & 0 \\ 0 & 1 \end{pmatrix}. \tag{10.82}$$

For $\mathbf{J}_{+\pi/4}$ and $\mathbf{J}_{-\pi/4}$, we may consider the orthogonal pair:

$$\frac{1}{\sqrt{2}}\begin{pmatrix} 1 \\ 1 \end{pmatrix} \quad \text{and} \quad \frac{1}{\sqrt{2}}\begin{pmatrix} 1 \\ -1 \end{pmatrix}.$$

From

$$\frac{1}{\sqrt{2}}\begin{pmatrix} J_{11} & J_{12} \\ J_{21} & J_{22} \end{pmatrix}\begin{pmatrix} 1 \\ 1 \end{pmatrix} = \frac{1}{\sqrt{2}}\begin{pmatrix} 1 \\ 1 \end{pmatrix}, \tag{10.83}$$

$$\frac{1}{\sqrt{2}}\begin{pmatrix} J_{11} & J_{12} \\ J_{21} & J_{22} \end{pmatrix}\begin{pmatrix} 1 \\ -1 \end{pmatrix} = \begin{pmatrix} 0 \\ 0 \end{pmatrix}, \tag{10.84}$$

we obtain

$$\mathbf{J}_{+\pi/4} = \frac{1}{2}\begin{pmatrix} 1 & 1 \\ 1 & 1 \end{pmatrix}. \tag{10.85}$$

Likewise, from

$$\frac{1}{\sqrt{2}}\begin{pmatrix} J_{11} & J_{12} \\ J_{21} & J_{22} \end{pmatrix}\begin{pmatrix} 1 \\ 1 \end{pmatrix} = \begin{pmatrix} 0 \\ 0 \end{pmatrix}, \tag{10.86}$$

$$\frac{1}{\sqrt{2}}\begin{pmatrix} J_{11} & J_{12} \\ J_{21} & J_{22} \end{pmatrix}\begin{pmatrix} 1 \\ -1 \end{pmatrix} = \frac{1}{\sqrt{2}}\begin{pmatrix} 1 \\ -1 \end{pmatrix}, \tag{10.87}$$

we obtain

$$\mathbf{J}_{-\pi/4} = \frac{1}{2}\begin{pmatrix} 1 & -1 \\ -1 & 1 \end{pmatrix}. \tag{10.88}$$

For a circular polarizer, we may consider the orthogonal pair:

$$\frac{1}{\sqrt{2}}\begin{pmatrix} 1 \\ -i \end{pmatrix} \quad \text{and} \quad \frac{1}{\sqrt{2}}\begin{pmatrix} 1 \\ i \end{pmatrix}.$$

From

$$\frac{1}{\sqrt{2}}\begin{pmatrix} J_{11} & J_{12} \\ J_{21} & J_{22} \end{pmatrix}\begin{pmatrix} 1 \\ -i \end{pmatrix} = \frac{1}{\sqrt{2}}\begin{pmatrix} 1 \\ -i \end{pmatrix}, \tag{10.89}$$

$$\frac{1}{\sqrt{2}}\begin{pmatrix} J_{11} & J_{12} \\ J_{21} & J_{22} \end{pmatrix}\begin{pmatrix} 1 \\ i \end{pmatrix} = \begin{pmatrix} 0 \\ 0 \end{pmatrix}, \tag{10.90}$$

we obtain

$$\mathbf{J}_R = \frac{1}{2}\begin{pmatrix} 1 & i \\ -i & 1 \end{pmatrix}. \tag{10.91}$$

Likewise, from

$$\frac{1}{\sqrt{2}}\begin{pmatrix} J_{11} & J_{12} \\ J_{21} & J_{22} \end{pmatrix}\begin{pmatrix} 1 \\ -i \end{pmatrix} = \begin{pmatrix} 0 \\ 0 \end{pmatrix}, \tag{10.92}$$

$$\frac{1}{\sqrt{2}}\begin{pmatrix} J_{11} & J_{12} \\ J_{21} & J_{22} \end{pmatrix}\begin{pmatrix} 1 \\ i \end{pmatrix} = \frac{1}{\sqrt{2}}\begin{pmatrix} 1 \\ i \end{pmatrix}, \tag{10.93}$$

we obtain

$$\mathbf{J}_L = \frac{1}{2}\begin{pmatrix} 1 & -i \\ i & 1 \end{pmatrix}. \tag{10.94}$$

10.12. CONVERSION FROM JONES CALCULUS TO MUELLER CALCULUS

For a nondepolarizing optical element, the Jones matrix and the Mueller matrix are equivalent. Unlike the Mueller matrix, the Jones matrix uses complex elements. Because one phase is typically arbitrary and is set to zero, a Jones matrix has seven independent real parameters. Consequently, a nondepolarizing Mueller matrix has seven independent parameters.

A Jones matrix $\mathbf{J}$ can be transformed into an equivalent Mueller matrix $\mathbf{M}$ as follows:

$$\mathbf{M} = \mathbf{U}(\mathbf{J} \otimes \mathbf{J}^*)\mathbf{U}^{-1}. \tag{10.95}$$

Here, $\mathbf{U}$ is the Jones–Mueller transformation matrix

$$\mathbf{U} = \frac{1}{\sqrt{2}}\begin{pmatrix} 1 & 0 & 0 & 1 \\ 1 & 0 & 0 & -1 \\ 0 & 1 & 1 & 0 \\ 0 & -i & i & 0 \end{pmatrix}, \tag{10.96}$$

and $\otimes$ represents the Krönecker tensor product. The Krönecker tensor product of $\mathbf{A}$ and $\mathbf{B}$ is defined as

$$\mathbf{A} \otimes \mathbf{B} = \begin{pmatrix} A(1,1)\mathbf{B} & A(1,2)\mathbf{B} & \ldots & A(1,n)\mathbf{B} \\ A(2,1)\mathbf{B} & A(2,2)\mathbf{B} & \ldots & A(2,n)\mathbf{B} \\ \ldots & \ldots & \ldots & \ldots \\ A(m,1)\mathbf{B} & A(m,2)\mathbf{B} & \ldots & A(m,n)\mathbf{B} \end{pmatrix}, \tag{10.97}$$

where m and n represent the dimensions of $\mathbf{A}$.

A Jones vector **E** can be transformed into a Stokes vector **S** using

$$\mathbf{S} = \sqrt{2}\mathbf{U}\langle \mathbf{E} \otimes \mathbf{E}^* \rangle, \quad (10.98)$$

where ensemble averaging is carried out when the light is quasimonochromatic instead of monochromatic.

Example 10.6. Given

$$\mathbf{A} = \begin{pmatrix} a_{11} & a_{12} \\ a_{21} & a_{22} \end{pmatrix} \quad \text{and} \quad \mathbf{B} = \begin{pmatrix} b_{11} & b_{12} \\ b_{21} & b_{22} \end{pmatrix},$$

calculate the Krönecker tensor product.

$$\mathbf{A} \otimes \mathbf{B} = \begin{pmatrix} a_{11}b_{11} & a_{11}b_{12} & a_{12}b_{11} & a_{12}b_{12} \\ a_{11}b_{21} & a_{11}b_{22} & a_{12}b_{21} & a_{12}b_{22} \\ a_{21}b_{11} & a_{21}b_{12} & a_{22}b_{11} & a_{22}b_{12} \\ a_{21}b_{21} & a_{21}b_{22} & a_{22}b_{21} & a_{22}b_{22} \end{pmatrix}. \quad (10.99)$$

10.13. DEGREE OF POLARIZATION IN OCT

After completely polarized monochromatic light (DOP = 1) is scattered multiple times in a scattering medium, the reemitted light generally becomes partially polarized (DOP < 1) unless the area of the detector is much smaller than the average size of the speckle grains. OCT, however, measures the amplitude—rather than the intensity—of backscattered light. As a result, only the part of the backscattered light that is coherent with the reference beam is detected, which leads to a DOP of unity as explained below.

In OCT, the interference signal I_{AC} received by a detector of finite area can be considered as the sum of the interference contributions from all points on the detector:

$$\begin{aligned} I_{AC} &\propto \mathbf{E}_r^{T*} \cdot \mathbf{E}_{s1} + \mathbf{E}_r^{T*} \cdot \mathbf{E}_{s2} + \mathbf{E}_r^{T*} \cdot \mathbf{E}_{s3} + \cdots \\ &= \mathbf{E}_r^{T*} \cdot (\mathbf{E}_{s1} + \mathbf{E}_{s2} + \mathbf{E}_{s3} + \cdots) \quad (10.100) \\ &= \mathbf{E}_r^{T*} \cdot \mathbf{E}_s. \end{aligned}$$

Here, $\mathbf{E}_r$ represents the Jones vector of the reference beam, which is assumed to be uniform across the detection plane; $\mathbf{E}_{si}$ ($i = 1, 2, \ldots$) represents the Jones vector of the coherent backscattered wave reaching point i on the detector; $\mathbf{E}_s$ represents the Jones vector of equivalent total coherent backscattered light; and the dot product represents the interference effect. The projection of each $\mathbf{E}_{si}$ onto $\mathbf{E}_r$ is summed into I_{AC}. Equivalently, $\mathbf{E}_{si}$ can be vector summed to $\mathbf{E}_s$, whose projection onto $\mathbf{E}_r$ contributes to I_{AC}. If all $\mathbf{E}_{si}$ components share the same polarization state, $\mathbf{E}_s$ has the same polarization state. Otherwise, $\mathbf{E}_s$ has a net apparent polarization state. In either case, E_s has a unique polarization state with a DOP of unity. In an intensity-based detection system, by contrast, the intensities of the backscattered optical fields that reach the various points of the detector are added; thus, unless all E_{si} components share the same polarization state, the DOP is less than unity.

10.14. SERIAL MUELLER OCT

In serial Mueller OCT discussed in this section, 16 independent OCT measurements are acquired to determine the Mueller matrix. Only five measurements, however, are required in Mueller OCT, as is discussed in the following section. From Chapter 9, we know that an interference signal for sample light from a given depth has a peak amplitude I_{AC0} given by

$$I_{AC0} \propto \sqrt{I_{s,A}(l_s) I_{r,A}}. \tag{10.101}$$

Here, $I_{r,A}$ denotes the intensity of the reference beam, which has polarization state A; $I_{s,A}$ denotes the intensity of the sample beam projected onto polarization state A. Thus, we have

$$I_{s,A} \propto \frac{I_{AC0}^2}{I_{r,A}}. \tag{10.102}$$

A serial Mueller OCT system that can measure the Mueller matrix of a scattering medium is shown in Figure 10.2. The light from the source, an SLD (superluminescent diode), has a center wavelength of 850 nm and a FWHM bandwidth of 26 nm. After passing through the polarizer, the lightbeam has a power of 0.4 mW. After passing through the half-wave and the quarter-wave plates, the lightbeam is split by a nonpolarizing beamsplitter. The sample beam is focused into the sample by an objective lens. The reference beam passes through a variable-wave plate and is then reflected back. The reflected beams from the reference and the sample arms are coupled into a single-mode fiber and detected by a photodiode. The spatial resolution is $\sim$10 μm.

Four different incident polarization states—H, V, $+\pi/4$, and R—are achieved by rotating the half-wave and the quarter-wave plates in the source arm. For each incident polarization state, the variable-wave plate in the reference arm is adjusted to sequentially achieve polarization states H, V, $+\pi/4$, and R for the round-trip reference beam. Thus, a total of 16 polarization-sensitive OCT images are acquired. Then, four Stokes vectors are computed using Eq. (10.21) and then further processed to construct a Mueller matrix using Eqs. (10.38) and (10.39).

10.15. PARALLEL MUELLER OCT

In parallel Mueller OCT discussed in this section, a Jones matrix is first measured by OCT and then converted into a Mueller matrix. Usually, the coordinates for the Jones vector follow the light propagation direction. For example, a reflection

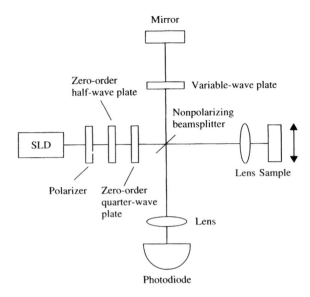

Figure 10.2. Schematic of a serial Mueller OCT system.

Jones matrix converts the Jones vector of the incident light expressed in the forward coordinates (z axis aligned with the direction of incidence) into the Jones vector of the reflected light expressed in the backward coordinates (z axis aligned with the direction of reflection). In this section, however, we use the forward coordinates for the Jones vectors of both the incident and the reflected lightbeams.

An OCT configuration that can measure the Jones matrix of a scattering medium with parallel channels is shown in Figure 10.3. Since at least two independent incident polarization states are required to fully measure a Jones matrix, two SLD light sources are used to provide horizontal polarization $(1\ \ 0)^T$ and vertical polarization $(0\ \ 1)^T$, respectively. The two sources—each of which has a center wavelength of 850 nm and a FWHM bandwidth of 26 nm—are amplitude-modulated at 3 and 3.5 kHz, respectively, for encoded parallel detection. The two source beams are merged by a polarizing beamsplitter, filtered by a spatial filter, and then split into the reference and the sample arms by a nonpolarizing beamsplitter. The sample beam passes through a quarter-wave plate with its fast axis inclined at $+\pi/4$ with respect to the x axis and then is focused into the sample by an objective lens (focal length $f = 15$ mm and NA $= 0.25$). Each source delivers about 0.2 mW of power to the sample. At the sample surface, the Jones vectors of the sample beam are $(1\ \ i)^T$ and $(1\ \ -i)^T$ for the two sources. The reference arm consists of a quarter-wave plate with its fast axis inclined at $+\pi/8$ with respect to the x axis, a lens, and a mirror. After the reference beam passes through the quarter-wave plate twice, the incident horizontal and vertical polarizations are converted into $+\pi/4$ polarization $(1\ \ 1)^T$ and $-\pi/4$ polarization $(1\ \ -1)^T$, respectively. Then, the reference and the sample beams

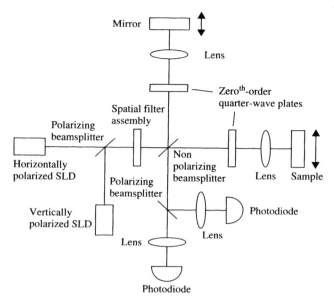

Figure 10.3. Schematic of a parallel Mueller OCT system.

are combined by the nonpolarizing beamsplitter. The combined light is split into horizontal and vertical components by a polarizing beamsplitter; each component is coupled into a single-mode fiber through an objective lens and then detected by a photodiode. A data-acquisition board, sampling at 50 kHz/channel, digitizes the two signals. The scan speed of the reference arm is 0.5 mm/s, generating a Doppler frequency of about 1.2 kHz. The carrier frequencies—the beat and the sum frequencies between the Doppler frequency and the modulation frequencies of the light source—are 1.8, 2.3, 4.2, and 4.7 kHz.

For singly backscattered light, the incident Jones vector $\mathbf{E}_{in}$ in the sample arm is converted to the detected Jones vector $\mathbf{E}_{out}$ by

$$\mathbf{E}_{out} = \mathbf{J}_{NBS}\mathbf{J}_{QB}(\mathbf{J}_{SB}\mathbf{J}_M\mathbf{J}_{SI})\mathbf{J}_{QI}\mathbf{E}_{in}. \tag{10.103}$$

Here, $\mathbf{J}_{QI}$ and $\mathbf{J}_{QB}$ are the Jones matrices of the quarter-wave plate in the incident and the backscattered directions, respectively; $\mathbf{J}_{SI}$ and $\mathbf{J}_{SB}$ are the Jones matrices of the sample in the incident and the backscattered directions, respectively; $\mathbf{J}_M$ is the Jones matrix of the backscattering, which functions as a mirror reflection; $\mathbf{J}_{NBS}$ is the Jones matrix of the reflecting surface of the nonpolarizing beamsplitter. According to the convention of the coordinates used in this section, both $\mathbf{J}_M$ and $\mathbf{J}_{NBS}$ are equal to the identity matrix:

$$\mathbf{J}_M = \mathbf{J}_{NBS} = \begin{pmatrix} 1 & 0 \\ 0 & 1 \end{pmatrix}. \tag{10.104}$$

The combined round-trip Jones matrix of the scattering medium $\mathbf{J}_{S2}$ is given by

$$\mathbf{J}_{S2} = \mathbf{J}_{SB}\mathbf{J}_M\mathbf{J}_{SI}. \tag{10.105}$$

The overall round-trip Jones matrix $\mathbf{J}_T$ is given by

$$\mathbf{J}_T = \mathbf{J}_{NBS}\mathbf{J}_{QB}\mathbf{J}_{S2}\mathbf{J}_{QI}. \tag{10.106}$$

Substituting Eqs. (10.105) and (10.106) into (10.103), we obtain

$$\mathbf{E}_{out} = \mathbf{J}_T \mathbf{E}_{in}, \tag{10.107}$$

or

$$\begin{pmatrix} E_H^o \\ E_V^o \end{pmatrix} = \begin{pmatrix} J_{T11} & J_{T12} \\ J_{T21} & J_{T22} \end{pmatrix} \begin{pmatrix} E_H^i \\ E_V^i \end{pmatrix}. \tag{10.108}$$

The Jones reversibility theorem states that the Jones matrices of an optical element for backward and forward light propagations—$\mathbf{J}_{BWD}$ and $\mathbf{J}_{FWD}$, respectively—are transposition-symmetric if the same coordinates are used for the Jones vectors:

$$\mathbf{J}_{BWD} = \mathbf{J}_{FWD}^T. \tag{10.109}$$

Thus, we have

$$\mathbf{J}_{SB} = \mathbf{J}_{SI}^T, \tag{10.110}$$

$$\mathbf{J}_{QB} = \mathbf{J}_{QI}^T, \tag{10.111}$$

which together lead to

$$\mathbf{J}_{S2} = \mathbf{J}_{S2}^T, \tag{10.112}$$

$$\mathbf{J}_T = \mathbf{J}_T^T. \tag{10.113}$$

According to these symmetry relations, the number of independent real parameters in $\mathbf{J}_{S2}$ or $\mathbf{J}_T$ reduces from seven to five, which means that only five real independent measurements are required to measure a Jones or Mueller matrix in OCT.

As discussed in Chapter 9, multiple-scattered light can contribute to OCT signals. In the presence of multiple-scattering contributions, Eq. (10.112) still holds as long as each photon path is reversible (the probabilities for photons to travel along the same path but in opposite directions are equal). This condition is met, for example, in single-mode optical-fiber-based OCT systems, where light delivery and detection share the same area and angular distribution. Apart from a constant factor, $\mathbf{J}_{S2}$ is the sum of the Jones matrices for all possible paths:

$$\mathbf{J}_{S2} = \sum_k w_k (\mathbf{J}_{Fk} + \mathbf{J}_{Rk}). \tag{10.114}$$

Here, w denotes the weight of a path; subscripts F and R denote the forward and the reversed directions of propagation, respectively; k denotes the index of a path. From the Jones reversibility theorem, we have

$$\mathbf{J}_{Rk} = \mathbf{J}_{Fk}^T, \mathbf{J}_{Fk} = \mathbf{J}_{Rk}^T. \tag{10.115}$$

Substituting Eqs. (10.115) into Eq. (10.114), we reach Eq. (10.112) again.

For two light sources of different polarization states, Eq. (10.108) leads to

$$\begin{pmatrix} E_{H1}^o & E_{H2}^o \\ E_{V1}^o & E_{V2}^o \end{pmatrix} = \begin{pmatrix} J_{T11} & J_{T12} \\ J_{T21} & J_{T22} \end{pmatrix} \begin{pmatrix} E_{H1}^i & E_{H2}^i e^{i\beta} \\ E_{V1}^i & E_{V2}^i e^{i\beta} \end{pmatrix}. \tag{10.116}$$

Here, subscripts 1 and 2 for the electric fields denote sources 1 and 2, respectively; β is a phase difference related to the difference in the spectral characteristics of the two light sources; and β is zero if the two characteristics are identical. Inverting Eq. (10.116) yields

$$\begin{pmatrix} J_{T11} & J_{T12} \\ J_{T21} & J_{T22} \end{pmatrix} = \begin{pmatrix} E_{H1}^o & E_{H2}^o \\ E_{V1}^o & E_{V2}^o \end{pmatrix} \begin{pmatrix} E_{H1}^i & E_{H2}^i e^{i\beta} \\ E_{V1}^i & E_{V2}^i e^{i\beta} \end{pmatrix}^{-1}. \tag{10.117}$$

The inverse matrix exists and is given by

$$\begin{pmatrix} E_{H1}^i & E_{H2}^i e^{i\beta} \\ E_{V1}^i & E_{V2}^i e^{i\beta} \end{pmatrix}^{-1} = \frac{1}{D} \begin{pmatrix} E_{V2}^i e^{i\beta} & -E_{H2}^i e^{i\beta} \\ -E_{V1}^i & E_{H1}^i \end{pmatrix}, \tag{10.118}$$

if the determinant D is nonzero:

$$D = \begin{vmatrix} E_{H1}^i & E_{H2}^i e^{i\beta} \\ E_{V1}^i & E_{V2}^i e^{i\beta} \end{vmatrix} = e^{i\beta} \begin{vmatrix} E_{H1}^i & E_{H2}^i \\ E_{V1}^i & E_{V2}^i \end{vmatrix} \neq 0. \tag{10.119}$$

This condition simply means that the two light sources have independent polarization states.

Substituting Eq. (10.118) into Eq. (10.117) yields

$$\mathbf{J}_T = \frac{1}{D} \begin{pmatrix} E_{H1}^o & E_{H2}^o \\ E_{V1}^o & E_{V2}^o \end{pmatrix} \begin{pmatrix} E_{V2}^i e^{i\beta} & -E_{H2}^i e^{i\beta} \\ -E_{V1}^i & E_{H1}^i \end{pmatrix}. \tag{10.120}$$

242 MUELLER OPTICAL COHERENCE TOMOGRAPHY

On the basis of the transposition symmetry of $\mathbf{J}_T$, β can be eliminated. Substituting Eq. (10.120) into Eq. (10.113) yields

$$e^{i\beta}\left(E_{H1}^o E_{H2}^i + E_{V1}^o E_{V2}^i\right) = E_{H2}^o E_{H1}^i + E_{V2}^o E_{V1}^i, \qquad (10.121)$$

which can be solved for β unless

$$E_{H1}^o E_{H2}^i + E_{V1}^o E_{V2}^i = 0. \qquad (10.122)$$

If the two incident polarization states are orthogonal and either one happens to be an eigenpolarization state of the sample, Eq. (10.122) holds (see Problem 10.1). This drawback can be overcome by using two nonorthogonal source polarization states, for example, a horizontal polarization state and a $+\pi/4$ polarization state. Once $\mathbf{J}_T$ is found, $\mathbf{J}_{S2}$ can then be determined by solving Eq. (10.106).

A piece of porcine tendon is imaged by this system. The Mueller matrix image is shown in Figure 10.4; each element of the Mueller matrix is a 2D image. Some of the images present periodical stripes presumably due to the birefringence of the collagen fibers in the porcine tendon. Since it is free of polarization effects, image M_{00} presents no such periodicity but shows backscattering contrast instead.

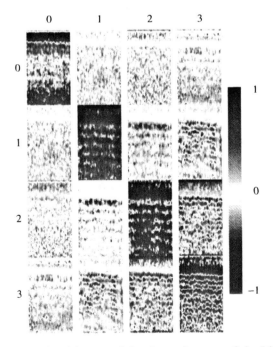

Figure 10.4. Two-dimensional images, 0.5 × 1 mm in area, of the Mueller matrix of a piece of porcine tendon. Each image except M_{00} is normalized by M_{00} pixel-by-pixel.

PROBLEMS

10.1 Substantiate the interpretation of Eq. (10.122) mathematically.

10.2 Given that

$$\mathbf{S}_H^o = \begin{pmatrix} 1 \\ 1 \\ 0 \\ 0 \end{pmatrix}, \quad \mathbf{S}_V^o = \begin{pmatrix} 1 \\ -1 \\ 0 \\ 0 \end{pmatrix}, \quad \mathbf{S}_{+\pi/4}^o = \begin{pmatrix} 1 \\ 0 \\ 0 \\ 1 \end{pmatrix}, \quad \mathbf{S}_R^o = \begin{pmatrix} 1 \\ 0 \\ -1 \\ 0 \end{pmatrix},$$

construct the Mueller matrix $\mathbf{M}$.

10.3 Show that the eigenvalues of the six polarizers given in Example 10.5 are equal to 0 or 1 and that the eigenvectors are orthogonal for each pair.

10.4 Find the Jones matrix for a linear retarder oriented at an angle with respect to the x axis.

10.5 (a) Show that a linear polarizer inclined at $-\pi/4$ with respect to the x axis, followed by a quarter-wave plate, leads to a circular polarizer. (b) Find the eigenvalues and eigenvectors for this matrix. (c) Show that the eigenvectors are nonorthogonal and allow both right and left circularly polarized beams to pass through. Swap the two optical components and derive the new Jones matrix of the system and show whether the two optical components are permutable.

10.6 Construct the Jones matrix for a polarizing element that converts a $\pm\pi/4$ linear polarization into a circular polarization but maintains the horizontal and the vertical polarizations.

10.7 Use the Jones representations to show that a quarter-wave plate can be used to convert linear polarization into circular polarization and vice versa.

10.8 Use Jones vectors to prove that the superposition of two equal-amplitude circularly polarized beams with opposite handedness and a phase difference can be used to form a linear polarization state of an arbitrary orientation. In practice, circular birefringence in either optically active media, such as glucose or magnetooptical media, can be used to produce the phase difference.

10.9 Prove that the Jones or the Mueller matrix of a rotator is independent of the orientation of the rotator.

10.10 Derive Eqs. (10.6), (10.8), (10.10), and (10.11).

10.11 Use MATLAB to produce a movie that shows an electric field vector tracing out the polarization ellipse and its special cases.

244 MUELLER OPTICAL COHERENCE TOMOGRAPHY

10.12 From the six measurements, I_H, I_V, $I_{+\pi/4}$, $I_{-\pi/4}$, I_R, and I_L, assuming that four are acquired to construct the Stokes vector, derive the number of valid choices.

10.13 Prove Eq. (10.22).

10.14 Prove Eq. (10.23).

10.15 When a linear polarizer is rotated a full circle in front of natural light (unpolarized light), draw the locus of the transmitted light polarization on the Poincaré sphere in MATLAB in pseudo-3D.

10.16 Prove Eqs. (10.27), (10.30), and (10.32).

10.17 Prove Eq. (10.33).

10.18 Prove Eqs. (10.55) and (10.57)–(10.59).

10.19 Use Eq. (10.71) to form the Jones matrix for a quarter-wave plate.

10.20 Prove conversion Eqs. (10.95) and (10.98).

10.21 Assume that a quarter-wave plate oriented at 45° with respect to the x axis is measured by the system in Figure 10.3. Derive the Jones vector of the lightbeam after each polarizing element, starting from the source. First, use right-handed coordinates whose z axis always follows the propagation direction of the lightbeam. Revise the Jones reversibility theorem in this convention. Then, use right-handed coordinates whose z axis is always in the propagation direction of the lightbeams incident on the reference mirror and the sample as in Section 10.14.

10.22 Extend polarization-difference imaging (see Chapter 8) to birefringent scattering media using the Stokes vector.

READING

Born M and Wolf E (1999): *Principles of Optics: Electromagnetic Theory of Propagation, Interference and Diffraction of Light*, Cambridge Univ. Press, New York. (See Sections 10.3–10.5, above.)

Chipman RA (1995): Polarimetry, in *Handbook of Optics*, Bass M. and Optical Society of America, eds., McGraw-Hill, New York. Vol. II, Chapter 22. (See Sections 10.4–10.6, above.)

Collett E (1993): *Polarized Light: Fundamentals and Applications*, Marcel Dekker, New York. (See Sections 10.3–10.6 and 10.8–10.11, above.)

Jiao SL, Yao G, and Wang LHV (2000): Depth-resolved two-dimensional Stokes vectors of backscattered light and Mueller matrices of biological tissue measured with optical coherence tomography, *Appl. Opt.* **39**(34): 6318–6324. (See Sections 10.7 and 10.13, above.)

Jiao SL and Wang LHV (2002): Two-dimensional depth-resolved Mueller matrix of biological tissue measured with double-beam polarization-sensitive optical coherence tomography, *Opt. Lett.* **27**(2): 101–103. (See Section 10.15, above.)

Vansteenkiste N, Vignolo P, and Aspect A (1993): Optical reversibility theorems for polarization—application to remote-control of polarization, *J. Opt. Soc. Am. A* **10**(10): 2240–2245. (See Section 10.12, above.)

Yao G and Wang LHV (1999): Two-dimensional depth-resolved Mueller matrix characterization of biological tissue by optical coherence tomography, *Opt. Lett.* **24**(8): 537–539. (See Section 10.14, above.)

FURTHER READING

Brosseau C (1998): *Fundamentals of Polarized Light: A Statistical Optics Approach*, Wiley, New York.

Bueno JM and Campbell MCW (2002): Confocal scanning laser ophthalmoscopy improvement by use of Mueller-matrix polarimetry, *Opt. Lett.* **27**(10): 830–832.

Cameron BD, Rakovic MJ, Mehrubeoglu M, Kattawar GW, Rastegar S, Wang LHV, and Cote GL (1998): Measurement and calculation of the two-dimensional backscattering Mueller matrix of a turbid medium, *Opt. Lett.* **23**(7): 485–487.

Cense B, Chen TC, Park BH, Pierce MC, and de Boer JF (2004): Thickness and birefringence of healthy retinal nerve fiber layer tissue measured with polarization-sensitive optical coherence tomography, *Invest. Ophthalmol. Visual Sci.* **45**(8): 2606–2612.

de Boer JF 1997, Milner TE, vanGemert MJC, and Nelson JS (1997): Two-dimensional birefringence imaging in biological tissue by polarization-sensitive optical coherence tomography, *Opt. Lett.* **22**(12): 934–936.

de Boer JF 1999, Milner TE, and Nelson JS (1999): Determination of the depth-resolved Stokes parameters of light backscattered from turbid media by use of polarization-sensitive optical coherence tomography, *Opt. Lett.* **24**(5): 300–302.

de Boer JF 1999, Srinivas SM, Park BH, Pham TH, Chen ZP, Milner TE, and Nelson JS (1999): Polarization effects in optical coherence tomography of various biological tissues, *IEEE J. Select. Topics Quantum Electron.* **5**(4): 1200–1204.

de Boer JF 2002 and Milner TE (2002): Review of polarization sensitive optical coherence tomography and Stokes vector determination, *J. Biomed. Opt.* **7**(3): 359–371.

Everett MJ, Schoenenberger K, Colston BW, and Da Silva LB (1998): Birefringence characterization of biological tissue by use of optical coherence tomography, *Opt. Lett.* **23**(3): 228–230.

Eyal A and Zadok A (2005): Optical noise induced by Gaussian sources in Stokes parameter measurements, *J. Opt. Soc. Am. A* **22**(4): 662–671.

Gil JJ and Bernabeu E (1987): Obtainment of the polarizing and retardation parameters of a nondepolarizing optical-system from the polar decomposition of its Mueller matrix, *Optik* **76**(2): 67–71.

Guo SG, Zhang J, Wang L, Nelson JS, and Chen ZP (2004): Depth-resolved birefringence and differential optical axis orientation measurements with fiber-based polarization-sensitive optical coherence tomography, *Opt. Lett.* **29**(17): 2025–2027.

Hecht E (2002): *Optics*, Addison-Wesley, Reading, MA.

Hitzenberger CK, Gotzinger E, Sticker M, Pircher M, and Fercher AF (2001): Measurement and imaging of birefringence and optic axis orientation by phase resolved polarization sensitive optical coherence tomography, *Opt. Express* **9**(13): 780–790.

Jiao SL, Yao G, and Wang LHV (2000): Depth-resolved two-dimensional Stokes vectors of backscattered light and Mueller matrices of biological tissue measured with optical coherence tomography, *Appl. Opt.* **39**(34): 6318–6324.

Jiao SL and Wang LHV (2002): Jones-matrix imaging of biological tissues with quadruple-channel optical coherence tomography, *J. Biomed. Opt.* **7**(3): 350–358.

Jiao SL and Wang LHV (2002): Two-dimensional depth-resolved Mueller matrix of biological tissue measured with double-beam polarization-sensitive optical coherence tomography, *Opt. Lett.* **27**(2): 101–103.

Jiao SL, Yu WR, Stoica G, and Wang LHV (2003): Contrast mechanisms in polarization-sensitive Mueller-matrix optical coherence tomography and application in burn imaging, *Appl. Opt.* **42**(25): 5191–5197.

Jiao SL, Yu WR, Stoica G, and Wang LHV (2003): Optical-fiber-based Mueller optical coherence tomography, *Opt. Lett.* **28**(14): 1206–1208.

Jiao SL, Todorovic M, Stoica G, and Wang LHV (2005): Fiber-based polarization-sensitive Mueller matrix optical coherence tomography with continuous source polarization modulation, *Appl. Opt.* **44**(26): 5463–5467.

Kemp NJ, Zaatari HN, Park J, Rylander HG, and Milner TE (2005): Form-biattenuance in fibrous tissues measured with polarization-sensitive optical coherence tomography (PS-OCT), *Opt. Express* **13**(12): 4611–4628.

Li J, Yao G, and Wang LHV (2002): Degree of polarization in laser speckles from turbid media: Implications in tissue optics, *J. Biomed. Opt.* **7**(3): 307–312.

Liu B, Harman M, and Brezinski ME (2005): Variables affecting polarization-sensitive optical coherence tomography imaging examined through the modeling of birefringent phantoms, *J. Opt. Soc. Am. A* **22**(2): 262–271.

Makita S, Yasuno Y, Endo T, Itoh M, and Yatagai T (2005): Jones matrix imaging of biological samples using parallel-detecting polarization-sensitive Fourier domain optical coherence tomography, *Opt. Rev.* **12**(2): 146–148.

Makita S, Yasuno Y, Sutoh Y, Itoh M, and Yatagai T (2003): Polarization-sensitive spectral interferometric optical coherence tomography for human skin imaging, *Opt. Rev.* **10**(5): 366–369.

Moreau J, Loriette V, and Boccara AC (2003): Full-field birefringence imaging by thermal-light polarization-sensitive optical coherence tomography. I. Theory, *Appl. Opt.* **42**(19): 3800–3810.

Moreau J, Loriette V, and Boccara AC (2003): Full-field birefringence imaging by thermal-light polarization-sensitive optical coherence tomography. II. Instrument and results, *Appl. Opt.* **42**(19): 3811–3818.

Park BH, Pierce MC, Cense B, and de Boer JF (2004): Jones matrix analysis for a polarization-sensitive optical coherence tomography system using fiber-optic components, *Opt. Lett.* **29**(21): 2512–2514.

Park BH, Pierce MC, Cense B, and de Boer JF (2005): Optic axis determination accuracy for fiber-based polarization-sensitive optical coherence tomography, *Opt. Lett.* **30**(19): 2587–2589.

Pierce MC, Park BH, Cense B, and de Boer JF (2002): Simultaneous intensity, birefringence, and flow measurements with high-speed fiber-based optical coherence tomography, *Opt. Lett.* **27**(17): 1534–1536.

Pircher M, Goetzinger E, Leitgeb R, and Hitzenberger CK (2004): Transversal phase resolved polarization sensitive optical coherence tomography, *Phys. Med. Biol.* **49**(7): 1257–1263.

Ren HW, Ding ZH, Zhao YH, Miao JJ, Nelson JS, and Chen ZP (2002): Phase-resolved functional optical coherence tomography: Simultaneous imaging of insitu tissue structure, blood flow velocity, standard deviation, birefringence, and Stokes vectors in human skin, *Opt. Lett.* **27**(19): 1702–1704.

Roth JE, Kozak JA, Yazdanfar S, Rollins AM, and Izatt JA (2001): Simplified method for polarization-sensitive optical coherence tomography, *Opt. Lett.* **26**(14): 1069–1071.

Saxer CE, de Boer JF, Park BH, Zhao YH, Chen ZP, and Nelson JS (2000): High-speed fiber-based polarization-sensitive optical coherence tomography of in vivohuman skin, *Opt. Lett.* **25**(18): 1355–1357.

Schoenenberger K, Colston BW, Maitland DJ, Da Silva LB, and Everett MJ (1998): Mapping of birefringence and thermal damage in tissue by use of polarization-sensitive optical coherence tomography, *Appl. Opt.* **37**(25): 6026–6036.

Todorovic M, Jiao SL, and Wang LHV (2004): Determination of local polarization properties of biological samples in the presence of diattenuation by use of Mueller optical coherence tomography, *Opt. Lett.* **29**(20): 2402–2404.

Vansteenkiste N, Vignolo P, and Aspect A (1993): Optical reversibility theorems for polarization—application to remote-control of polarization, *J. Opt. Soc. Am. A* **10**(10): 2240–2245.

Yang Y, Wu L, Feng YQ, and Wang RK (2003): Observations of birefringence in tissues from optic-fibre-based optical coherence tomography, *Meas. Sci. Technol.* **14**(1): 41–46.

Yao G and Wang LHV (2000): Propagation of polarized light in turbid media: Simulated animation sequences, *Opt. Express* **7**(5): 198–203.

Yasuno Y, Sutoh Y, Makita S, Itoh M, and Yatagai T (2003): Polarization sensitive spectral interferometric optical coherence tomography for biological samples, *Opt. Rev.* **10**(5): 498–500.

Yasuno Y, Makita S, Endo T, Itoh M, Yatagai T, Takahashi M, Katada C, and Mutoh M (2004): Polarization-sensitive complex Fourier domain optical coherence tomography for Jones matrix imaging of biological samples, *Appl. Phys. Lett.* **85**(15): 3023–3025.

Zhang J, Guo SG, Jung WG, Nelson JS, and Chen ZP (2003): Determination of birefringence and absolute optic axis orientation using polarization-sensitive optical coherence tomography with PM fibers, *Opt. Express* **11**(24): 3262–3270.

Zhang J, Jung WG, Nelson JS, and Chen ZP (2004): Full range polarization-sensitive Fourier domain optical coherence tomography, *Opt. Express* **12**(24): 6033–6039.

CHAPTER 11

Diffuse Optical Tomography

11.1. INTRODUCTION

The term *diffuse optical tomography* (DOT) refers to the optical imaging of biological tissue in the diffusive regime. Since it has a $1/e$ penetration depth on the order of 0.5 cm, NIR light around 700-nm wavelength can penetrate several centimeters into biological tissue. As a result, DOT can image the human breast and brain. Image reconstruction in DOT involves both the forward and the inverse problems. The forward problem usually uses the diffusion equation to predict the distribution of reemitted light on the basis of presumed parameters for both the light source and the object. The inverse problem uses the forward problem to reconstruct the distributions of the optical properties of the object from a measured data set. Since the inverse problem is ill-posed, recovering imaging information from diffuse photons remains a challenge. As a rule of thumb, the spatial resolution of DOT is on the order of 20% of the imaging depth; hence, DOT is a low-resolution imaging technology. Nevertheless, DOT provides valuable rapid functional imaging at low cost.

11.2. MODES OF DIFFUSE OPTICAL TOMOGRAPHY

In a DOT system, sources and detectors are placed around the object to be imaged in various geometric configurations. Common geometric configurations, suited for different applications, fall into planar transmission, planar reflection, and cylindrical reemission. Most anatomical sites can be imaged in the planar reflection configuration. Human limbs as well as small animals can be imaged in either the planar transmission or the cylindrical configuration. Human breasts can be imaged in all three configurations. Generally, while one source illuminates the object, all detectors measure the reemitted light. This process is repeated with each source to complete a measurement data set; subsequently, images are reconstructed by computer.

Biomedical Optics: Principles and Imaging, by Lihong V. Wang and Hsin-i Wu
Copyright © 2007 John Wiley & Sons, Inc.

TABLE 11.1. Modes of DOT with Idealized Parameters[a]

Mode	Source Light $\Phi_s(\vec{r}', t')$	Reemitted Light $\Phi_m(\vec{r}, t; \vec{r}', t')$
Time domain	Impulse: $\delta(\vec{r}')\delta(t')$	Time-resolved: $\Phi_m(\vec{r}, t; \vec{r}', t')$
Frequency domain	Amplitude-modulated: $\delta(\vec{r}')[D_s + A_s \cos(\omega t' + \phi_s)]$	Amplitude-modulated: $D_m(\vec{r}; \vec{r}') + A_m(\vec{r}; \vec{r}') \cos(\omega t + \phi_m(\vec{r}; \vec{r}'))$
Direct current (DC)	DC: $D_s \delta(\vec{r}')$	DC: $D_m(\vec{r}; \vec{r}')$

[a] Symbol key: Φ—fluence rate; D—DC; A—AC amplitude; ω—angular frequency of AC; ϕ—phase of AC; $\vec{r}'$—source location; t'—source time; $\vec{r}$—detection location; t—detection time; subscripts s and m represent source and measurement, respectively

According to the type of signal, DOT is usually classified into three modes (Table 11.1): time domain, frequency domain, and direct current (DC). In all three modes, the reemitted light has the same general form as the source light since the system is linear and time-invariant. In the time-domain mode, the source light is ultra-short-pulsed (typically a few picoseconds wide), and the reemitted light pulses are broadened. In the frequency-domain mode, the source light intensity is amplitude-modulated sinusoidally at typically hundreds of MHz, and the reemitted light modulation has reduced modulation depth (AC amplitude/DC). In the DC mode, the source light is usually time-invariant, but it is sometimes modulated at low frequency (e.g., kHz) to improve the SNR or to encode the source. Such low-frequency modulation, however, does not attain the benefits of the frequency-domain mode.

The time-domain and the frequency-domain modes are mathematically related via the Fourier transformation. If measured at many frequencies (including DC) in a sufficiently broad bandwidth, frequency-domain signals can be converted to the time domain using the inverse Fourier transformation. Therefore, the time-domain mode is mathematically equivalent to the combination of the frequency-domain and DC modes. Further, the DC mode is a zero-frequency special case of the frequency-domain mode.

Among the three modes, the time-domain mode is the most information-rich but the slowest in data acquisition and the most expensive. The full extent of the information content, however, is yet to be fully explored. The frequency-domain mode, which typically operates at only a single modulation frequency, contains less information than does the time-domain mode but is faster and less expensive. In addition, it provides better SNR by means of narrowband detection. The DC mode contains no direct information about time of flight and hence has the most limited capability for separating absorption from scattering in a heterogeneous medium; it is the fastest and the least expensive, however.

In the following sections, experimental systems representative of all three modes are introduced. An in-depth description of the frequency-domain mode follows since this mode is the most mature and dominant to date.

11.3. TIME-DOMAIN SYSTEM

In the time-domain mode, temporal responses to an ultrashort laser pulse are measured around the scattering object to be imaged. Each temporal response represents reemitted light intensity as a function of time; each actually equals the convolution of three functions: the temporal PSF due to light propagation in the medium, the pulse profile of the laser beam, and the impulse response of the detection system. The responses can be temporally resolved using a streak camera (see Chapter 8) or a time-correlated single-photon counting system. The latter has the advantages of larger detection area, better temporal linearity, lower cost, and higher dynamic range over the former; however, it has the disadvantage of slower data acquisition.

A 32-channel imaging system, based on time-correlated single-photon counting, was constructed at University College London (UCL) (Figure 11.1). A laser provides picosecond light pulses with tunable wavelength. After going through a beamsplitter, the lightbeam is attenuated to a preset intensity by a neutral-density filter. After further passing through a shutter and a fiber coupler, the lightbeam is coupled to a 1 × 32 fiber switch, which selects one of the 32 source fibers

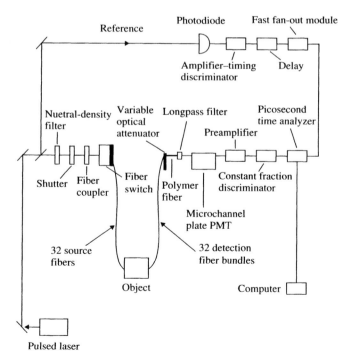

Figure 11.1. Schematic of the UCL time-resolved DOT system. For clarity, only one source fiber and one detection channel are shown.

at a time. The source fibers are routed to distributed locations on the surface of the object.

Reemitted photons are simultaneously detected by 32 detection channels, each with a large-diameter fiber bundle. Since the 32 detection positions are distributed around the object, the received optical signals have vastly different intensities. In other words, some detection fibers—typically those that are close to the source fiber—receive strong optical signals, while other detection fibers—typically the ones that are far away from the source fiber—receive weak optical signals. Since the single-photon counting technique requires that either one photon or none be measured per laser pulse, stronger optical signals are attenuated by computer-controlled variable optical attenuators to extend the dynamic range of the measurement. Then, a longpass filter reduces the ambient light that has shorter wavelengths than the signal light.

In each detection channel, the filtered signal light is delivered to the photocathode of a microchannel plate–photomultiplier tube (MCP-PMT), which converts the optical pulse into an analog electronic pulse. After preamplification, the analog pulse is shaped by a constant fraction discriminator into a logic pulse with high timing accuracy. In parallel, part of the main beam is split off by a beamsplitter to provide a reference lightbeam. The reference optical pulse is converted by a photodiode into an electronic pulse. The electronic pulse is also preamplified and shaped into a logic pulse by a combined amplifier–timing discriminator unit. Each reference logic pulse is then time-delayed and converted through a fast fan-out module into 32 outputs. Each output is connected to a picosecond time analyzer. Each picosecond time analyzer compares the signal logic pulse with the delayed reference logic pulse to measure the time of flight of each individual photon. The detection cycle is repeated with many laser pulses. At the end, the times of flight of the individual photons detected by each channel build a histogram, which represents the associated temporal spread function.

Once histograms are acquired for a source location, the fiber switch shifts the source light to the next source fiber. This process is repeated through all 32 source fibers to complete the measurement data set. Then, image reconstruction is performed by computer.

11.4. DIRECT-CURRENT SYSTEM

A DC system based on frequency-division multiplexing was built using 32 lasers and 32 detectors at Massachusetts General Hospital (MGH) (Figure 11.2). Half of the lasers operate at 690-nm wavelength and the other half at 830 nm. The lasers are encoded with 32 frequencies from a master clock; these frequencies are distributed uniformly between 6.4 and 12.6 kHz. The laser outputs are fiberoptically coupled to 16 paired positions around the object to be imaged. Reemitted light is collected through 32 channels of optical fibers and detected by 32 avalanche photodiodes (APDs) in parallel. The electronic output of each APD is bandpass-filtered. The filtered signal is amplified by a programmable-gain stage, which

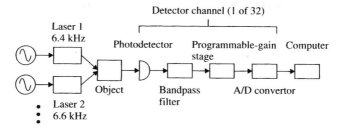

Figure 11.2. Schematic representation of the MGH DC imaging system.

matches the signal amplitude to the input range of the analog-to-digital (A/D) converter. The amplified signal is digitized at 45 kHz by the A/D converter and subsequently transferred to a computer. Then, the digitally encoded signal is decoded by the computer on the basis of the modulation frequencies to recover the reemitted optical signal components originating from each source simultaneously. Since all sources and detectors function concurrently, the data-acquisition rate is high. The system is well suited to observing rapid physiological phenomena.

11.5. FREQUENCY-DOMAIN SYSTEM

Before describing a frequency-domain imaging system, we first introduce a single-channel sensing system (Figure 11.3). The output power of a laser diode is modulated by an AC signal at a radiofrequency f (e.g., 200 MHz) from a function generator. The modulated light is delivered to the object through an optical

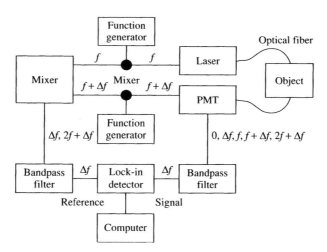

Figure 11.3. Schematic representation of a single-channel frequency-domain sensing system.

fiber. Reemitted light is detected by a PMT through another optical fiber. If the PMT has a constant gain, the output electronic signal will have frequency f and can be digitized by an A/D converter. As required by the Nyquist criterion, the A/D sampling frequency must be greater than $2f$.

The sampling frequency can be reduced by heterodyne detection, which is applicable to narrowband signals. For heterodyne detection, a local oscillator of frequency $f + \Delta f$—where Δf is typically a fraction of f (e.g., tens of kHz)—is produced by another function generator to modulate the gain of the PMT. Because it is proportional to the product of the input light power and the gain, the output of the PMT has multiple frequencies: 0, Δf, f, $f + \Delta f$, and $2f + \Delta f$. The Δf component is selected by a bandpass filter and then digitized by an A/D converter. The lower frequency Δf results in less lengthy data.

The amount of data can be further reduced by lock-in detection, which is applicable to single-frequency signals. For lock-in detection, a reliable reference of frequency Δf is produced by another heterodyne channel. An electronic mixer, which multiplies the two input signals, mixes the outputs from the two function generators to produce a signal of two frequencies: Δf and $2f + \Delta f$. The Δf component is selected by a bandpass filter to be the reference. The lock-in detector inputs the signal and the reference and outputs the amplitude and the phase of the signal.

The principle of dual-phase lock-in detection (Figure 11.4), also referred to as *IQ detection* ("I" for in-phase and "Q" for quadrature), is described as follows.

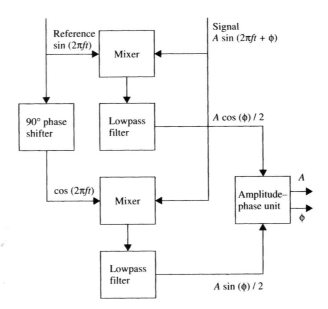

Figure 11.4. Block diagram of a dual-phase lock-in detection system.

Both the signal and the reference oscillate at frequency f. The signal is represented by $A\sin(2\pi ft + \phi)$, where A denotes the amplitude, ϕ denotes the phase, and t denotes time. The reference is represented simply by $\sin(2\pi ft + \phi)$ without taking the amplitude into account. The system consists primarily of two mixers, two lowpass filters, and a 90° phase shifter. In the first channel, the mixer mixes the signal and the reference to produce a signal consisting of a DC and a second-harmonic $2f$ component. The first lowpass filter passes the DC component and rejects the second-harmonic component, where the DC component can be represented by $S_I = \frac{1}{2}A\cos\phi$ apart from a constant factor. In the second channel, the reference signal is first phase shifted by 90° to produce a signal proportional to $\cos(\omega t)$. Then, the second mixer followed by the second low-pass filter produces a DC component that can be represented by $S_Q = \frac{1}{2}A\sin\phi$. Finally, the amplitude–phase unit outputs the two unknowns—A and ϕ—based on the two DC components:

$$A = 2\sqrt{S_I^2 + S_Q^2}, \tag{11.1}$$

$$\phi = \arctan\frac{S_Q}{S_I}. \tag{11.2}$$

A scanning frequency-domain imaging system was built at Dartmouth College (Figure 11.5). Unlike the system in Figure 11.3, this system automatically scans both the light incidence position and the light detection position around the object using 32 large-core fiberoptic bundles—16 for light delivery and 16 for light

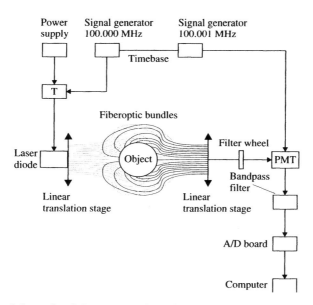

Figure 11.5. Schematic of the Dartmouth College frequency-domain imaging system.

detection. Two signal generators—sharing the same timebase—produce signals of radiofrequency $f_1 = 100.000$ MHz and $f_2 = 100.001$ MHz, respectively. A bias-T mixes the DC current from a power supply and the radiofrequency current of f_1. A laser, whose output light power is modulated by the output of the bias-T, provides 800-nm NIR light. A linear translation stage scans to one of the 16 light delivery fiberoptic bundles to receive the modulated laser light. The selected fiberoptic bundle delivers the light to the object at one of the 16 positions. Another linear translation stage scans to each of the 16 light detection fiberoptic bundles to gather reemitted light from every detection position. A filter wheel—made of neutral-density filters of various optical densities—attenuates stronger optical signals more strongly to compress the range of the received optical signals. A PMT detects the attenuated light and outputs a signal that has various frequencies. A bandpass filter passes the $f_2 - f_1$ component. An A/D board digitizes the filtered signal and then transfers the digital data to a computer for data processing. This detection process is repeated through all 16 light delivery fiberoptic bundles to complete the data acquisition.

11.6. FREQUENCY-DOMAIN THEORY: BASICS

As shown in Chapter 5, the time-resolved diffusion equation is

$$\frac{\partial \Phi(\vec{r}, t)}{c \partial t} + \mu_a \Phi(\vec{r}, t) - \nabla \cdot [D \, \nabla \Phi(\vec{r}, t)] = S(\vec{r}, t). \tag{11.3}$$

Here, Φ denotes the fluence rate, c denotes the speed of light in the scattering medium, μ_a denotes the absorption coefficient, D denotes the diffusion coefficient, and S denotes the source power density. Sometimes, the diffusivity D'—defined by $D' = cD$—is used instead of D.

For monochromatic light, Eq. (11.3) can be rewritten as

$$\frac{\partial U(\vec{r}, t)}{\partial t} + c\mu_a U(\vec{r}, t) - c\nabla \cdot [D \, \nabla U(\vec{r}, t)] = q(\vec{r}, t), \tag{11.4}$$

where U denotes the photon density and q denotes the photon density source strength. The following relations are used in the conversion:

$$\Phi(\vec{r}, t) = h\nu \, U(\vec{r}, t) c, \tag{11.5}$$

$$S(\vec{r}, t) = h\nu \, q(\vec{r}, t), \tag{11.6}$$

where $h\nu$ denotes the photon energy.

We first examine the spatial impulse response in an infinite homogeneous scattering medium. A point photon density source is denoted by the following phasor expression:

$$q(\vec{r}, t) = [A + B \exp(-i\omega t)]\delta(\vec{r}). \tag{11.7}$$

Here, ω denotes the angular frequency and A and $|B|$ denote the DC and the AC source amplitudes, respectively. The ratio $|B|/A$, termed the *modulation depth*, ranges between 0 and 1 because $q \geq 0$ must hold. From $B = |B|\exp(-i\phi_B)$, we have

$$B\exp(-i\omega t) = |B|\exp(-i\omega t - i\phi_B). \tag{11.8}$$

Of course, it is the real part of a phasor expression that represents the actual oscillation:

$$\text{Re}\{|B|\exp(-i\omega t - i\phi_B)\} = |B|\cos(\omega t + \phi_B). \tag{11.9}$$

Because the diffusion equation is linear, we assume the solution to be as follows:

$$U(\vec{r}, t) = U_{\text{DC}}(\vec{r}) + U_{\text{AC}}(\vec{r})\exp(-i\omega t), \tag{11.10}$$

where $U_{\text{AC}}(\vec{r})$ is complex. For the DC part, we have

$$c\mu_a U_{\text{DC}}(\vec{r}) - cD\,\nabla^2 U_{\text{DC}}(\vec{r}) = A\delta(\vec{r}). \tag{11.11}$$

The solution is given by

$$U_{\text{DC}}(\vec{r}) = A\frac{\exp(-\mu_{\text{eff}}r)}{4\pi cDr}, \tag{11.12}$$

where $\mu_{\text{eff}} = \sqrt{\mu_a/D}$.

For the AC part, we have

$$-i\omega U_{\text{AC}}(\vec{r}) + c\mu_a U_{\text{AC}}(\vec{r}) - cD\,\nabla^2 U_{\text{AC}}(\vec{r}) = B\delta(\vec{r}). \tag{11.13}$$

This equation can be reformulated to a Helmholtz wave equation:

$$(\nabla^2 + k^2)U_{\text{AC}}(\vec{r}) = -B\frac{\delta(\vec{r})}{cD}, \tag{11.14}$$

where

$$k^2 = \frac{-c\mu_a + i\omega}{cD}. \tag{11.15}$$

The solution is given by

$$U_{\text{AC}}(\vec{r}) = B\frac{\exp(ikr)}{4\pi cDr}, \tag{11.16}$$

which represents a photon-density wave, where k is the propagation constant.

Substituting Eqs. (11.12) and (11.16) into Eq. (11.10), we obtain the overall response:

$$U(\vec{r}, t) = \frac{A \exp(-\mu_{\text{eff}} r) + B \exp(ikr - i\omega t)}{4\pi c D r}. \tag{11.17}$$

When $\omega \to 0$, the AC solution should approach the DC solution. If $\omega = 0$, Eq. (11.15) becomes

$$k^2 = \frac{-\mu_a}{D} = -\mu_{\text{eff}}^2.$$

As a result, Eq. (11.16) reduces to

$$U_{\text{AC}}(\vec{r}) = B \frac{\exp(-\mu_{\text{eff}} r)}{4\pi c D r}, \tag{11.18}$$

which is identical to Eq. (11.12) except for the amplitudes.

Propagation constant k can be separated into real and imaginary parts:

$$k_r = \text{Re}\{k\} = \left(\frac{c^2 \mu_a^2 + \omega^2}{c^2 D^2}\right)^{1/4} \sin\left(\frac{1}{2} \arctan \frac{\omega}{c \mu_a}\right), \tag{11.19}$$

$$k_i = \text{Im}\{k\} = \left(\frac{c^2 \mu_a^2 + \omega^2}{c^2 D^2}\right)^{1/4} \cos\left(\frac{1}{2} \arctan \frac{\omega}{c \mu_a}\right). \tag{11.20}$$

Substituting $k = k_r + i k_i$ into Eq. (11.16), we obtain

$$U_{\text{AC}}(\vec{r}) = B \exp(-k_i r) \frac{\exp(i k_r r)}{4\pi c D r}. \tag{11.21}$$

Factor $\exp(i k_r r)$ represents the phase delay of the photon-density wave, which approximately equals the product of the average time of flight of the photons and the angular frequency ω (see Problems 11.1–11.3); therefore, the phase of $U_{\text{AC}}(\vec{r})$ provides information about the effective path length that the diffuse photons have taken. The real part k_r can be converted to a wavelength by

$$\lambda = \frac{2\pi}{k_r}. \tag{11.22}$$

Factor $\exp(-k_i r)$ represents the attenuation of the photon-density wave due to diffusion and absorption in addition to the $1/r$ geometric decay. Even in the absence of absorption (i.e., $\mu_a = 0$), the photon-density wave is still strongly damped as a result of diffusion. Because they follow physical paths of various lengths, photons that reach the observation point arrive at different times. Therefore, photons in the peaks and troughs of a photon-density wave mingle as the wave propagates, which dampens the modulation depth of the overall response.

Diffuse photon-density waves can be used to measure optical properties. Once the complex k is measured, μ_a and D can be deduced. From Eqs. (11.19) and (11.20), we obtain

$$\mu_a = -\frac{\omega \operatorname{Re}\{k^2\}}{c \operatorname{Im}\{k^2\}}, \qquad (11.23)$$

$$D = \frac{\sqrt{c^2 \mu_a^2 + \omega^2}}{c|k|^2}. \qquad (11.24)$$

Diffuse photon-density waves possess many of the common wave characteristics, such as reflection, refraction, diffraction, and dispersion. For example, Snell's law is applicable to diffuse photon-density waves:

$$\frac{\sin \theta_i}{\sin \theta_t} = \frac{\lambda_i}{\lambda_t}. \qquad (11.25)$$

Here, θ_i and θ_t denote the angles of incidence and transmission, respectively; λ_i and λ_t denote the wavelengths of the diffuse photon-density waves in the incident and transmitted media, respectively.

Because of its long wavelength (cm scale), a photon-density wave typically provides imaging in the near field, where the spatial resolution is related to the SNR rather than the wavelength. In far-field imaging, spatial resolution based on the Rayleigh criterion is related to $\sim \lambda/2$; yet, it can be improved to "superresolution" by a factor related to the SNR as well. Ultimately, the resolution in either case is limited by the SNR.

One must distinguish between an optical wave and a photon-density wave. The former is an electromagnetic vector wave, whereas the latter is a photon-density scalar wave. Note also that the latter is based on the former. Furthermore, their wavelengths and attenuation mechanisms are different.

Example 11.1. Illustrate the null plane between two photon-density sources that are $180°$ out of phase.

A representative MATLAB source code is shown below:

```
c = 3e8/1.37;     %m/s
mua = 0.1E2;      %/m
mus = 10E2; %(/m) mus = mus', g = 0
f = 200E6; %Hz

D = 1/(3*(mua+mus));   %m
k = sqrt((-c*mua + i*2*pi*f)/(c*D));      %wave vector

disp(['Absorption coeff. mua (/cm) = ', num2str(mua*1E-2)])
disp(['Reduced scattering coeff. mus'' (/cm) = ', num2str(mus*1E-2)])
disp(['Frequency f (MHz) = ', num2str(f*1E-6)])
disp(['Wavelength (cm) = ', num2str(2*pi/real(k)*1E2)])
```

```
disp(['Decay const (cm) = ', num2str(1/imag(k)*1E2)])

xs = 1E-2; %(m) sources at (xs, 0) & (-xs, 0)
yd = 3E-2; %(m) detector at (xd, yd)
xd = (-3:0.02:3)*1E-2; %m

r1 = sqrt((xd - xs).^2 + yd.^2); % distance b/t -src & detector
U1 = -exp(i*k*r1)./(4*pi*c*D*r1); % negative src

r2 = sqrt((xd + xs).^2 + yd.^2); % distance b/t +src & detector
U2 = exp(i*k*r2)./(4*pi*c*D*r2);

figure(1)
subplot(3,1,1)
plot([-xs xs]*1E2, [0, 0], '*', [0], [3], 'o', [0 0], [0 3])
text(-xs*1E2, 0.5, '+Source')
text(+xs*1E2, 0.5, '-Source')
text(0, 2.5, 'Scanning Detector')
axis([-3 3 0 3])
xlabel('Source & Detector Positions (cm)')
ylabel('y (cm)')
title('Null line')

subplot(3,1,2)
plot(xd*1E2, abs(U1+U2))
xlabel('Detector position (cm)')
ylabel('Amplitude')

subplot(3,1,3)
plot(xd*1E2, unwrap(angle(U1+U2))*180/pi)
xlabel('Detector position (cm)')
ylabel('Angle (deg)')
grid

figure(2) % optional
r = (1:1:20)*1E-2;
U = exp(i*k*r)./(4*pi*c*D*r);
subplot(2,1,1)
semilogy(r*1E2, abs(U))
xlabel('Source-detector distance (cm)')
ylabel('Amplitude')
title('Propagation of PDW')

subplot(2,1,2)
plot(r*1E2, unwrap(angle(U))*180/pi)
xlabel('Source-detector distance (cm)')
ylabel('Angle (deg)')
grid
```

The text output in MATLAB is shown below:

```
Absorption coeff. mua (/cm) = 0.01
Reduced scattering coeff. mus' (/cm) = 10
Frequency f (MHz) = 200
```

```
Wavelength (cm) = 7.382
Decay const (cm) = 0.9878
```

The graphical output in MATLAB is shown in Figure 11.6. The detector is scanned along the $y = 3$ line. Because the two sources are 180° out of phase, the phase difference between any two points that are symmetric about the null plane must be 180°.

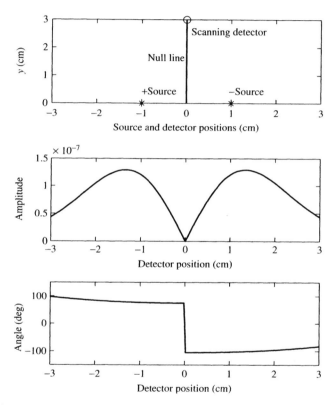

Figure 11.6. Illustration of the null plane—a null line in 2D space—in the antiphase dual-source response.

11.7. FREQUENCY-DOMAIN THEORY: LINEAR IMAGE RECONSTRUCTION

In this section, we illustrate frequency-domain image reconstruction in an infinite medium using a simple linear inverse theory. The object to be imaged is assumed to have absorption contrast only. The absorption coefficient $\mu_a(\vec{r})$ is expressed as

$$\mu_a(\vec{r}) = \mu_{a0}(\vec{r}) + \delta\mu_a(\vec{r}), \tag{11.26}$$

where $\mu_{a0}(\vec{r})$ represents the background absorption coefficient and $\delta\mu_a(\vec{r})$ represents the differential absorption coefficient of the heterogeneities relative to the background. In the forward problem, the perturbation in photon density is calculated from presumed $\mu_{a0}(\vec{r})$ and $\delta\mu_a(\vec{r})$ values. In the inverse problem, $\delta\mu_a(\vec{r})$ is calculated from the measured data.

For simplicity, we define D as $1/(3\mu'_s)$ in this section so that D does not depend on μ_a. We start by rewriting Eq. (11.14) as

$$(\nabla^2 + k^2)U_{AC}(\vec{r}, \vec{r}_s) = -B\frac{\delta(\vec{r} - \vec{r}_s)}{cD}, \qquad (11.27)$$

where $\vec{r}_s$ denotes the location of the photon-density source. Substituting Eq. (11.26) into Eq. (11.15), we obtain

$$k^2 = k_0^2 + O(\vec{r}), \qquad (11.28)$$

where

$$k_0^2 = \frac{-c\mu_{a0} + i\omega}{cD}, \qquad (11.29)$$

$$O(\vec{r}) = -\frac{\delta\mu_a(\vec{r})}{D}. \qquad (11.30)$$

Let

$$U_{AC}(\vec{r}, \vec{r}_s) = U_0(\vec{r}, \vec{r}_s) + U_{SC}(\vec{r}, \vec{r}_s), \qquad (11.31)$$

where U_0 represents the AC photon density in a homogeneous medium that has the background optical properties and U_{SC} represents the differential AC photon density due to the heterogeneities.

Substituting Eq. (11.31) into Eq. (11.27) yields

$$[\nabla^2 + k_0^2 + O(\vec{r})][U_0(\vec{r}, \vec{r}_s) + U_{SC}(\vec{r}, \vec{r}_s)] = -B\frac{\delta(\vec{r} - \vec{r}_s)}{cD}. \qquad (11.32)$$

The diffusion equation for U_0 is given by

$$(\nabla^2 + k_0^2)U_0(\vec{r}, \vec{r}_s) = -B\frac{\delta(\vec{r} - \vec{r}_s)}{cD}. \qquad (11.33)$$

Taking the difference between Eqs. (11.32) and (11.33), we obtain

$$(\nabla^2 + k_0^2)U_{SC}(\vec{r}, \vec{r}_s) = -O(\vec{r})[U_0(\vec{r}, \vec{r}_s) + U_{SC}(\vec{r}, \vec{r}_s)]. \qquad (11.34)$$

If $\delta\mu_a(\vec{r}) \ll \mu_{a0}(\vec{r})$, we assume that the Born approximation, $U_{SC}(\vec{r}, \vec{r}_s) \ll U_0(\vec{r}, \vec{r}_s)$, is valid; hence, Eq. (11.34) becomes

$$(\nabla^2 + k_0^2)U_{SC}(\vec{r}, \vec{r}_s) = -O(\vec{r})U_0(\vec{r}, \vec{r}_s), \qquad (11.35)$$

which can be solved by the following Green function method. For a point source in an infinite medium, Eq. (11.35) becomes

$$(\nabla^2 + k_0^2) G(\vec{r} - \vec{r}_s) = -\delta(\vec{r} - \vec{r}_s), \quad (11.36)$$

where G, referred to as a *Green function*, is given by

$$G(\vec{r} - \vec{r}_s) = \frac{\exp(ik_0|\vec{r} - \vec{r}_s|)}{4\pi |\vec{r} - \vec{r}_s|}. \quad (11.37)$$

The solution to Eq. (11.33) has a similar form:

$$U_0(\vec{r}, \vec{r}_s) = \frac{B}{cD} \frac{\exp(ik_0|\vec{r} - \vec{r}_s|)}{4\pi |\vec{r} - \vec{r}_s|}. \quad (11.38)$$

In response to the general forcing function on the right-hand side of Eq. (11.35), we have the following solution to the forward problem, based on Green's theorem:

$$U_{\text{sc}}(\vec{r}, \vec{r}_s) = \int U_0(\vec{r}', \vec{r}_s) O(\vec{r}') G(\vec{r} - \vec{r}') \, d\vec{r}'. \quad (11.39)$$

Because G is shift-invariant, this expression is actually equivalent to a convolution. The physical meaning of the Green function method is graphically represented in Figure 11.7. As can be seen, $U_0(\vec{r}', \vec{r}_s)$ represents the light propagated from the source to a point inside the object; $U_0(\vec{r}', \vec{r}_s) O(\vec{r}')$ serves as a new source for further light propagation to the detector; the propagation is described by $G(\vec{r} - \vec{r}')$.

To demonstrate the inverse problem for image reconstruction, we discretize Eq. (11.39) in an xy plane:

$$U_{\text{SC}}(\vec{r}_j, \vec{r}_{si}) = \sum_m \sum_n U_0(\vec{r}'_{mn}, \vec{r}_{si}) O(\vec{r}'_{mn}) G(\vec{r}_j - \vec{r}'_{mn}) h^3. \quad (11.40)$$

Here, i is an index of the source, j is an index of the detector, m and n are indices of the coordinates of the perturbations, and h is the size of each grid

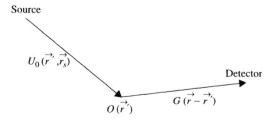

Figure 11.7. Illustration of the Green function method.

element (for simplicity, assumed to be cubical). Although the sources, perturbations, and detectors are restricted to the plane of interest, photons still propagate in 3D space.

We rewrite Eq. (11.40) as

$$U_{SC}(\vec{r}_j, \vec{r}_{si}) = \sum_m \sum_n W_{ij,mn} \delta\mu_a(\vec{r}'_{mn}), \qquad (11.41)$$

where

$$W_{ij,mn} = -U_0(\vec{r}'_{mn}, \vec{r}_{si}) G(\vec{r}_j - \vec{r}'_{mn}) \frac{h^3}{D}. \qquad (11.42)$$

To compress each pair of indices to a single index (analogous to counting the squares in a chessboard sequentially), we let

$$i' = i + jN_i, \qquad (11.43)$$

$$m' = m + nN_m, \qquad (11.44)$$

where N_i is the range of i (the number of values represented by i) and N_m is the range of m. Thus, Eq. (11.41) becomes

$$U_{SC}(i') = \sum_{m'} W_{i',m'} \delta\mu_a(\vec{r}'_{m'}), \qquad (11.45)$$

which can be rewritten in matrix form with separated real and imaginary parts:

$$\begin{bmatrix} \text{Re}\{U_{SC}(i')\} \\ \text{Im}\{U_{SC}(i')\} \end{bmatrix} = \begin{bmatrix} \text{Re}\{W_{i'm'}\} \\ \text{Im}\{W_{i'm'}\} \end{bmatrix} [\delta\mu_a(m')], \qquad (11.46)$$

or

$$[U_{SC}] = [W][\delta\mu_a]. \qquad (11.47)$$

Of course, we can also use complex matrices directly. For matrix $[W]$, the number of rows N_r equals twice the range of i' or twice the number of measurements, and the number of columns N_c equals the range of m' or the number of volume elements (voxels). Matrix $[W]$, which is generally nonsquare, can be inverted to solve for $[\delta\mu_a]$:

$$[\delta\mu_a] = [W]^{-1}[U_{SC}]. \qquad (11.48)$$

Here, $[\delta\mu_a]$ represents the image that we are pursuing. Matrix $[W]$ is referred to as the *Jacobian matrix*; it is also referred to as the *sensitivity matrix* because it relates the changes in the measurements to the perturbations in the optical properties. The forward problem provides matrix $[W]$. For regular boundaries such as semiinfinite, cylindrical, and spherical ones, $[W]$ can be computed by analytical methods. In general, however, $[W]$ must be computed by numerical methods.

Singular-value decomposition (SVD) can be used to invert $[W]$. First, $[W]$ is decomposed into three matrices as follows:

$$[W]_{N_r,N_c} = [U]_{N_r,N_r}[\text{diag}(w_i)]_{N_r,N_c}[V]^T_{N_c,N_c}. \tag{11.49}$$

The middle matrix is diagonal; the subscripts describe the matrix dimensions; $[U]$ and $[V]$ have the following orthogonality:

$$[U]^T[U] = [I], [V]^T[V] = [I], \tag{11.50}$$

where $[I]$ denotes identity matrix. Then, inversion leads to

$$[W]^{-1} = \{[V]^T\}^{-1}[\text{diag}(w_i)]^{-1}[U]^{-1} = [V]\left[\text{diag}\left(\frac{1}{w_i}\right)\right][U]^T. \tag{11.51}$$

To avoid overflow, we set $1/w_j$ to zero when w_j is less than a preset threshold. Alternatively, we can use the following smoothing algorithm:

$$w'_j = w_j + \frac{\sigma}{w_j}, \tag{11.52}$$

where σ is a free parameter. Both methods trade accuracy for stability.

Iterative methods, such as the algebraic reconstruction technique (ART) or the simultaneous iterative reconstruction technique (SIRT) (Appendix 11A), may be used to solve Eq. (11.47) as well. The advantage of an iterative method over the preceding matrix inversion method is that the former permits hard constraints. For instance, because the reconstructed absorption coefficient should always be nonnegative, a negative value in any voxel can be set to zero at the end of each iteration cycle.

Up until this point, only an infinite medium has been considered. With a finite medium, an extrapolated boundary condition can be used (see Chapter 5). The Green function approach takes the following form instead of Eq. (11.39):

$$U_{\text{SC}}(\vec{r}) = \int_V U_0(\vec{r}', \vec{r}_s) O(\vec{r}') G(\vec{r}, \vec{r}') d\vec{r}' + \frac{1}{4\pi}$$
$$\times \int_S \left[G(\vec{r}, \vec{r}') \frac{\partial U_{\text{SC}}}{\partial n'} - U_{\text{SC}}(\vec{r}') \frac{\partial G}{\partial n'} \right] dS. \tag{11.53}$$

Here, S denotes the extrapolated boundary, V denotes the volume enclosed by S, and n' denotes the outward surface normal.

Since U_{SC} is approximately zero on the extrapolated boundary, the surface integral of the second term in the brackets in Eq. (11.53) vanishes. If we choose a Green's function that satisfies the homogeneous Dirichlet boundary condition (namely, G is zero on the extrapolated boundary), the surface integral of the first term in the brackets in Eq. (11.53) vanishes as well. Therefore, Eq. (11.53) becomes

$$U_{SC}(\vec{r}) = \int_V U_0(\vec{r}',\vec{r}_s) O(\vec{r}') G(\vec{r},\vec{r}') d\vec{r}'. \qquad (11.54)$$

It is important to note that the G here is different from the G for an infinite medium. The Green functions that satisfy the homogeneous Dirichlet boundary condition for some regular geometric shapes (e.g., semiinfinite, spherical, cylindrical spaces) are analytically available. Green's functions for more complex geometric shapes, however, can be computed only using numerical methods. Because no perturbation of optical properties exists outside the real boundary, the volume integration outside the original object is zero and, therefore, V can be reduced to the actual volume of the object.

Example 11.2. Derive Eq. (11.39).

Since the system is linear, a linear operator $\mathcal{L}$ can be used to represent the system as follows:

$$G(\vec{r}-\vec{r}_s) = \mathcal{L}\{\delta(\vec{r}-\vec{r}_s)\}, \qquad (11.55)$$

$$U_{SC}(\vec{r},\vec{r}_s) = \mathcal{L}\{O(\vec{r})U_0(\vec{r},\vec{r}_s)\}. \qquad (11.56)$$

A general forcing function $O(\vec{r})U_0(\vec{r},\vec{r}_s)$ can be expanded using $\delta(\vec{r}-\vec{r}_s)$

$$O(\vec{r})U_0(\vec{r},\vec{r}_s) = \int O(\vec{r}')U_0(\vec{r}',\vec{r}_s)\delta(\vec{r}-\vec{r}') d\vec{r}', \qquad (11.57)$$

which is based on the sifting property of the delta function (as if a point were sifted out of a function).

Substituting Eq. (11.57) into Eq. (11.56) and operating $\mathcal{L}$ on the $\vec{r}$-dependent quantities only, we obtain

$$U_{SC}(\vec{r},\vec{r}_s) = \mathcal{L}\{O(\vec{r})U_0(\vec{r},\vec{r}_s)\} = \mathcal{L}\left\{\int O(\vec{r}')U_0(\vec{r}',\vec{r}_s)\delta(\vec{r}-\vec{r}') d\vec{r}'\right\}$$

$$= \int O(\vec{r}')U_0(\vec{r}',\vec{r}_s)\mathcal{L}\{\delta(\vec{r}-\vec{r}')\} d\vec{r}' \qquad (11.58)$$

$$= \int O(\vec{r}')U_0(\vec{r}',\vec{r}_s) G(\vec{r}-\vec{r}') d\vec{r}'.$$

Example 11.3. Implement the linear inverse algorithm in C.

The following data structure is defined (the source code is available on the Web at ftp://ftp.wiley.com/public/sci_tech_med/biomedical_optics):

```
typedef struct {
  double f; /* frequency (Hz). */
  double c; /* speed of light in medium (cm/s). */
  double h; /* grid size (cm), dx = dy = dz = h. */
  int N; /* NxN grid. */
  int gap; /* gap between src/det & boundary (grid). */
  double mua0; /* mua of background (/cm). */
  double mus0; /* mus' of background (/cm). */

  int obj_x, obj_y, obj_size; /* object location and size in grid. */
  double dmua; /* delta mua of object (/cm). */
} ParamStru;
```

For the following parameters, the reconstructed image is shown in Figure 11.8:

```
void SetParam(ParamStru *par)
{
  par->f = 500e6;
  par->c = 3E10 / 1.37;
  par->h = 0.2;
  par->N = 20; /* number of voxels in each direction. */
  par->gap = 3;
  par->mua0 = 0.1;
  par->mus0 = 10;

  par->obj_x = 12;
  par->obj_y = 6;
  par->obj_size = 2;
  par->dmua = 0.01;
}
```

If the frequency is set to zero, the AC photon-density wave reduces to DC. The corresponding reconstructed image is shown in Figure 11.9.

11.8. FREQUENCY-DOMAIN THEORY: GENERAL IMAGE RECONSTRUCTION

In this section, we formulate and solve a general frequency-domain imaging problem in an infinite medium. The goal is to map both the absorption coefficient $\mu_a(\vec{r})$ and the reduced scattering coefficient $\mu'_s(\vec{r})$ from the measured photon-density distribution.

11.8.1. Problem Formulation

The starting point is again the time-dependent diffusion equation for a single frequency. From Eq. (11.4), the AC component of the photon-density wave $[U_{AC}(\vec{r})]$

268 DIFFUSE OPTICAL TOMOGRAPHY

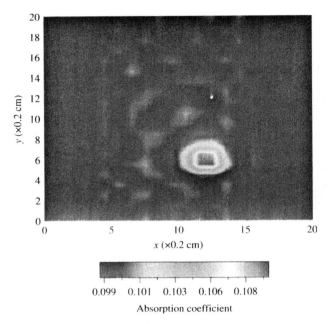

Figure 11.8. Reconstructed image based on simulated frequency-domain data in an infinite medium.

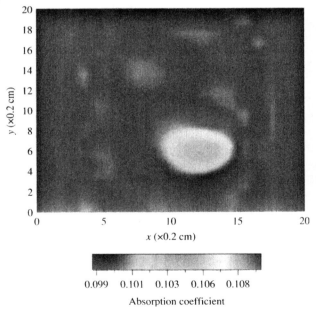

Figure 11.9. Reconstructed image based on simulated DC data in an infinite medium.

from a point source $B\delta(\vec{r} - \vec{r}_s)$ satisfies

$$c\nabla \cdot [D(\vec{r})\nabla U_{AC}(\vec{r})] - [c\mu_a(\vec{r}) - i\omega]U_{AC}(\vec{r}) = -B\delta(\vec{r} - \vec{r}_s). \quad (11.59)$$

Since $\mu'_s(\vec{r})$ is implicit in this equation, we reconstruct $D(\vec{r})$ and $\mu_a(\vec{r})$, from which $\mu'_s(\vec{r})$ can be obtained.

As in the previous section, $D(\vec{r})$ and $\mu_a(\vec{r})$ are decomposed as

$$D(\vec{r}) = D_0 + \delta D(\vec{r}), \quad (11.60)$$

$$\mu_a(\vec{r}) = \mu_{a0} + \delta\mu_a(\vec{r}), \quad (11.61)$$

where D_0 and μ_{a0} represent the constant background optical properties and $\delta\mu_a(\vec{r})$ and $\delta D(\vec{r})$ represent the differential optical properties of the heterogeneities relative to the background. We duplicate Eq. (11.31) here:

$$U_{AC}(\vec{r}, \vec{r}_s) = U_0(\vec{r}, \vec{r}_s) + U_{SC}(\vec{r}, \vec{r}_s). \quad (11.62)$$

Substituting Eqs. (11.60)–(11.62) into Eq. (11.59), we obtain a differential equation for U_{SC}, the solution of which is based on the Green function method:

$$U_{SC}(\vec{r}_d, \vec{r}_s) = -\int \frac{\delta\mu_a(\vec{r})}{D_0} G_0(\vec{r}_d, \vec{r}) U_{AC}(\vec{r}, \vec{r}_s) \, d\vec{r}$$
$$+ \int \frac{\delta D(\vec{r})}{D_0} \nabla G_0(\vec{r}_d, \vec{r}) \cdot \nabla U_{AC}(\vec{r}, \vec{r}_s) \, d\vec{r}. \quad (11.63)$$

Here, $\vec{r}_d$ denotes the location of the detector, $\vec{r}$ denotes a position within the object to be imaged, and G_0 denotes the Green function associated with the diffusion equation for a homogeneous medium that has the background optical properties. The integrals presented above are 3D over the entire object. The first integral is what we obtained in the previous section from perturbation $\delta\mu_a$. The second integral is the contribution from perturbation δD.

Experimentally, we measure a quantity directly related to U_{AC}. If the background optical properties are known, we can compute U_0. Next, we subtract U_0 from U_{AC} to yield U_{SC}. Then, from Eq. (11.63), we solve for $\delta\mu_a$ and δD, which represent two images of different contrasts. Although $\delta\mu_a$ and δD appear in Eq. (11.63) in apparent linear form, this imaging problem is intrinsically nonlinear because U_{AC} is an implicit function of $\delta\mu_a$ and δD.

Example 11.4. Derive Eq. (11.63).

Substituting Eqs. (11.60) and (11.61) into Eq. (11.59), we obtain

$$cD_0\nabla^2 U_{AC} + c\nabla \cdot [\delta D \nabla U_{AC}] - [c(\mu_{a0} + \delta\mu_a) - i\omega]U_{AC} = -B\delta(\vec{r} - \vec{r}_s). \quad (11.64)$$

In a homogeneous medium with the background optical properties, Eq. (11.64) becomes

$$cD_0\nabla^2 U_0 - (c\mu_{a0} - i\omega)U_0 = -B\delta(\vec{r} - \vec{r}_s). \tag{11.65}$$

Subtracting Eq. (11.65) from Eq. (11.64) and using Eq. (11.62), we obtain

$$\nabla^2 U_{SC} + k_0^2 U_{SC} = \frac{\delta\mu_a}{D_0} U_{AC} - \frac{1}{D_0}\nabla\cdot(\delta D\nabla U_{AC}), \tag{11.66}$$

where k_0 is given by Eq. (11.29). Equation (11.66) is a Helmholtz equation and can be solved using the Green function method.

The Green function that satisfies differential equation

$$\nabla^2 G_0 + k_0^2 G_0 = -\delta(\vec{r}_d - \vec{r}) \tag{11.67}$$

is given by

$$G_0(\vec{r}_d, \vec{r}) = \frac{\exp(ik_0|\vec{r}_d - \vec{r}|)}{4\pi|\vec{r}_d - \vec{r}|}. \tag{11.68}$$

Thus, the solution to Eq. (11.66) is

$$U_{SC}(\vec{r}_d, \vec{r}_s) = -\int \frac{\delta\mu_a(\vec{r})}{D_0} G_0(\vec{r}_d, \vec{r}) U_{AC}(\vec{r}, \vec{r}_s)\, d\vec{r}$$
$$+ \frac{1}{D_0}\int \nabla\cdot[\delta D(\vec{r})\nabla U_{AC}(\vec{r}, \vec{r}_s)] G_0(\vec{r}_d, \vec{r})\, d\vec{r}. \tag{11.69}$$

The second integral on the right-hand side is rewritten as

$$\frac{1}{D_0}\int \nabla\cdot[\delta D(\vec{r})\nabla U_{AC}(\vec{r}, \vec{r}_s)] G_0(\vec{r}_d, \vec{r})\, d\vec{r}$$
$$= \frac{1}{D_0}\int \nabla\delta D(\vec{r})\cdot\nabla U_{AC}(\vec{r}, \vec{r}_s) G_0(\vec{r}_d, \vec{r})\, d\vec{r} \tag{11.70}$$
$$+ \frac{1}{D_0}\int \delta D(\vec{r})\nabla^2 U_{AC}(\vec{r}, \vec{r}_s) G_0(\vec{r}_d, \vec{r})\, d\vec{r}.$$

According to Green's second identity

$$\int (u\nabla^2 v + \nabla u\cdot\nabla v)\, d\vec{r} = \oint_S u\nabla v\cdot\hat{n}\, dS, \tag{11.71}$$

where S is an arbitrary surface that encloses the volume to be imaged and $\hat{n}$ is the surface normal, the second integral on the right-hand side of Eq. (11.70) is expressed as

$$\frac{1}{D_0}\int \delta D(\vec{r})\nabla^2 U_{AC}(\vec{r},\vec{r}_s)G_0(\vec{r}_d,\vec{r})\,d\vec{r}$$
$$=\frac{1}{D_0}\int [\delta D(\vec{r})G_0(\vec{r}_d,\vec{r})]\nabla^2 U_{AC}(\vec{r},\vec{r}_s)\,d\vec{r}$$
$$=\frac{1}{D_0}\oint_S [\delta D(\vec{r})G_0(\vec{r}_d,\vec{r})]\nabla U_{AC}(\vec{r},\vec{r}_s)\cdot\hat{n}\,dS \quad (11.72)$$
$$-\frac{1}{D_0}\int \nabla[\delta D(\vec{r})G_0(\vec{r}_d,\vec{r})]\cdot\nabla U_{AC}(\vec{r},\vec{r}_s)\,d\vec{r}.$$

As S approaches infinity, the surface integral in Eq. (11.72) vanishes, which leads to

$$\frac{1}{D_0}\int \delta D(\vec{r})\nabla^2 U_{AC}(\vec{r},\vec{r}_s)G_0(\vec{r}_d,\vec{r})\,d\vec{r}$$
$$=-\frac{1}{D_0}\int \nabla[\delta D(\vec{r})G_0(\vec{r}_d,\vec{r})]\cdot\nabla U_{AC}(\vec{r},\vec{r}_s)\,d\vec{r} \quad (11.73)$$
$$=-\frac{1}{D_0}\int \delta D(\vec{r})[\nabla G_0(\vec{r}_d,\vec{r})\cdot\nabla U_{AC}(\vec{r},\vec{r}_s)]\,d\vec{r}$$
$$-\frac{1}{D_0}\int G_0(\vec{r}_d,\vec{r})[\nabla\delta D(\vec{r})\cdot\nabla U_{AC}(\vec{r},\vec{r}_s)]\,d\vec{r}.$$

The second term on the right-hand side of Eq. (11.73) cancels out the first term on the right-hand side of Eq. (11.70); thus, Eq. (11.69) becomes

$$U_{SC}(\vec{r}_d,\vec{r}_s)=-\int \frac{\delta\mu_a(\vec{r})}{D_0}G_0(\vec{r}_d,\vec{r})U_{AC}(\vec{r},\vec{r}_s)\,d\vec{r}$$
$$+\frac{1}{D_0}\int \delta D(\vec{r})[\nabla G_0(\vec{r}_d,\vec{r})\cdot\nabla U_{AC}(\vec{r},\vec{r}_s)]\,d\vec{r}, \quad (11.74)$$

which is Eq. (11.63).

11.8.2. Linearized Problem

Although nonlinear with respect to the optical properties, the solution given by Eq. (11.63) can be linearized when the heterogeneity is weak. In this case, the Born approximation $U_{SC}\ll U_0$ is assumed; hence, U_{AC} on the right-hand side of Eq. (11.63) can be replaced by U_0, which can be computed if μ_{a0} and D_0 are known. Consequently, Eq. (11.63) can be linearized and discretized to

$$U_{SC}(\vec{r}_d, \vec{r}_s) = \sum_{j=1}^{N_V} [W_{a,j} \delta\mu_a(\vec{r}_j) + W_{s,j} \delta D(\vec{r}_j)]. \quad (11.75)$$

Here, the summation is over all N_V voxels within the object to be imaged; $W_{a,j}$ and $W_{s,j}$ represent weights

$$W_{a,j} = -\frac{G_0(\vec{r}_d, \vec{r}_j) U_0(\vec{r}_j, \vec{r}_s) \Delta x \, \Delta y \, \Delta z}{D_0}, \quad (11.76)$$

$$W_{s,j} = \frac{\nabla G_0(\vec{r}_d, \vec{r}_j) \cdot \nabla U_0(\vec{r}_j, \vec{r}_s) \Delta x \, \Delta y \, \Delta z}{D_0}, \quad (11.77)$$

where Δx, Δy, and Δz represent the sizes of the grid elements along the x, y, and z directions, respectively.

We rewrite Eq. (11.75) in the following matrix form

$$[W_{a,ji}, W_{s,ji}] \begin{bmatrix} \delta\mu_a(\vec{r}_j) \\ \delta D(\vec{r}_j) \end{bmatrix} = [U_{SC}(\vec{r}_{di}, \vec{r}_{si})] \quad (11.78)$$

or

$$[W][\delta x] = [U_{SC}]. \quad (11.79)$$

Here, subscript i is the index of the measurement with the source at position $\vec{r}_{si}$ and the detector at position $\vec{r}_{di}$; subscript j is the index of the position within the object to be imaged. If N_S source positions and N_D detector positions exist in the image acquisition, $N_M = N_S \times N_D$ measurements exist. Measurement column vector $[U_{SC}]$ has N_M elements; column vector $[\delta x]$ has $2N_V$ elements since vectors $[\delta\mu_a]$ and $[\delta D]$ are concatenated; thus, matrix $[W]$ has dimensions $N_M \times 2N_V$. Unknown vector $[\delta x]$ can be solved from Eq. (11.79) using various mathematical methods.

11.8.3. Nonlinear Problem

When the perturbation is not small, the image reconstruction is nonlinear and is usually solved iteratively with the following steps:

1. Assume the initial optical properties.
2. Solve the forward problem.
3. Calculate the error and check the convergence. If the error is sufficiently small, terminate the loop. Otherwise, continue to the next step.
4. Set up the inverse problem to update the optical properties.

5. Solve the inverse problem.
6. Update the optical properties and return to step 2.

In step 1, the initial optical properties are assumed. Usually, a homogeneous distribution of average optical properties is judiciously selected.

In step 2, the forward problem is solved with the current optical properties to calculate the diffuse photon density U_C at all detection locations for each source position. The finite-element or the finite-difference method can provide a solution to the forward problem on the basis of the diffusion equation. For finite objects, boundary conditions must be imposed.

In step 3, the χ^2 error is typically calculated as follows:

$$\chi^2 = \sum_{i=1}^{N_M} \left[\frac{U_C(\vec{r}_{di}, \vec{r}_{si}) - U_M(\vec{r}_{di}, \vec{r}_{si})}{\sigma_i} \right]^2, \qquad (11.80)$$

where U_M denotes the measured diffuse photon density and σ_i denotes the ith measurement error. If $\chi^2 < \varepsilon$, where ε is a predefined small quantity, the problem has converged, and the looping is terminated. Otherwise, the looping proceeds to step 4.

In step 4, an inverse problem is set up to update the optical properties. Rather than another random assumption, an optimal update of the optical properties is computed by accounting for the difference between U_C and U_M. Since the goal is to reach U_M from the current U_C, we expand U_M to the first order by a Taylor series in matrix form:

$$[U_M] = [U_C] + \left[\frac{\partial U_C}{\partial \mu_a} \right] [\Delta \mu_a] + \left[\frac{\partial U_C}{\partial D} \right] [\Delta D], \qquad (11.81)$$

where

$$\left[\frac{\partial U_C}{\partial \mu_a} \right]_{ij} = -\frac{\Delta x \Delta y \Delta z}{D_0} G_0(\vec{r}_{di}, \vec{r}_j) U_C(\vec{r}_j, \vec{r}_{si}), \qquad (11.82)$$

$$\left[\frac{\partial U_C}{\partial D} \right]_{ij} = \frac{\Delta x \Delta y \Delta z}{D_0} \nabla G_0(\vec{r}_{di}, \vec{r}_j) \cdot \nabla U_C(\vec{r}_j, \vec{r}_{si}). \qquad (11.83)$$

Here, vectors $[U_M]$ and $[U_C]$ have dimension N_M; vectors $[\Delta \mu_a]$ and $[\Delta D]$ have dimension N_V and denote the differential updates for μ_a and D, respectively; matrices $[\partial U_C / \partial \mu_a]$ and $[\partial U_C / \partial D]$ have dimensions $N_M \times N_V$. From Eq. (11.81), the inverse problem can be formulated as

$$[J] \begin{bmatrix} \Delta \mu_a \\ \Delta D \end{bmatrix} = [U_M - U_C]. \qquad (11.84)$$

The Jacobian matrix is

$$[J] = \begin{bmatrix} \dfrac{\partial U_C}{\partial \mu_a} \\ \dfrac{\partial U_C}{\partial D} \end{bmatrix}. \tag{11.85}$$

The inverse problem in Eq. (11.84) is a linearized problem, which is analogous to Eq. (11.78).

In step 5, Eq. (11.84) is solved for $[\Delta\mu_a]$ and $[\Delta D]$ with mathematical techniques to be discussed in the following section.

In step 6, the optical properties are updated with $[\Delta\mu_a]$ and $[\Delta D]$. The looping then returns to step 2.

Although the entire problem is nonlinear, each iteration cycle is linear. Thus, the nonlinear problem is solved by a series of linear steps. To understand this concept by analogy, draw a parabola with the minimum slightly above the abscissa and the opening facing up. Try to reach the minimum from a higher point on the curve by searching along the local tangent, which is a linear step. Iterate the linear search to approach the neighborhood of the minimum.

11.8.4. Inverse Method

Step 5 is an inverse mathematical problem. The Jacobian matrix can be constructed explicitly with either the direct or the adjoint method. In the direct method, the forward problem calculates the derivatives of the matrix elements. We write the forward problem in operator form

$$\{A\}\{U_C\} = \{S\},$$

where $\{A\}$ denotes the operator and $\{S\}$ denotes the source. The following equations are solved for derivatives $\{\partial U_C/\partial \mu_a\}$ and $\{\partial U_C/\partial D\}$:

$$\{A\}\left\{\dfrac{\partial U_C}{\partial \mu_a}\right\} = \left\{\dfrac{\partial S}{\partial \mu_a}\right\} - \left\{\dfrac{\partial A}{\partial \mu_a}\right\}\{U_C\}, \tag{11.86}$$

$$\{A\}\left\{\dfrac{\partial U_C}{\partial D}\right\} = \left\{\dfrac{\partial S}{\partial D}\right\} - \left\{\dfrac{\partial A}{\partial D}\right\}\{U_C\}. \tag{11.87}$$

In the adjoint method, the forward problem $\{A\}\{U_C\} = \{S\}$ is solved for U_C. Next, the adjoint equation $\{A'\}\{U_C\} = \{S'\}$ is solved for Green's function G_0 in response to a point source at a detector position. Then, the Jacobian matrix is computed by Eqs. (11.82) and (11.83).

To solve Eq. (11.84), we must invert a nonsquare Jacobian matrix of dimensions $N_M \times 2N_V$. As a result of photon diffusion, the Jacobian matrix is ill-conditioned (nearly singular); hence, direct matrix inversion is unreliable. Usually, the Jacobian matrix is first multiplied by its transpose to form a square matrix; accordingly, Eq. (11.84) becomes

$$[J]^T[J]\begin{bmatrix}\Delta\mu_a \\ \Delta D\end{bmatrix} = [J]^T[U_M - U_C]. \tag{11.88}$$

However, square matrix $[J]^T[J]$ is still ill-conditioned. A regularization technique is usually used to improve stability at the expense of image quality as follows:

$$([J]^T[J] + \eta_r[C_r]^T[C_r])\begin{bmatrix}\Delta\mu_a \\ \Delta D\end{bmatrix} = [J]^T[U_M - U_C]. \tag{11.89}$$

Here, η_r is the regularization parameter, which can be adjusted to control the stability of the inversion; $[C_r]$ is the regularizing operator (sometimes simply the identity matrix). This regularized equation is usually solved using the conjugate-gradient method.

APPENDIX 11A. ART AND SIRT

If a unique solution exists for a set of linear equations, $\mathbf{Ax} = \mathbf{b}$, where $\mathbf{A}$ is a matrix with elements a_{jk}, $\mathbf{x}$ is an n-tuple vector of unknowns, and $\mathbf{b}$ is an n-tuple vector of measurements; then $\mathbf{A}$ must be a nonsingular square matrix of dimensions $n \times n$. The goal is to solve for $\mathbf{x}$ iteratively.

In the algebraic reconstruction technique (ART), the search typically starts from the origin; the iterative equation for element i of $\mathbf{x}$ is

$$(x_i^p)_j = (x_i^p)_{j-1} - \frac{\sum_{k=1}^n a_{jk}(x_k^p)_{j-1} - b_j}{\sum_{k=1}^n a_{jk}a_{jk}} a_{ji}. \tag{11.90}$$

Here, p is the index of iteration; j and k are the row and column indices of $\mathbf{A}$; and $i, j, k = 1, 2, \ldots, n$. For brevity, $(x_k^p)_0$ denotes the final search point before the current iteration.

The first two movements of the iteration for $n = 2$ are illustrated in Figure 11.10, where the two lines represent $\sum_{k=1}^n a_{1k}x_k = b_1$ and $\sum_{k=1}^n a_{2k}x_k = b_2$. In move 1, the search point moves from the current position (the origin, in this figure) perpendicularly to line 1 and reaches intersection $(\mathbf{x}^p)_{j=1}$. In move 2, the search point moves from the current position perpendicularly to line 2 and reaches intersection $(\mathbf{x}^p)_{j=2}$. These two moves complete the first iteration cycle for $n = 2$. The cycle is iterated until convergence is reached.

If there are more equations than unknowns, the problem is overdetermined, and no unique solution exists. In this case, the solution of the ART oscillates in

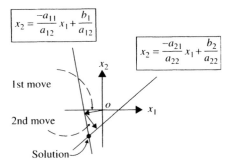

Figure 11.10. Illustration of ART.

the vicinity of the "true solution." If there are more unknowns than equations, the solution converges to a subspace, for instance, a line for $n = 2$.

The simultaneous iterative reconstruction technique (SIRT) is a variant of the ART. In each iteration cycle, n search points are first found by movements toward all the "lines" from the final search point before the current iteration:

$$(x_i^p)_j = (x_i^p)_0 - \frac{\sum_{k=1}^{n} a_{jk}(x_i^p)_0 - b_j}{\sum_{k=1}^{n} a_{jk} a_{jk}} a_{ji}, \quad j = 1, 2, \ldots, n. \tag{11.91}$$

Then, the final search point for this iteration cycle is the average of all n search points:

$$x_i^p = \frac{1}{n} \sum_{j=1}^{n} (x_i^p)_j. \tag{11.92}$$

The SIRT in general yields better images than the ART does but has a slower convergence speed.

PROBLEMS

11.1 (a) On the basis of the diffusion theory for an infinite medium, derive the mean time of flight between the observation point and the isotropic point source

$$\langle t \rangle = \frac{r^2}{2(D + r\sqrt{\mu_a D})c} = \frac{r^2}{2Dc[1 + (r/\delta)]},$$

where c is the speed of light in the scattering medium, r is the distance between the observation point and the source, and δ is the penetration depth. **(b)** Relate the mean path length of flight $\langle p \rangle$ with $\langle t \rangle$. **(c)** Show that $\langle p \rangle = (r^2/2D) \propto r^2$ if $r \ll \delta$ and $\langle p \rangle = (r\delta/2D) \propto r$ given $r \gg \delta$. *Hints*: Use fluence rate

$$\Phi(\vec{r}, t) = \frac{c}{(4\pi Dct)^{3/2}} \exp\left(-\frac{r^2}{4Dct} - \mu_a ct\right)$$

and current density $R = -D\partial\Phi/\partial r$. Define

$$\langle t \rangle_\Phi = \frac{\int t\Phi\, dt}{\int \Phi\, dt} \quad \text{and} \quad \langle t \rangle_R = \frac{\int tR\, dt}{\int R\, dt}.$$

11.2 Define the differential path length as $p_d = -d\ln R/d\mu_a$, where R is the current density. We have $R = -D(\partial/\partial r)\Phi(\vec{r})$, where Φ is the steady-state fluence rate in response to an isotropic point source in an infinite medium: $\Phi(\vec{r}) = \exp(-\mu_{eff}r)/(4\pi Dr)$.

(a) Show that

$$p_d = \frac{1}{2}\frac{r^2[1 + 3\mu_a D]}{D + \sqrt{\mu_a D}r}.$$

(b) Relate p_d to $\langle p \rangle$ in Problem 11.1.
(c) Show that $R \propto \exp\left(-\frac{1}{2}\mu_{eff}^2 r^2\right)$ when $r \ll \delta$ and $R \propto \exp(-\mu_{eff}r)$ when $r \gg \delta$.

11.3 A photon-density wave is generated from an isotropic point source in an infinite medium.
(a) Show that the phase at distance r from the source is

$$\Psi(r) = \arctan\left[\frac{Ar\sin(\tau/2)}{1 + Ar\cos(\tau/2)}\right] - Ar\sin(\tau/2),$$

where $A = [(\mu_a c)^2 + \omega^2]^{1/4}/\sqrt{Dc}$ and $\tau = \arctan[\omega/(\mu_a c)]$.
(b) Plot $\Psi(r)$ and $\omega\langle t \rangle$ versus $\omega \in [10, 100]$, where $\langle t \rangle$ is as defined in Problem 11.1. [*Hint*: Start with fluence rate $\Phi(r, \omega)$ and current density $R = -D(\partial/\partial r)\Phi(r, \omega)$.]

11.4 Use diffusion theory for an infinite scattering medium to plot the output as a function of time in MATLAB to graphically illustrate the three imaging modalities.

11.5 Derive Eqs. (11.19) and (11.20).

11.6 Show that

$$k_r = \sqrt{\frac{\mu_a}{2D}\left[\sqrt{1+\left(\frac{\omega}{c\mu_a}\right)^2}-1\right]} \quad \text{and}$$

$$k_i = \sqrt{\frac{\mu_a}{2D}\left[\sqrt{1+\left(\frac{\omega}{c\mu_a}\right)^2}+1\right]}. \quad (11.93)$$

11.7 (a) Calculate the wavelength of a photon-density wave with the following parameters: frequency $f = 200$ MHz, $\mu_a = 0.1$ cm^{-1}, $\mu_s' = 10$ cm^{-1}, index of refraction $n = 1.37$. (b) Plot the wavelength and k_i as a function of f in the range 100–1000 MHz.

11.8 (a) Duplicate the figures presented in Example 11.1. Known parameters include $\mu_a = 0.1$ cm^{-1}, $\mu_s' = 10$ cm^{-1}, and frequency $f = 200$ MHz. (b) Find the location of the maximal amplitude.

11.9 Generate a movie in MATLAB to demonstrate the propagation of a photon-density wave.

11.10 Show Snell's law for photon-density waves.

11.11 Derive the following diffuse photon-density wave image formation formula for a spherical refracting surface: $\lambda_{in}/S_{in} + \lambda_{out}/S_{out} = (\lambda_{in} - \lambda_{out})/R$. Here, λ_{in} is the wavelength in the incident medium, λ_{out} is the wavelength in the transmitted medium, S_{in} is the distance of object, S_{out} is the distance of image, and R is the radius of curvature.

11.12 Derive the Rytov approximation counterpart of the Born approximation solution. In the Rytov approximation, the photon-density distribution is expressed as the product—instead of the sum—of the incident (homogeneous) and scattered (heterogeneous) parts: $U(\vec{r},\vec{r}_s) = \exp[u_0(\vec{r},\vec{r}_s) + u_{SC}(\vec{r},\vec{r}_s)]$, where $U_0(\vec{r},\vec{r}_s) = \exp[u_0(\vec{r},\vec{r}_s)]$. Show that the solution is

$$u_{SC}(\vec{r}_d,\vec{r}_s) = -\frac{1}{U_0(\vec{r}_d,\vec{r}_s)}\int \frac{\delta\mu_a(\vec{r})}{D}G(\vec{r}-\vec{r}_d)U_0(\vec{r},\vec{r}_s)\,d\vec{r}.$$

11.13 Derive the Green function for a photon-density wave in a semiinfinite scattering medium. The isotropic point source is one transport mean free path below the surface. Use the extrapolated boundary condition.

11.14 Derive the Green function for a photon-density wave in a slab scattering medium. The isotropic point source is one transport mean free path below the surface. Use the extrapolated boundary condition.

11.15 Estimate the relative changes in the amplitude and the phase of a photon-density wave when a small absorber within a breast is moved by a small distance. Use realistic parameters for the estimation.

11.16 Rewrite the inverse algorithm in Section 11.7 in MATLAB or C/C++. Explore whether the real part of Eq. (11.46) is sufficient to provide an image.

11.17 Rewrite the inverse algorithm in Section 11.7 in MATLAB or C/C++ using ART and SIRT, respectively.

READING

Chance B, Kang K, He L, Weng J, and Sevick E (1993): Highly sensitive object location in tissue models with linear in-phase and antiphase multielement optical arrays in one and 2 dimensions, *Proc. Natl. Acad. Sci. USA* **90**(8): 3423–3427. (See Section 11.6, above.)

O'Leary MA (1996): *Imaging with Diffuse Photon Density Waves*, Ph.D. thesis, Univ. Pennsylvania, Philadelphia. (See Sections 11.5–11.7, above.)

Pogue BW, Testorf M, McBride T, Osterberg U, and Paulsen K (1997): Instrumentation and design of a frequency domain diffuse optical tomography imager for breast cancer detection, *Opt. Express* **1**(13): 391–403. (See Section 11.5, above.)

Schmidt FEW, Fry ME, Hillman EMC, Hebden JC, and Delpy DT (2000): A 32-channel time-resolved instrument for medical optical tomography, *Rev. Sci. Instrum.* **71**(1): 256–265. (See Section 11.3, above.)

Yodh AG and Boas DA (2003): Functional imaging with diffusing light, in *Biomedical Photonics Handbook*, Vo-Dinh T, ed., CRC Press, Boca Raton, FL, pp. 21.1–21.45. (See Sections 11.4 and 11.8, above.)

FURTHER READING

Aronson R (1995): Boundary-conditions for diffusion of light, *J. Opt. Soc. Am. A* **12**(11): 2532–2539.

Arridge SR (1999): Optical tomography in medical imaging, *Inverse Problems* **15**(2): R41–R93.

Boas DA, Brooks DH, Miller EL, DiMarzio CA, Kilmer M, Gaudette RJ, and Zhang Q (2001): Imaging the body with diffuse optical tomography, *IEEE Signal Process. Mag.* **18**(6): 57–75.

Boas DA, Gaudette T, Strangman G, Cheng XF, Marota JJA, and Mandeville JB (2001): The accuracy of near infrared spectroscopy and imaging during focal changes in cerebral hemodynamics, *Neuroimage* **13**(1): 76–90.

Boas DA, O'Leary MA, Chance B, and Yodh AG (1997): Detection and characterization of optical inhomogeneities with diffuse photon density waves: A signal-to-noise analysis, *Appl. Opt.* **36**(1): 75–92.

Cerussi AE, Berger AJ, Bevilacqua F, Shah N, Jakubowski D, Butler J, Holcombe RF, and Tromberg BJ (2001): Sources of absorption and scattering contrast for near-infrared optical mammography, *Acad. Radiol.* **8**(3): 211–218.

Cerussi AE, Jakubowski D, Shah N, Bevilacqua F, Lanning R, Berger AJ, Hsiang D, Butler J, Holcombe RF, and Tromberg BJ (2002): Spectroscopy enhances the information content of optical mammography, *J. Biomed. Opt.* **7**(1): 60–71.

Chance B, Anday E, Nioka S, Zhou S, Hong L, Worden K, Li C, Murray T, Ovetsky Y, Pidikiti D, and Thomas R (1998): A novel method for fast imaging of brain function, non-invasively, with light, *Opt. Express* **2**(10): 411–423.

Colak SB, van der Mark MB, Hooft GW, Hoogenraad JH, van der Linden ES, and Kuijpers FA (1999): Clinical optical tomography and NIR spectroscopy for breast cancer detection, *IEEE J. Select. Topics Quantum Electron.* **5**(4): 1143–1158.

Dorn O (1998): A transport-backtransport method for optical tomography, *Inverse Problems* **14**(5): 1107–1130.

Durduran T, Choe R, Culver JP, Zubkov L, Holboke MJ, Giammarco J, Chance B, and Yodh AG (2002): Bulk optical properties of healthy female breast tissue, *Phys. Med. Biol.* **47**(16): 2847–2861.

Durian DJ and Rudnick J (1997): Photon migration at short time and distances and in cases of strong absorption, *J. Opt. Soc. Am. A* **14**(1): 235–245.

Fantini S, Franceschini MA, and Gratton E (1994): Semi-infinite-geometry boundary-problem for light migration in highly scattering media—a frequency-domain study in the diffusion-approximation, *J. Opt. Soc. Am. B* **11**(10): 2128–2138.

Fantini S, Franceschini MA, Gaida G, Gratton E, Jess H, Mantulin WW, Moesta KT, Schlag PM, and Kaschke M (1996): Frequency-domain optical mammography: Edge effect corrections, *Med. Phys.* **23**(1): 149–157.

Fantini S, Walker SA, Franceschini MA, Kaschke M, Schlag PM, and Moesta KT (1998): Assessment of the size, position, and optical properties of breast tumors in vivoby noninvasive optical methods, *Appl. Opt.* **37**(10): 1982–1989.

Franceschini MA, Moesta KT, Fantini S, Gaida G, Gratton E, Jess H, Mantulin WW, Seeber M, Schlag PM, and Kaschke M (1997): Frequency-domain techniques enhance optical mammography: Initial clinical results, *Proc. Natl. Acad. Sci. USA* **94**(12): 6468–6473.

Franceschini MA, Toronov V, Filiaci ME, Gratton E, and Fantini S (2000): On-line optical imaging of the human brain with 160-ms temporal resolution, *Opt. Express* **6**(3): 49–57.

Gratton G, Fabiani M, Friedman D, Franceschini MA, Fantini S, Corballis P, and Gratton E (1995): Rapid changes of optical-parameters in the human brain during a tapping task, *J. Cogn. Neurosci.* **7**(4): 446–456.

Graves EE, Ripoll J, Weissleder R, and Ntziachristos V (2003): A submillimeter resolution fluorescence molecular imaging system for small animal imaging, *Med. Phys.* **30**(5): 901–911.

Grosenick D, Moesta KT, Wabnitz H, Mucke J, Stroszczynski C, Macdonald R, Schlag PM, and Rinneberg H (2003): Time-domain optical mammography: Initial clinical results on detection and characterization of breast tumors, *Appl. Opt.* **42**(16): 3170–3186.

Grosenick D, Wabnitz H, Rinneberg HH, Moesta KT, and Schlag PM (1999): Development of a time-domain optical mammography and first in vivo applications, *Appl. Opt.* **38**(13): 2927–2943.

Hawrysz DJ and Sevick-Muraca EM (2000): Developments toward diagnostic breast cancer imaging using near-infrared optical measurements and fluorescent contrast agents, *Neoplasia* **2**(5): 388–417.

Hebden JC, Veenstra H, Dehghani H, Hillman EMC, Schweiger M, Arridge SR, and Delpy DT (2001): Three-dimensional time-resolved optical tomography of a conical breast phantom, *Appl. Opt.* **40**(19): 3278–3287.

Hielscher AH, Klose AD, and Hanson KM (1999): Gradient-based iterative image reconstruction scheme for time-resolved optical tomography, *IEEE Trans. Med. Imaging* **18**(3): 262–271.

Klose AD and Hielscher AH (1999): Iterative reconstruction scheme for optical tomography based on the equation of radiative transfer, *Med. Phys.* **26**(8): 1698–1707.

Li XD, Durduran T, Yodh AG, Chance B, and Pattanayak DN (1997): Diffraction tomography for biochemical imaging with diffuse-photon density waves, *Opt. Lett.* **22**(8): 573–575.

Licha K, Riefke B, Ntziachristos V, Becker A, Chance B, and Semmler W (2000): Hydrophilic cyanine dyes as contrast agents for near-infrared tumor imaging: Synthesis, photophysical properties and spectroscopic in vivocharacterization, *Photochem. Photobiol.* **72**(3): 392–398.

McBride TO, Pogue BW, Gerety ED, Poplack SB, Osterberg UL, and Paulsen KD (1999): Spectroscopic diffuse optical tomography for the quantitative assessment of hemoglobin concentration and oxygen saturation in breast tissue, *Appl. Opt.* **38**(25): 5480–5490.

Ntziachristos V, Ma XH, and Chance B (1998): Time-correlated single photon counting imager for simultaneous magnetic resonance and near-infrared mammography, *Rev. Sci. Instrum.* **69**(12): 4221–4233.

Ntziachristos V, Yodh AG, Schnall M, and Chance B (2000): Concurrent MRI and diffuse optical tomography of breast after indocyanine green enhancement, *Proc. Natl. Acad. Sci. USA* **97**(6): 2767–2772.

O'Leary MA, Boas DA, Chance B, and Yodh AG (1992): Refraction of diffuse photon density waves, *Phys. Rev. Lett.* **69**(18): 2658–2661.

O'Leary MA, Boas DA, Li XD, Chance B, and Yodh AG (1996): Fluorescence lifetime imaging in turbid media, *Opt. Lett.* **21**(2): 158–160.

Pogue BW, Patterson MS, Jiang H, and Paulsen KD (1995): Initial assessment of a simple system for frequency-domain diffuse optical tomography, *Phys. Med. Biol.* **40**(10): 1709–1729.

Pogue BW, Poplack SP, McBride TO, Wells WA, Osterman KS, Osterberg UL, and Paulsen KD (2001): Quantitative hemoglobin tomography with diffuse near-infrared spectroscopy: Pilot results in the breast, *Radiology* **218**(1): 261–266.

Reynolds JS, Troy TL, Mayer RH, Thompson AB, Waters DJ, Cornell KK, Snyder PW, and Sevick-Muraca EM (1999): Imaging of spontaneous canine mammary tumors using fluorescent contrast agents, *Photochem. Photobiol.* **70**(1): 87–94.

Shah N, Cerussi A, Eker C, Espinoza J, Butler J, Fishkin J, Hornung R, and Tromberg B (2001): Noninvasive functional optical spectroscopy of human breast tissue, *Proc. Natl. Acad. Sci. USA* **98**(8): 4420–4425.

Yodh A and Chance B (1995): Spectroscopy and imaging with diffusing light, *Phys. Today* **48**(3): 34–40.

CHAPTER 12

Photoacoustic Tomography

12.1. INTRODUCTION

The term *photoacoustic tomography* (PAT) refers to imaging that is based on the photoacoustic effect. Although the photoacoustic effect was first reported on by Alexander Graham Bell in 1880, PAT was invented only after the advent of ultrasonic transducers, computers, and lasers. In PAT, the object is usually irradiated by a short-pulsed laser beam. Some of the light is absorbed by the object and partially converted into heat. The heat is then further converted to a pressure rise via thermoelastic expansion. The pressure rise is propagated as an ultrasonic wave, which is referred to as a *photoacoustic wave*. The photoacoustic wave is detected by ultrasonic transducers and is used by a computer to form an image.

12.2. MOTIVATION FOR PHOTOACOUSTIC TOMOGRAPHY

The motivation for PAT is to combine the contrast of optical absorption with the spatial resolution of ultrasound for deep imaging in the optical quasidiffusive or diffusive regime. Optical absorption is desirable because of its high sensitivity to molecules such as oxygenated and deoxygenated hemoglobin. In Table 12.1, PAT is compared with optical coherence tomography (OCT; see Chapters 9 and 10), diffuse optical tomography (DOT; see Chapter 11), and ultrasonography (US). Because of the strong optical scattering, pure optical imaging in biological tissue has either shallow imaging depth or low spatial resolution. Pure ultrasonic imaging can provide better resolution than pure optical imaging in the optical quasidiffusive or diffusive regime because ultrasonic scattering is two to three orders of magnitude weaker than optical scattering. Ultrasonic imaging, however, detects only mechanical properties and has weak contrast in early-stage tumors.

PAT overcomes the limitations of existing pure optical and pure ultrasonic imaging; its contrast is based on optical absorption in the photoacoustic excitation phase, whereas its resolution is derived from ultrasonic detection in the

---

Biomedical Optics: Principles and Imaging, by Lihong V. Wang and Hsin-i Wu
Copyright © 2007 John Wiley & Sons, Inc.

TABLE 12.1. Comparison of Optical Coherence Tomography (OCT), Diffuse Optical Tomography (DOT), Ultrasonography (US) Operating at 5 MHz, and Photoacoustic Tomography (PAT)

Property	OCT	DOT	US	PAT
Contrast	Good	Excellent	Poor for early cancers	Excellent
Imaging depth	Poor (~1 mm)	Good (~50 mm)	Excellent and scalable (~60 mm)	Good and scalable
Resolution	Excellent (~0.01 mm)	Poor (~5 mm)	Excellent and scalable (~0.3 mm)	Excellent and scalable
Speckle artifacts	Strong	None	Strong	None
Scattering coefficient	Strong (~10 mm^{-1})	Strong (~10 mm^{-1})	Weak (~0.03 mm^{-1})	Mixed[a]

[a] Strong for the excitation light and weak for the photoacoustic wave.

photoacoustic emission phase. The image resolution, as well as the maximum imaging depth, is scalable with ultrasonic frequency within the reach of diffuse photons. As ultrasonic center frequency and bandwidth increase, spatial resolution improves at the expense of imaging depth. In addition, PAT provides images that are devoid of speckle artifacts, which are conspicuous in both OCT and US images.

12.3. INITIAL PHOTOACOUSTIC PRESSURE

Two important timescales exist in laser heating:

1. The *thermal relaxation time*, which characterizes the thermal diffusion, is estimated by

$$\tau_{th} = \frac{d_c^2}{\alpha_{th}}, \tag{12.1}$$

where α_{th} is the thermal diffusivity (m^2/s) and d_c is the characteristic dimension of the heated region (the dimension of the structure of interest or the decay constant of the optical energy deposition, whichever is smaller).

2. The *stress relaxation time*, which characterizes the pressure propagation, is given by

$$\tau_s = \frac{d_c}{v_s}, \tag{12.2}$$

where v_s is the speed of sound (~1480 m/s in water).

If the laser pulsewidth is much shorter than τ_{th}, the excitation is said to be in thermal confinement, and heat conduction is negligible during the laser pulse. Likewise, if the laser pulsewidth is much shorter than τ_s, the excitation is said to be in stress confinement, and stress propagation is negligible during the laser pulse.

On laser excitation, the fractional volume expansion dV/V can be expressed as

$$\frac{dV}{V} = -\kappa p + \beta T. \tag{12.3}$$

Here, κ denotes the isothermal compressibility ($\sim 5 \times 10^{-10}$ Pa^{-1} for water or soft tissue); β denotes the thermal coefficient of volume expansion ($\sim 4 \times 10^{-4}$ K^{-1} for muscle); p and T denote the changes in pressure (Pa) and temperature (K), respectively. The isothermal compressibility κ can be expressed as

$$\kappa = \frac{C_P}{\rho v_s^2 C_V}. \tag{12.4}$$

Here, ρ denotes the mass density (~ 1000 kg/m^3 for water and soft tissue); C_P and C_V [~ 4000 J/(kg K) for muscle] denote the specific heat capacities at constant pressure and volume, respectively.

If the laser excitation is in both the thermal and stress confinements, the fractional volume expansion is negligible and the local pressure rise p_0 immediately after the laser pulse can be derived from Eq. (12.3):

$$p_0 = \frac{\beta T}{\kappa}, \tag{12.5}$$

which can be rewritten as

$$p_0 = \frac{\beta}{\kappa \rho C_V} \eta_{th} A_e. \tag{12.6}$$

Here, A_e is the specific optical absorption (J/m^3) and η_{th} is the percentage that is converted into heat. We define the Grueneisen parameter (dimensionless) as

$$\Gamma = \frac{\beta}{\kappa \rho C_V} = \frac{\beta v_s^2}{C_P}. \tag{12.7}$$

For water and diluted aqueous solutions, Γ can be estimated by the following empirical formula:

$$\Gamma_w(T_0) = 0.0043 + 0.0053 T_0, \tag{12.8}$$

where T_0 is the temperature in degrees Celsius. At body temperature, $\Gamma_w(37°\text{C}) = 0.20$. From Eq. (12.7), Eq. (12.6) becomes

$$p_0 = \Gamma \eta_{th} A_e \tag{12.9}$$

or

$$p_0 = \Gamma \eta_{th} \mu_a F. \tag{12.10}$$

Here, μ_a is the optical absorption coefficient and F is the optical fluence (J/cm^2).

Example 12.1. Show that the dimensions of the energy density and the pressure are the same.

$$J/m^3 = N\ m/m^3 = N/m^2 = Pa.$$

Note that 1 bar = 10^5 Pa.

Example 12.2. Given $d_c = 0.15$ cm and 15 µm, compute τ_{th} and τ_s in soft tissue.

With the typical properties of soft tissue, Eqs. (12.1) and (12.2) predict the following values:
For $d_c = 0.15$ cm:

$$\tau_{th} = \frac{(0.15\ \text{cm})^2}{1.3 \times 10^{-3}\ \text{cm}^2/\text{s}} = 17\ \text{s},$$

$$\tau_s = \frac{0.15\ \text{cm}}{0.15\ \text{cm}/\mu\text{s}} = 1\ \mu\text{s}.$$

For $d_c = 15$ µm:

$$\tau_{th} = \frac{(15 \times 10^{-4}\ \text{cm})^2}{1.3 \times 10^{-3}\ \text{cm}^2/\text{s}} = 17 \times 10^{-4}\ \text{s},$$

$$\tau_s = \frac{15 \times 10^{-4}\ \text{cm}}{0.15\ \text{cm}/\mu\text{s}} = 0.01\ \mu\text{s}.$$

Example 12.3. Estimate the temperature and the pressure rises on a short-pulse laser excitation of soft tissue at body temperature with a fluence of 10 mJ/cm^2. Assume $\mu_a = 0.1$ cm^{-1}.

$$A_e = 0.1\ \text{cm}^{-1} \times 10\ \text{mJ/cm}^2 = 1\ \text{mJ/cm}^3,$$

$$T = \frac{A_e}{\rho C_V} = \frac{1\ \text{mJ/cm}^3}{1\ \text{g/cm}^3 \times 4\ \text{J g}^{-1}\text{K}^{-1}} = 0.25\ \text{mK},$$

$$p_0 = \Gamma A_e = 0.20 \times 10\ \text{mbar} = 2\ \text{mbar}.$$

The results also indicate that for each mK (millikelvin) temperature rise, an 8-mbar pressure rise is produced.

12.4. GENERAL PHOTOACOUSTIC EQUATION

The photoacoustic wave generation and propagation in an inviscid medium is described by the following general photoacoustic equation (see Example 12.4):

$$\left(\nabla^2 - \frac{1}{v_s^2}\frac{\partial^2}{\partial t^2}\right) p(\vec{r}, t) = -\frac{\beta}{\kappa v_s^2}\frac{\partial^2 T(\vec{r}, t)}{\partial t^2}, \tag{12.11}$$

where $p(\vec{r}, t)$ denotes the acoustic pressure at location $\vec{r}$ and time t and T denotes the temperature rise. The left-hand side of this equation describes the wave propagation, whereas the right-hand side represents the source term.

In thermal confinement, the thermal equation becomes

$$\rho C_V \frac{\partial T(\vec{r}, t)}{\partial t} = H(\vec{r}, t). \tag{12.12}$$

Here, $H(\vec{r}, t)$ is the heating function defined as the thermal energy converted per unit volume and per unit time; it is related to the specific optical power deposition A_p by $H = \eta_{th} A_p$ and to the optical fluence rate Φ by $H = \eta_{th} \mu_a \Phi$. Substituting Eq. (12.12) into Eq. (12.11), we obtain the following less general photoacoustic equation:

$$\left(\nabla^2 - \frac{1}{v_s^2}\frac{\partial^2}{\partial t^2}\right) p(\vec{r}, t) = -\frac{\beta}{C_P}\frac{\partial H}{\partial t}. \tag{12.13}$$

The source term is related to the first time derivative of H. Therefore, time-invariant heating does not produce a pressure wave; only time-variant heating does.

Sometimes, velocity potential ϕ_v—which is related to p as follows—is used:

$$p = -\rho \frac{\partial \phi_v}{\partial t}. \tag{12.14}$$

Substituting Eq. (12.14) into Eq. (12.13) yields the following equation, which avoids the time derivative of H:

$$\left(\nabla^2 - \frac{1}{v_s^2}\frac{\partial^2}{\partial t^2}\right) \phi_v = \frac{\beta}{\rho C_P} H. \tag{12.15}$$

Example 12.4. Derive the photoacoustic equation shown in Eq. (12.11).

The two basic equations responsible for photoacoustic generation are the thermal expansion equation (generalized Hooke's law),

$$\nabla \cdot \vec{\xi}(\vec{r}, t) = -\kappa p(\vec{r}, t) + \beta T(\vec{r}, t) \tag{12.16}$$

and the linear inviscid force equation (the equation of motion)

$$\rho \frac{\partial^2}{\partial t^2}\vec{\xi}(\vec{r},t) = -\nabla p(\vec{r},t), \tag{12.17}$$

where vector $\vec{\xi}$ denotes the medium displacement. The left-hand side of Eq. (12.16) represents the fractional volume expansion, while the right-hand side represents the two factors related to the volume expansion. The left-hand side of Eq. (12.17) represents the mass density times the acceleration, and the right-hand side represents the force applied per unit volume. Thus, Eq. (12.17) is an incarnation of Newton's second law. The reader can reduce the two equations above to their 1D counterparts to understand their physical meanings more clearly.

Taking the divergence of Eq. (12.17), we obtain

$$\rho \frac{\partial^2}{\partial t^2}[\nabla \cdot \vec{\xi}(\vec{r},t)] = -\nabla^2 p(\vec{r},t). \tag{12.18}$$

Substituting Eq. (12.16) into Eq. (12.18) leads to Eq. (12.11), where $v_s = 1/\sqrt{\rho\kappa}$ is used. A more detailed derivation of the acoustic wave equation is shown in Appendix 12A.

12.5. GENERAL FORWARD SOLUTION

The general photoacoustic equation shown in Eq. (12.11) can be solved by the Green function approach (see Appendix 12B). *Green's function* is defined here as the response to a spatial and temporal impulse source term

$$\left(\nabla^2 - \frac{1}{v_s^2}\frac{\partial^2}{\partial t^2}\right)G(\vec{r},t;\vec{r}',t') = -\delta(\vec{r}-\vec{r}')\delta(t-t'), \tag{12.19}$$

where $\vec{r}'$ and t' denote the source location and time, respectively. In infinite space, where no boundary exists, Green's function is given by

$$G(\vec{r},t;\vec{r}',t') = \frac{\delta\left(t - t' - \frac{|\vec{r}-\vec{r}'|}{v_s}\right)}{4\pi|\vec{r}-\vec{r}'|}, \tag{12.20}$$

which represents an impulse diverging spherical wave. The following reciprocity relation holds:

$$G(\vec{r},t;\vec{r}',t') = G(\vec{r}',-t';\vec{r},-t). \tag{12.21}$$

To see this relationship more clearly, one observes $G(\vec{r},t;\vec{r}',0) = G(\vec{r}',0;\vec{r},-t)$ by setting $t' = 0$.

GENERAL FORWARD SOLUTION **289**

The physical meaning of Green's function should be interpreted carefully. A spatial delta function in the source term of the photoacoustic equation simply represents a point acoustic source. A temporal delta function in the source term, however, is translated into a step heating function or a ramp temperature rise for the following reason; the source term of the photoacoustic equation is proportional to the first time derivative of the heating function in thermal confinement or the second time derivative of the temperature in general. In other words, Green's function represents the response of a point absorber to step heating, rather than impulse heating.

Applying the Green function approach to Eq. (12.11) yields

$$p(\vec{r},t) = \int_{-\infty}^{t^+} dt' \int d\vec{r}' G(\vec{r},t;\vec{r}',t') \frac{\beta}{\kappa v_s^2} \frac{\partial^2 T(\vec{r}',t')}{\partial t'^2}, \quad (12.22)$$

which represents the pressure in response to an arbitrary source. Substituting Eq. (12.20) into Eq. (12.22) leads to

$$p(\vec{r},t) = \frac{\beta}{4\pi\kappa v_s^2} \int d\vec{r}' \frac{1}{|\vec{r}-\vec{r}'|} \left. \frac{\partial^2 T(\vec{r}',t')}{\partial t'^2} \right|_{t'=t-|\vec{r}-\vec{r}'|/v_s}. \quad (12.23)$$

In thermal confinement, substituting Eq. (12.12) into Eq. (12.23) yields

$$p(\vec{r},t) = \frac{\beta}{4\pi C_P} \int d\vec{r}' \frac{1}{|\vec{r}-\vec{r}'|} \left. \frac{\partial H(\vec{r}',t')}{\partial t'} \right|_{t'=t-|\vec{r}-\vec{r}'|/v_s}. \quad (12.24)$$

or

$$p(\vec{r},t) = \frac{\beta}{4\pi C_P} \frac{\partial}{\partial t} \int d\vec{r}' \frac{1}{|\vec{r}-\vec{r}'|} H\left(\vec{r}', t - \frac{|\vec{r}-\vec{r}'|}{v_s}\right). \quad (12.25)$$

If the heating function can be decomposed as $H(\vec{r}',t') = H_s(\vec{r}') H_t(t')$, Eq. (12.25) can be further simplified to

$$p(\vec{r},t) = \frac{\beta}{4\pi C_P} \frac{\partial}{\partial t} \int d\vec{r}' \frac{H_s(\vec{r}')}{|\vec{r}-\vec{r}'|} H_t\left(t - \frac{|\vec{r}-\vec{r}'|}{v_s}\right). \quad (12.26)$$

If $H_t(t') = \delta(t')$, Eq. (12.26) provides the delta heating response of an arbitrary absorbing object as

$$p(\vec{r},t) = \frac{\beta}{4\pi C_P} \frac{\partial}{\partial t} \int d\vec{r}' \frac{H_s(\vec{r}')}{|\vec{r}-\vec{r}'|} \delta\left(t - \frac{|\vec{r}-\vec{r}'|}{v_s}\right) \quad (12.27)$$

or

$$p(\vec{r},t) = \frac{\partial}{\partial t} \left[\frac{\beta}{4\pi C_P} \frac{1}{v_s t} \int d\vec{r}' H_s(\vec{r}') \delta\left(t - \frac{|\vec{r}-\vec{r}'|}{v_s}\right) \right], \quad (12.28)$$

where the quantity within the square brackets is the step heating response. From Eq. (12.9), the initial pressure response due to delta heating can be expressed as

$$p_0(\vec{r}) = \Gamma H_s(\vec{r}'). \tag{12.29}$$

Using Eqs. (12.7) and (12.29), we rewrite Eq. (12.28) as

$$p(\vec{r}, t) = \frac{1}{4\pi v_s^2} \frac{\partial}{\partial t}\left[\frac{1}{v_s t} \int d\vec{r}' p_0(\vec{r}') \delta\left(t - \frac{|\vec{r} - \vec{r}'|}{v_s}\right)\right]. \tag{12.30}$$

Example 12.5. Derive Eq. (12.20) from Eq. (12.19).

The following Fourier transformations are used:

$$g(\vec{k}, \omega) = \iint G(\vec{r}, t; \vec{r}', t') \exp[-i\vec{k} \cdot (\vec{r} - \vec{r}')] \exp[i\omega(t - t')] d\vec{r}\, dt, \tag{12.31}$$

$$1 = \iint \delta(\vec{r} - \vec{r}') \delta(t - t') \exp[-i\vec{k} \cdot (\vec{r} - \vec{r}')]$$
$$\times \exp[i\omega(t - t')] d\vec{r}\, dt. \tag{12.32}$$

Taking the Fourier transformation of both sides of Eq. (12.19) yields

$$g(\vec{k}, \omega) = \frac{1}{k^2 - \omega^2/v_s^2}. \tag{12.33}$$

Substituting this equation into the following inverse Fourier transformation

$$G(\vec{r}, t; \vec{r}', t') = \frac{1}{(2\pi)^4} \iint g(\vec{k}, \omega) \exp[i\vec{k} \cdot (\vec{r} - \vec{r}')] \exp[-i\omega(t - t')] d\omega\, d\vec{k}, \tag{12.34}$$

we obtain

$$G(\vec{r}, t; \vec{r}', t') = \frac{1}{16\pi^4} \iint \frac{1}{k^2 - \omega^2/v_s^2} \exp[i\vec{k} \cdot (\vec{r} - \vec{r}')]$$
$$\times \exp[-i\omega(t - t')] d\omega\, d\vec{k}. \tag{12.35}$$

The integral on the right-hand side involves singularities at $k = \pm\omega/v_s$, but it can be evaluated by Cauchy's contour integration:

$$\iint \frac{\exp(i\vec{k} \cdot \vec{\xi} - i\omega\tau)}{k^2 - \omega^2/v_s^2} d\omega\, d\vec{k}$$
$$= v_s \iint \frac{\exp(i\vec{k} \cdot \vec{\xi} - iv_s\tau\omega/v_s)}{(k + \omega/v_s)(k - \omega/v_s)} d\left(\frac{\omega}{v_s}\right) d\vec{k}$$

$$= 2\pi i v_s \int \exp(i\vec{k}\cdot\vec{\xi}) \frac{\exp(-iv_s k\tau) - \exp(iv_s k\tau)}{2k} d\vec{k}$$

$$= 2\pi v_s \int \exp(i\vec{k}\cdot\vec{\xi}) \frac{\sin(v_s k\tau)}{k} d\vec{k}. \tag{12.36}$$

Here, $\vec{\xi} = \vec{r} - \vec{r}'$ and $\tau = t - t'$. Since $d\vec{k} = 2\pi k^2 \sin\theta \, d\theta \, dk$ in spherical coordinates, we derive

$$\begin{aligned}
G(\vec{r}, t; \vec{r}', t') &= \frac{v_s}{4\pi^2} \int_0^\infty \int_0^\pi k \exp(ik\xi\cos\theta) \sin(v_s k\tau) \sin\theta \, d\theta \, dk \\
&= -\frac{v_s}{4\pi^2 \xi i} \int_0^\infty [\exp(-ik\xi) - \exp(ik\xi)] \sin(v_s k\tau) \, dk \\
&= \frac{v_s}{2\pi^2 \xi} \int_0^\infty \sin(k\xi) \sin(v_s k\tau) \, dk \\
&= \frac{1}{8\pi^2 \xi} \int_{-\infty}^\infty \left\{ \exp\left[iv_s k\left(\tau - \frac{\xi}{v_s}\right)\right] \right. \\
&\quad \left. - \exp\left[iv_s k\left(\tau + \frac{\xi}{v_s}\right)\right] \right\} d(v_s k) \\
&= \frac{1}{4\pi\xi} \left[\delta\left(\tau - \frac{\xi}{v_s}\right) - \delta\left(\tau + \frac{\xi}{v_s}\right)\right] \\
&= \frac{1}{4\pi|\vec{r}-\vec{r}'|}\left[\delta\left(t-t'-\frac{|\vec{r}-\vec{r}'|}{v_s}\right) - \delta\left(t-t'+\frac{|\vec{r}-\vec{r}'|}{v_s}\right)\right].
\end{aligned} \tag{12.37}$$

The second term on the right-hand side of this equation violates causality because the signal detected at distance $|\vec{r} - \vec{r}'|$ from $\vec{r}'$ can take place only when $t > t'$; thus, it must be dropped. As a result, we reach Eq. (12.20).

Example 12.6. From the scalar wave equation, generalize the reciprocity relation to a finite medium.

The Green function for a scalar wave equation satisfies the following differential equation:

$$\left(\nabla^2 - \frac{1}{v_s^2}\frac{\partial^2}{\partial t^2}\right) G(\vec{r}, t; \vec{r}', t') = -\delta(\vec{r}-\vec{r}')\delta(t-t'), \tag{12.38}$$

where location $\vec{r}'$ and time t' represent the source point and $\vec{r}$ and t denote the observation point. Time reversal of Eq. (12.38) leads to

$$\left(\nabla^2 - \frac{1}{v_s^2}\frac{\partial^2}{\partial t^2}\right) G(\vec{r}, -t; \vec{r}'', -t'') = -\delta(\vec{r}-\vec{r}'')\delta(t''-t). \tag{12.39}$$

292   PHOTOACOUSTIC TOMOGRAPHY

Multiplying Eq. (12.38) by $G(\vec{r}, -t; \vec{r}'', -t'')$, multiplying Eq. (12.39) by $G(\vec{r}, t; \vec{r}', t')$, subtracting the two, and integrating over the volume of investigation V and over time t from $-\infty$ to t_{max} with $t_{max} > \max(t', t'')$, we obtain

$$\int_{-\infty}^{t_{max}} dt \int_V d\vec{r} \bigg[G(\vec{r}, t; \vec{r}', t') \nabla^2 G(\vec{r}, -t; \vec{r}'', -t'')$$
$$- G(\vec{r}, -t; \vec{r}'', -t'') \nabla^2 G(\vec{r}, t; \vec{r}', t')$$
$$- \frac{1}{v_s^2} G(\vec{r}, t; \vec{r}', t') \frac{\partial^2}{\partial t^2} G(\vec{r}, -t; \vec{r}'', -t'') \qquad (12.40)$$
$$+ \frac{1}{v_s^2} G(\vec{r}, -t; \vec{r}'', -t'') \frac{\partial^2}{\partial t^2} G(\vec{r}, t; \vec{r}', t') \bigg]$$
$$= G(\vec{r}', -t'; \vec{r}'', -t'') - G(\vec{r}'', t''; \vec{r}', t').$$

Note that

$$G(\vec{r}, t; \vec{r}', t') \nabla^2 G(\vec{r}, -t; \vec{r}'', -t'') - G(\vec{r}, -t; \vec{r}'', -t'') \nabla^2 G(\vec{r}, t; \vec{r}', t')$$
$$= \nabla \cdot [G(\vec{r}, t; \vec{r}', t') \nabla G(\vec{r}, -t; \vec{r}'', -t'')$$
$$- G(\vec{r}, -t; \vec{r}'', -t'') \nabla G(\vec{r}, t; \vec{r}', t')] \qquad (12.41)$$

and

$$-G(\vec{r}, t; \vec{r}', t') \frac{\partial^2}{\partial t^2} G(\vec{r}, -t; \vec{r}'', -t'') + G(\vec{r}, -t; \vec{r}'', -t'') \frac{\partial^2}{\partial t^2} G(\vec{r}, t; \vec{r}', t')$$
$$= \frac{\partial}{\partial t} \bigg[-G(\vec{r}, t; \vec{r}', t') \frac{\partial}{\partial t} G(\vec{r}, -t; \vec{r}'', -t'')$$
$$+ G(\vec{r}, -t; \vec{r}'', -t'') \frac{\partial}{\partial t} G(\vec{r}, t; \vec{r}', t') \bigg]. \qquad (12.42)$$

Substituting Eqs. (12.41) and (12.42) into Eq. (12.41) and applying Green's theorem, we obtain

$$\int_{-\infty}^{t_{max}} dt \int_S d\vec{S} \cdot [G(\vec{r}, t; \vec{r}', t') \nabla G(\vec{r}, -t; \vec{r}'', -t'')$$
$$- G(\vec{r}, -t; \vec{r}'', -t'') \nabla G(\vec{r}, t; \vec{r}', t')]$$
$$+ \frac{1}{v_s^2} \int_V d\vec{r} \bigg[-G(\vec{r}, t; \vec{r}', t') \frac{\partial}{\partial t} G(\vec{r}, -t; \vec{r}'', -t'') \qquad (12.43)$$
$$+ G(\vec{r}, -t; \vec{r}'', -t'') \frac{\partial}{\partial t} G(\vec{r}, t; \vec{r}', t') \bigg]_{t=-\infty}^{t=t_{max}}$$
$$= G(\vec{r}', -t'; \vec{r}'', -t'') - G(\vec{r}'', t''; \vec{r}', t'),$$

where S is an arbitrary surface enclosing V. The first integral on the left-hand side vanishes because both Green functions satisfy the same homogeneous boundary condition (either G or its normal gradient is zero) on S. The second integral also vanishes because both $G(\vec{r}, t = -\infty; \vec{r}', t')$ and $G(\vec{r}, -t = -t_{max}; \vec{r}'', -t'')$, where $-\infty < t'$ and $-t_{max} < -t''$, as well as their time derivatives, must be zero as required by causality. Thus, Eq. (12.44) becomes

$$G(\vec{r}', -t'; \vec{r}'', -t'') = G(\vec{r}'', t''; \vec{r}', t'), \qquad (12.44)$$

which is the reciprocity relation.

12.6. DELTA-PULSE EXCITATION OF A SLAB

When a delta excitation pulse heats up a slab of thickness d (Figure 12.1), an initial pressure p_0 is first built up within the slab and then propagated outward in both the positive and the negative z directions. The pressure distribution can be derived by Eq. (12.30), where the integral over the volume is converted into an integral over a solid angle as shown in Figure 12.1. The solid angle is subtended by a section of the spherical shell of radius $v_s t$ from observation point $\vec{r} = (0, 0, z)$, where the section is the intersection with the slab.

We first consider an observation point on the positive z axis ($z \geq 0$). When the observation point is outside the slab ($z > d/2$), three cases are considered in terms of the propagation time:

When $v_s t < z - d/2$, the spherical shell does not touch the slab; thus, $p(z, t) = 0$.

When $z - d/2 \leq v_s t \leq z + d/2$, the spherical shell intersects the near edge of the slab at polar angle θ with respect to the negative z axis. The pressure

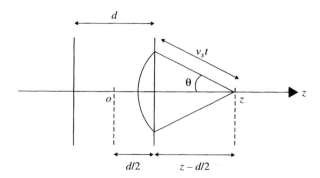

Figure 12.1. Diagram for a slab object.

distribution is calculated as follows:

$$\begin{aligned}
p(z,t) &= \frac{1}{4\pi v_s^2} \frac{\partial}{\partial t} \left[\frac{1}{v_s t} \int d\vec{r}'\, p_0 \delta\left(t - \frac{|\vec{r}-\vec{r}'|}{v_s}\right) \right] \\
&= \frac{p_0}{4\pi v_s^2} \frac{\partial}{\partial t} \left[\frac{1}{t} \int d\vec{r}'\, \delta(v_s t - |\vec{r}'-\vec{r}|) \right] \\
&= \frac{p_0}{4\pi v_s^2} \frac{\partial}{\partial t} \left[\frac{1}{t} \int_0^{2\pi} \int_0^{\theta} (v_s t)^2 \sin\theta\, d\theta\, d\phi \right] \\
&= \frac{p_0}{2} \frac{\partial}{\partial t}[t(1-\cos\theta)] = \frac{p_0}{2} \frac{\partial}{\partial t}\left[t\left(1 - \frac{z-d/2}{v_s t}\right) \right] = \frac{p_0}{2}.
\end{aligned} \qquad (12.45)$$

When $v_s t > z + d/2$, the spherical shell intersects the far edge of the slab at polar angle θ' as well, and we have

$$\begin{aligned}
p(z,t) &= \frac{p_0}{4\pi v_s^2} \frac{\partial}{\partial t}\left[\frac{1}{t} \int_0^{2\pi} \int_{\theta'}^{\theta} (v_s t)^2 \sin\theta\, d\theta\, d\phi \right] \\
&= \frac{p_0}{2} \frac{\partial}{\partial t}[t(\cos\theta' - \cos\theta)] \\
&= \frac{p_0}{2} \frac{\partial}{\partial t}\left[t\left(\frac{z+d/2}{v_s t} - \frac{z-d/2}{v_s t} \right) \right] = 0. \qquad (12.46)
\end{aligned}$$

When the observation point is inside the slab ($0 \le z \le d/2$), Eq. (12.45) can still be used.

When $v_s t < d/2 - z$, the upper limit for the integral over θ in Eq. (12.45) becomes π because the spherical shell is totally within the slab; as a result, $p(z,t) = p_0$.

When $d/2 - z \le v_s t \le d/2 + z$, the upper limit for the integral over θ in Eq. (12.45) becomes

$$\theta = \cos^{-1}\left(\frac{z-d/2}{v_s t}\right);$$

as a result, $p(z,t) = p_0/2$.

When $v_s t > d/2 + z$, as in Eq. (12.46), we have $p(z,t) = 0$.

For $z < 0$, the results are similar owing to symmetry.

In summary, the initial pressure distribution can be rewritten as

$$p_0(z) = p_0 U\left(z + \frac{d}{2}\right) U\left(-z + \frac{d}{2}\right), \qquad (12.47)$$

where U is the Heaviside function defined as

$$U(z) = \begin{cases} 1 & \text{for} \quad z \geq 0, \\ 0 & \text{for} \quad z < 0. \end{cases} \tag{12.48}$$

Consequently, the pressure distribution at any later time can be expressed as

$$p(z, t) = \frac{1}{2} p_0(z - v_s t) + \frac{1}{2} p_0(z + v_s t). \tag{12.49}$$

The first term on the right-hand side represents a right-propagating (along the $+z$ axis) plane wave and the second term, a left-propagating (along the $-z$ axis) plane wave. The physical meaning of Eq. (12.49) is interpreted as follows. Pressure p_0 is generated within the slab on delta heating. Immediately thereafter, p_0 is split into two plane waves, each having a magnitude of $p_0/2$ but propagating in opposite directions.

Example 12.7. Plot snapshots of the propagating pressure on the basis of Eq. (12.49).

The following MATLAB code produces Figure 12.2; the partial pressures in Eq. (12.49) are denoted by ppos (p_+) and pneg (p_-), respectively:

```
% Photoacoustic signal from a homogeneously heated slab
% Use SI units
clear all
vs = 1500;
p0 = 1;
d = 1E-3;
dhalf = d/2;
zmax = 2;
z = linspace(-zmax, zmax, 1000)*d;

figure(1)
clf

i_axis = 1;
for t = [0:1/2:1, 2]*dhalf/vs
    ppos = p0/2.*heaviside(z-(-dhalf+vs*t)).*heaviside(dhalf+vs*t-z);
    pneg = p0/2.*heaviside(z-(-dhalf-vs*t)).*heaviside(dhalf-vs*t-z);
    p = ppos + pneg;

    subplot(4, 2, i_axis, 'align')
    plot(z/d, ppos/p0, 'k-', z/d, pneg/p0, 'k--')
    tick = [.015 .025];
    set(0,'DefaultAxesTickLength',tick)
    title(['\itt\rm = ', num2str(vs*t/d), '\itxd rm/ itv_s'])
    axis([-zmax, zmax, 0, 1.1])
    if (i_axis == 7)
```

```
        xlabel('\itz\rm/\itd')
    end
    ylabel('Partial pressures/\it{p}\rm_0')
    if (i_axis == 1)
        legend('\itp\rm_+', '\itp\rm_-')
    end
    i_axis = i_axis + 1;

    subplot(4, 2, i_axis, 'align')
    plot(z/d, p/p0, 'k-')
    tick = [.015 .025];
    set(0,'DefaultAxesTickLength',tick)
    title(['\itt\rm = ', num2str(vs*t/d), '\itxd\rm/\itv_s'])
    axis([-zmax, zmax, 0, 1.1])
    if (i_axis == 8)
        xlabel('\itz\rm/\itd')
    end
    ylabel('Total pressure/\it{p}\rm_0')
    i_axis = i_axis + 1;
end
```

We can also make a movie showing the pressure propagation using the following MATLAB script:

```
% Photoacoustic signal from a homogeneously heated slab
% Use SI units

fig=figure(1);
set(fig,'DoubleBuffer','on'); % Flash-free rendering for animations

clear all
vs = 1500;
p0 = 1;
d = 1E-3;
dhalf = d/2;
zmax = 2;
z = linspace(-zmax, zmax, 1000)*d;

mov = avifile('Example07_PA_Slab.avi')

for t = [0:0.1:2]*dhalf/vs
    ppos = p0/2.*heaviside(z-(-dhalf+vs*t)).*heaviside(dhalf+vs*t-z);
    pneg = p0/2.*heaviside(z-(-dhalf-vs*t)).*heaviside(dhalf-vs*t-z);
    p = ppos + pneg;

    subplot(1, 2, 1)
    hold off;
    plot(z/d, ppos/p0, 'k-', z/d, pneg/p0, 'k--')
    grid
    axis([-zmax, zmax, 0, 1.1])
    xlabel('\itz/d')
    ylabel('Partial pressures/\itp\rm_0')
    legend('\itp\rm_+', '\itp\rm_-')
```

```
subplot(1, 2, 2)
hold off;
plot(z/d, p/p0, 'k-')
grid
axis([-zmax, zmax, 0, 1.1])
xlabel('z/d')
ylabel('Total pressure/\itp\rm_0')
pause(0.01)

    mov = addframe(mov,getframe(gcf));
end

mov = close(mov);
```

12.7. DELTA-PULSE EXCITATION OF A SPHERE

When a sphere of radius R_s is heated up with a delta pulse, an initial pressure p_0 is generated inside the sphere. As in the case of a slab, the pressure distribution can be derived from Eq. (12.30). However, the propagation involves spherical waves instead of plane waves. When the observation point is outside the sphere ($r > R_s$), three cases are considered according to the propagation time. Figure 12.3 shows part of a spherical shell of radius $v_s t$ centered at the observation point.

When $v_s t < r - R_s$, the spherical shell does not touch the heated spherical object; thus, $p(r, t) = 0$.

When $r - R_s \leq v_s t \leq r + R_s$, the spherical shell intersects the heated spherical object. Thus, the pressure distribution can be derived similarly as in Eq. (12.45):

$$p(r, t) = \frac{p_0}{2} \frac{\partial}{\partial t}[t(1 - \cos\theta)] = \frac{p_0}{2} \frac{\partial}{\partial t}\left[t\left(1 - \frac{(v_s t)^2 + r^2 - R_s^2}{2r v_s t}\right)\right]$$
$$= \frac{p_0}{2r}(r - v_s t). \qquad (12.50)$$

When $v_s t > r + R_s$, the spherical shell passes the far edge of the heated spherical object and no longer intersects with the heated spherical object; thus, $p(r, t) = 0$.

When the observation point is inside the sphere ($r \leq R_s$), the pressure distribution can be derived similarly:

When $v_s t < R_s - r$, the spherical shell is entirely enclosed by the heated spherical object, which means that $\theta = \pi$; as a result, $p(r, t) = p_0$.

When $R_s - r \leq v_s t \leq R_s + r$, the spherical shell emerges out of the heated spherical object; the pressure distribution can be derived similarly as in Eq. (12.50):

$$p(r, t) = \frac{p_0}{2r}(r - v_s t). \qquad (12.51)$$

298 PHOTOACOUSTIC TOMOGRAPHY

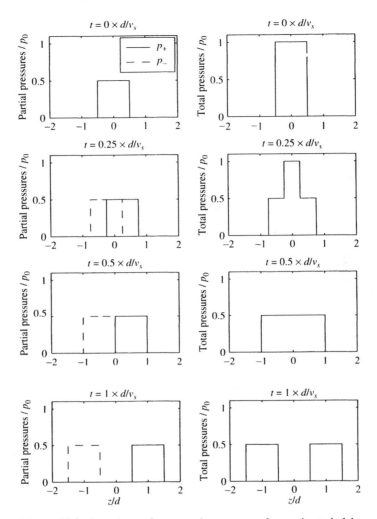

Figure 12.2. Snapshots of propagating pressure from a heated slab.

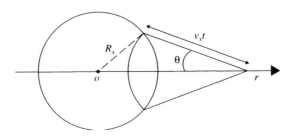

Figure 12.3. Diagram for a heated spherical object.

When $v_s t > R_s + r$, the spherical shell encloses the heated spherical object and no longer intersects with the heated spherical object; thus, $p(r, t) = 0$.

The results listed above are summarized using the Heaviside function U as

$$p(r, t) = p_0 \left[U(R_s - v_s t - r) + \frac{r - v_s t}{2r} U(r - |R_s - v_s t|) U(R_s + v_s t - r) \right]. \quad (12.52)$$

If we write the initial pressure as

$$p_0(r) = p_0 U(r) U(-r + R_s) \quad \text{for} \quad 0 \leq r < R_s, \quad (12.53)$$

we have

$$p(r, t) = \frac{r + v_s t}{2r} p_0(r + v_s t) + \frac{r - v_s t}{2r} p_0(-r + v_s t) + \frac{r - v_s t}{2r} p_0(r - v_s t). \quad (12.54)$$

The first term on the right-hand side represents a converging spherical wave; the second term represents a diverging spherical wave that originates from the initially converging wave propagating through the center; the third term represents a diverging spherical wave. On delta heating, an initial pressure p_0—which is constant across the entire heated sphere—is generated. This initial pressure is divided into two equal parts, each initiating a spherical wave. One travels inward

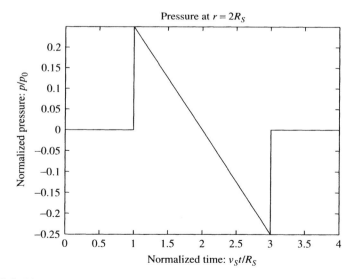

Figure 12.4. Bipolar (positive followed by negative) pressure profile from a heated sphere versus time.

as a converging spherical compression wave (first term), and the other travels outward as a diverging spherical compression wave (third term). When reaching the center of the heated spherical object, the converging spherical wave becomes a diverging spherical rarefaction wave (second term).

Example 12.8. Plot pressure versus time on the basis of Eq. (12.54).

The following MATLAB code produces Figure 12.4; the partial pressures in Eq. (12.54) are denoted by pin1, pinr, and pout, respectively:

```
% Photoacoustic signal from a homogeneously heated sphere
% Use SI units

clear all

vs = 1500;
p0 = 1;
Rs = 0.5E-3;

rd = 2*Rs; % Location of detector
t = linspace(0, (rd + 2*Rs)/vs, 1000);

figure(1)
clf

pin1 = p0/2*(1+vs*t./rd).*heaviside(rd+vs*t).*heaviside(Rs-rd-vs*t);
pinr = p0/2*(1-vs*t./rd).*heaviside(-rd+vs*t).*heaviside(Rs+rd-vs*t);
pout = p0/2*(1-vs*t./rd).*heaviside(rd-vs*t).*heaviside(Rs-rd+vs*t);
p = pin1 + pinr + pout;

plot(vs*t/Rs, p, 'k')
tick = [.015 .025];
set(0,'DefaultAxesTickLength',tick)
xlabel('Normalized time: \itv_s t\rm/\itR_s')
ylabel('Normalized pressure: \itp\rm/\itp\rm_0')
title('Pressure at \itr \rm= 2\itR_s')
```

Example 12.9. Plot snapshots of the propagating pressure on the basis of Eq. (12.54).

The following MATLAB code produces Figure 12.5; the partial pressures in Eq. (12.54) are denoted by pin1, pinr, and pout, respectively.

```
% Photoacoustic signal from a homogeneously heated sphere
% Use SI units

clear all
vs = 1500;
p0 = 1;
Rs = 0.5E-3;
rmax = 4;
```

```
rmin = 1E-3*Rs;
r = linspace(0, rmax*Rs, 1000) + rmin;

theta = linspace(-pi/2, pi/2);

figure(1)
clf

i_axis = 1;
for t = [0:1/2:1, 2]*Rs/vs
   pin1 = p0/2*(1+vs*t./r).*heaviside(r+vs*t).*heaviside(Rs-r-vs*t);
   pinr = p0/2*(1-vs*t./r).*heaviside(-r+vs*t).*heaviside(Rs+r-vs*t);
   pout = p0/2*(1-vs*t./r).*heaviside(r-vs*t).*heaviside(Rs-r+vs*t);
   p = pin1 + pinr + pout;

   subplot(4, 2, i_axis, 'align')
   hold off;
   plot(r/Rs, pin1/p0, 'k--', ...
      r/Rs, pinr/p0, 'k-.', ...
      r/Rs, pout/p0, 'k-', ...
      cos(theta), sin(theta), 'k-')
   tick = [.015 .025];
   set(0,'DefaultAxesTickLength',tick)
   title(['t = ', num2str(vs*t/Rs), 'x\itR_s/v_s'])
   axis equal;
   axis([0, rmax, -2, 2])
   ylabel('Partial pressures/\itp\rm_0')
   if (i_axis == 1)
      legend('p_{in1}', 'p_{inr}', 'p_{out}')
   end
   if (i_axis == 7)
      xlabel('\itr/R_s')
   end
   i_axis = i_axis + 1;

   subplot(4, 2, i_axis, 'align')
   hold off;
   plot(r/Rs, p/p0, 'k-', cos(theta), sin(theta), 'k-')
   tick = [.015 .025];
   set(0,'DefaultAxesTickLength',tick)
   title(['t = ', num2str(vs*t/Rs), 'x\itR_s/v_s'])
   axis equal;
   axis([0, rmax, -2, 2])
   ylabel('Total pressure/\itp\rm_0')
   if (i_axis == 8)
      xlabel('\itr/R_s')
   end
   i_axis = i_axis + 1;
end
```

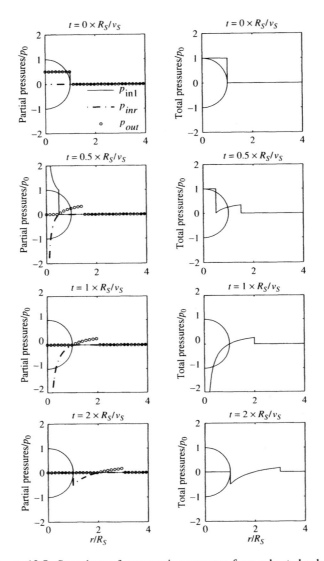

Figure 12.5. Snapshots of propagating pressure from a heated sphere.

12.8. FINITE-DURATION PULSE EXCITATION OF A THIN SLAB

Because the photoacoustic response in an infinite medium is both linear to the excitation power and timeshift-invariant, the response to a finite-duration excitation pulse $R(t)$ can be computed by convolution:

$$R(t) = \int_{-\infty}^{+\infty} dt' G(t-t') S(t') = \int_{-\infty}^{+\infty} dt' G(t') S(t-t'), \qquad (12.55)$$

where $G(t)$ denotes the response to a delta excitation pulse and $S(t)$ denotes the power of the pulse as a function of time.

For an excited slab, $G(t)$ is a top-hat function of time as shown in Figure 12.2. If the slab becomes so thin that the acoustic transit time across the slab is much less than the duration of the excitation pulse, $G(t)$ approaches a delta function apart from a constant factor:

$$G(t) \propto \delta(v_s t - z). \tag{12.56}$$

Substituting Eq. (12.56) into Eq. (12.55) leads to

$$R(t) \propto S(v_s t - z), \tag{12.57}$$

which means that the photoacoustic pressure is proportional to the excitation pulse. For example, if the pulse is Gaussian, that is,

$$S(t) = S_0 \exp\left[-\frac{1}{2}\frac{(t-t_0)^2}{\sigma^2}\right], \tag{12.58}$$

the photoacoustic pressure is Gaussian as well. Here, S_0 denotes the peak power, t_0 denotes the center timepoint corresponding to the peak power, and σ denotes the standard deviation.

12.9. FINITE-DURATION PULSE EXCITATION OF A SMALL SPHERE

Outside an excited sphere, $G(t)$ is a bipolar function of time as shown in Figure 12.4. If the sphere becomes so small that the acoustic transit time across the sphere is much less than the width of the excitation pulse, $G(t)$ approaches the derivative of a delta function apart from a constant factor:

$$G(t) \propto \frac{d}{dt}\delta(v_s t - r). \tag{12.59}$$

Substituting Eq. (12.59) into Eq. (12.55) leads to

$$R(t) \propto \frac{d}{dt}S(v_s t - r), \tag{12.60}$$

which means that for a small spherical object, the photoacoustic pressure is proportional to the time derivative of the excitation pulse.

12.10. DARK-FIELD CONFOCAL PHOTOACOUSTIC MICROSCOPY

Reflection-mode (or backward-mode) confocal photoacoustic microscopy (PAM) has been implemented using both dark-field pulsed-laser illumination and high-NA ultrasonic detection. In conventional dark-field transmission optical microscopy, an opaque disk is placed between the light source and the condenser lens so that ballistic light is rejected; as a result, only nonballistic light—which is scattered by the sample—is detected. In dark-field PAM, the excitation laser beam has a donut-shaped cross section; therefore, the photoacoustic signal from the tissue surface in the field of view is minimized.

In the system shown in Figure 12.6, a Q-switched pulsed Nd:YAG (neodymium: yttrium aluminum garnet) laser, operating at 532-nm wavelength, delivers 0.3 mJ of energy per pulse to the object through a 0.6-mm-diameter optical fiber. The laser beam has a 6.5-ns pulsewidth and a 10-Hz pulse repetition frequency. The lightbeam from the fiber output end is coaxially aligned with a focused ultrasonic transducer; both are mounted on a 3D mechanical translation stage. The ultrasonic transducer has a 50-MHz center frequency and a 70% nominal bandwidth. A concave acoustic lens with a 5.5-mm aperture diameter and a 5.6-mm working distance is attached to the ultrasonic transducer. This positive lens provides an NA of 0.44, which is considered large in ultrasonics. Light from the fiber is expanded by a conical lens and then focused onto the object through an optical condenser that has an NA of 1.1. The optical focal region overlaps with the focal spot of the ultrasonic transducer; thus, the optical dark-field illumination and ultrasonic detection are confocal.

Photoacoustic signals are received by the ultrasonic transducer, amplified by a low-noise amplifier, and then recorded digitally. For ultrasonic coupling, the ultrasonic transducer is immersed in water inside a plastic container; the bottom of the container has an opening that is sealed with a thin disposable polyethylene membrane; the object is first coated with ultrasound-coupling gel and then placed below the membrane.

On each laser-pulse excitation, the emitted photoacoustic wave is recorded as a function of time at each location of the ultrasonic transducer. The photoacoustic

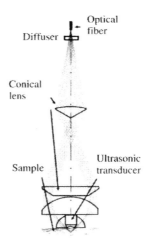

Figure 12.6. Schematic of a PAM system.

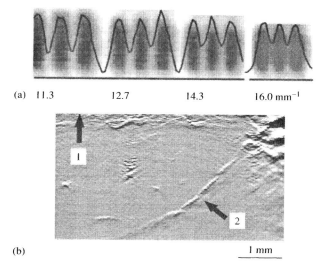

Figure 12.7. (a) Spatial resolution test with a bar chart placed in a tissue phantom (numbers below the images indicate the spatial modulation frequency); (b) imaging depth test with a black double-stranded cotton thread placed obliquely in the abdominal area of a rat (1 represents the skin surface; 2 represents the thread).

signal is converted into a 1D depth-resolved image (A-scan), based on the sound velocity in soft tissue (1.54 mm/μs). Raster scanning of the PAM probing head in a horizontal (xy) plane produces a 3D image.

Four PAM images of a Mylar USAF-1951 target taken through a 4-mm-thick layer of light-scattering tissue phantom made of 2% Intralipid® solution and 1% agar gel are shown in Figure 12.7a; the solid curves show the relative peak-to-peak amplitude of the received photoacoustic pressure across the bars on the target. The reduced scattering coefficient μ_s' of the phantom is 15 cm^{-1}. The thickness of the phantom equals six transport mean free paths. The modulation transfer function of the system is extracted from Figure 12.7a and extrapolated to its cutoff spatial frequency, producing an estimated lateral resolution of 45 μm. From another image of a 6-μm-diameter carbon fiber, the axial resolution is estimated to be ∼15 μm.

A PAM B-scan image of a black double-stranded cotton thread of 0.2 mm in diameter and 1.25 mm in pitch, which is inserted obliquely into the abdominal area of a sacrificed rat, is shown in Figure 12.7b. The thread is clearly visible in the image up to 3 mm in depth.

PAM images of the vasculature in the dorsal dermis (the upper lumbar area left of the vertebra) of a rat, with the hair removed using commercial hair remover lotion, is shown in Figure 12.8. Four in situ consecutive PAM B-scan images are obtained 0.2 mm apart laterally (Figure 12.8a). Each image is a gray-scale plot of the peak-to-peak amplitudes of the received photoacoustic signals; the vertical and the horizontal axes represent the depth from the skin surface and

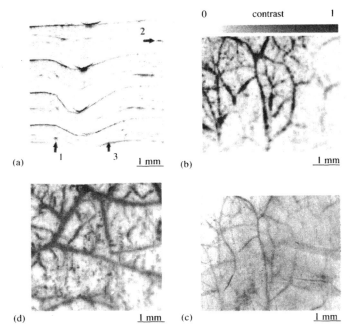

Figure 12.8. (a) Four in situ consecutive PAM B-scan images (1 and 2 represent vessels perpendicular to the imaging plane; 3 represents an in-plane vessel); (b) in situ maximum-amplitude projection PAM image taken from epidermal side; (c) photograph taken from dermal side using transmission illumination; (d) in vivo noninvasive maximum-amplitude projection PAM image taken from epidermal side.

the horizontal ultrasonic transducer position, respectively. The focal plane of the ultrasonic transducer is located at a depth of 1.2 mm. The PAM probing head is scanned 100 steps horizontally with a step size of 0.1 mm. The slightly inclined solid line in the upper part of each B-scan delineates the skin surface. The vessels marked by 1 and 2 are nearly perpendicular to the imaging plane, whereas the vessel marked by 3 is nearly parallel to the imaging plane.

An in situ maximum-amplitude projection PAM image on the skin surface (100×100 pixels, 0.1 mm step size), which plots the maximum peak-to-peak amplitude of each received photoacoustic signal (A-scan) within a 0.2–2-mm depth interval from the skin surface versus the ultrasonic transducer position on the tissue surface (x, y), is shown in Figure 12.8b. For comparison, a photograph (obtained using transmission light illumination) of the inner surface of the harvested skin is shown in Figure 12.8c. Good agreement in the vasculature is observed between the PAM image and the photograph. The photograph indicates that the major vessels are ~100 μm in diameter and the smaller vessels ~30 μm in diameter.

An in vivo maximum-amplitude projection PAM image of a similar area (100×100 pixels, 0.05 mm step size, and 0.5–3-mm depth interval) is shown in Figure 12.8d. Blood vessels with a density of up to a few counts per mm are observed.

PAM is capable of imaging optical absorption in biological tissue in the quasidiffusive regime, where the spatial resolution is determined primarily by the ultrasonic detection parameters instead of the optical excitation parameters. If the laser pulse is sufficiently short, a high-NA acoustic lens and a high-center-frequency ultrasonic transducer provide high lateral resolution while a wideband ultrasonic transducer provides high axial resolution.

Compared with bright-field illumination, dark-field illumination provides several advantages: (1) a large illumination area reduces the optical fluence on the tissue surface to less than 1 mJ/cm^2, which is well within the safety standards; (2) a large illumination area partially averages out the shadows of superficial heterogeneities in the image; and (3) dark-field illumination reduces the otherwise strong interference of the extraneous photoacoustic signal from the superficial paraxial area.

12.11. SYNTHETIC APERTURE IMAGE RECONSTRUCTION

Like optical waves, ultrasonic waves can be focused when passing through an acoustic lens, which is the basis of PAM as described in the previous section. Acoustic focusing can also be achieved synthetically by scanning a single-element ultrasonic transducer or using a multielement ultrasonic transducer. According to the principle of reciprocity, ultrasonic transmission and detection of the same acoustic lens are reciprocal.

In Figure 12.9, ultrasonic transmission focusing is illustrated with a five-element array transducer. If all elements are excited at the same time with the same voltage, an approximate plane ultrasonic wave is produced. If the elements are excited at different times, the produced ultrasonic wave can be focused to various points. In this illustration, the excitation of the outermost elements precedes that of the center element by $\Delta R/v_s$, where ΔR is the difference in radial distance to the desired focal point between the outermost elements and the center element. If similar proper delays are applied to all of the elements accordingly, ultrasonic pulses from these elements arrive at the focal point simultaneously; thus, the produced ultrasonic wave is focused to the desired point. Further, the focal point can be off the ultrasonic axis. If the focal point is set to infinity, an approximate plane wave is produced.

Like an optical grating, an ultrasonic array transducer produces the following far-field amplitude distribution (Figure 12.10):

$$H(\theta) = \frac{\sin(Nv_d)}{N \sin v_d} \frac{\sin v_e}{v_e}, \tag{12.61}$$

where

$$v_d = \frac{1}{2} k d_e \sin \theta \tag{12.62}$$

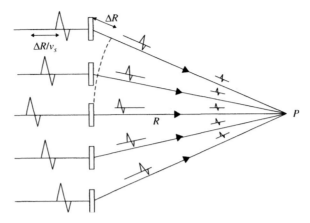

Figure 12.9. Ultrasonic transmission focusing by a multielement ultrasonic transducer.

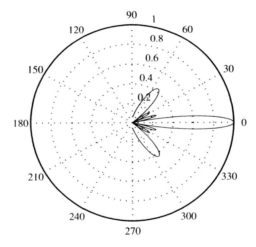

Figure 12.10. Acoustic-amplitude pattern produced by an ultrasonic array transducer versus the polar angle in a polar plot.

and

$$v_e = \frac{1}{2} k w_e \sin \theta. \tag{12.63}$$

Here, θ denotes the polar angle, N denotes the number of elements, k denotes the magnitude of the wavevector, d_e denotes the periodic distance between the elements, and w_e denotes the width of each element. Sidelobes, also termed *grating lobes*, appear at polar angles given by

$$\theta_g = \sin^{-1} \frac{m\lambda_a}{d_e}, \tag{12.64}$$

where m is a nonzero integer and λ_a is the acoustic wavelength. The angular width of the mainlobe or each grating lobe is given by

$$\Delta\theta_g = \sin^{-1} \frac{\lambda_a}{N d_e}. \tag{12.65}$$

Grating lobes deteriorate the lateral resolution in imaging but can be minimized using various designs. For example, the width of each element can be enlarged to reduce the magnitude of the grating lobes relative to the mainlobe, the excitation pulses can be shortened as much as possible, or the spacing between elements can be minimized or randomized.

Synthetic-aperture detection, also referred to as *delay-and-sum detection* or *beamforming*, can be applied to imaging. This detection scheme is reciprocal to the aforementioned transmission focusing. For each focal point, the image signal S can be calculated using

$$S(t) = \sum_i S_i(t + \Delta t_i). \tag{12.66}$$

Here, S_i is the signal from the ith ultrasonic transducer element; Δt_i is the time delay for the ith transducer element, which can be calculated similarly as in Figure 12.9.

12.12. GENERAL IMAGE RECONSTRUCTION

In this section, we consider general image reconstruction for an infinite acoustically homogeneous medium. The initial photoacoustic pressure excited by pulse $\delta(t)$ equals $p_0(\vec{r}) = \Gamma(\vec{r}) H_s(\vec{r})$ [Eq. (12.29)]. The acoustic pressure $p(\vec{r}, t)$ at position $\vec{r}$ and time t, initiated by source $p_0(\vec{r})$, satisfies the following photoacoustic wave equation [see Eq. (12.13)]:

$$\left(\nabla^2 - \frac{1}{v_s^2} \frac{\partial^2}{\partial t^2}\right) p(\vec{r}, t) = -\frac{p_0(\vec{r})}{v_s^2} \frac{d\delta(t)}{dt}. \tag{12.67}$$

Three detection configurations are considered. As shown in Figure 12.11, the detection surface is represented by S_0. For the planar geometry, if another planar surface S_0' parallel to S_0 is added, the combination of S_0' and S_0 encloses the source $p_0(\vec{r})$. For convenience, we write $S = S_0 + S_0'$ for the planar geometry and $S = S_0$ for the cylindrical or spherical geometry.

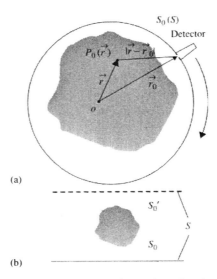

Figure 12.11. (a) During measurement, an ultrasonic point detector at position $\vec{r}_0$ on surface S_0 receives photoacoustic signals emitted from source $p_0(\vec{r})$—during image reconstruction, a quantity related to the measurement at position $\vec{r}_0$ projects backward via a spherical surface centered at $\vec{r}_0$; (b) in the planar geometry, another surface S_0' is combined with S_0 to enclose the entire source.

The following Fourier transformation pair is used to convert pressure between the time and frequency domains:

$$\tilde{F}(k) = \int_{-\infty}^{+\infty} F(\bar{t}) \exp(ik\bar{t}) \, d\bar{t}, \tag{12.68}$$

$$F(\bar{t}) = \frac{1}{2\pi} \int_{-\infty}^{+\infty} \tilde{F}(k) \exp(-ik\bar{t}) \, dk, \tag{12.69}$$

where $\bar{t} = v_s t$ and $k = \omega/v_s$ (ω is the angular frequency).

According to Green's theorem, the spectrum of the measured pressure $p(\vec{r}_0, \bar{t})$ is given by

$$\tilde{p}(\vec{r}_0, k) = -ik \int_V d\vec{r} \, \tilde{G}_k^{(\text{out})}(\vec{r}, \vec{r}_0) p_0(\vec{r}). \tag{12.70}$$

Here, V is a volume enclosing the entire source $p_0(\vec{r})$; $\tilde{G}_k^{(\text{out})}(\vec{r}, \vec{r}_0)$ is a Green function representing a monochromatic diverging spherical wave:

$$\tilde{G}_k^{(\text{out})}(\vec{r}, \vec{r}_0) = \frac{\exp(ik|\vec{r} - \vec{r}_0|)}{4\pi |\vec{r} - \vec{r}_0|}. \tag{12.71}$$

The acoustic pressure $\tilde{p}(\vec{r}, k)$ inside S can be computed by

$$\tilde{p}(\vec{r}, k) = \int_S dS \, \tilde{p}^*(\vec{r}_0, k)[2\hat{n}_0^s \cdot \nabla_0 \tilde{G}_k^{(\text{out})}(\vec{r}, \vec{r}_0)], \tag{12.72}$$

where $*$ indicates complex conjugation—equivalent to time reversal (see Problem 12.1), ∇_0 denotes the gradient on $\vec{r}_0$, and $\hat{n}_0^s$ denotes the normal vector of S pointing inward. The term in the square brackets indicates dipole radiation. Since $p_0(\vec{r}) = p(\vec{r}, \bar{t} = 0)$, taking the inverse Fourier transformation of Eq. (12.72) leads to the following backprojection formula:

$$p_0(\vec{r}) = \frac{1}{\pi} \int_{-\infty}^{+\infty} dk \int_S dS \, \tilde{p}^*(\vec{r}_0, k)[\hat{n}_0^s \cdot \nabla_0 \tilde{G}_k^{(\text{out})}(\vec{r}, \vec{r}_0)]. \tag{12.73}$$

For the planar geometry, if S is replaced by S_0, the right-hand side yields $p_0(\vec{r})/2$ instead. Since the reconstructed pressure is real, Eq. (12.73) can be rewritten as

$$p_0(\vec{r}) = \frac{1}{\pi} \int_S dS \int_{-\infty}^{+\infty} dk \, \tilde{p}(\vec{r}_0, k)[\hat{n}_0^s \cdot \nabla_0 \tilde{G}_k^{(\text{in})}(\vec{r}, \vec{r}_0)], \tag{12.74}$$

where $\tilde{G}_k^{(\text{in})}(\vec{r}, \vec{r}_0)$ is a Green function corresponding to a monochromatic converging spherical wave:

$$\tilde{G}_k^{(\text{in})}(\vec{r}, \vec{r}_0) = \frac{\exp(-ik|\vec{r} - \vec{r}_0|)}{4\pi|\vec{r} - \vec{r}_0|}. \tag{12.75}$$

A rigorous proof of Eq. (12.74) for the three common detection geometric configurations is given in the references for this section.

Employing $\nabla_0 \tilde{G}_k^{(\text{in})}(\vec{r}, \vec{r}_0) = -\nabla \tilde{G}_k^{(\text{in})}(\vec{r}, \vec{r}_0)$ and inverse-Fourier-transforming $\tilde{p}(\vec{r}_0, k)$, we can rewrite Eq. (12.74) in the time domain

$$p_0(\vec{r}) = -\frac{2}{\Omega_0} \nabla \cdot \int_{S_0} \hat{n}_0^s \, dS_0 \left[\frac{p(\vec{r}_0, \bar{t})}{\bar{t}} \right]_{\bar{t} = |\vec{r} - \vec{r}_0|}, \tag{12.76}$$

where Ω_0 is the solid angle subtended by the entire surface S_0 with respect to the reconstruction point $\vec{r}$ inside S_0. We have $\Omega_0 = 2\pi$ for the planar geometry and $\Omega_0 = 4\pi$ for the spherical or cylindrical geometry.

Further, we can rewrite Eq. (12.76) in a backprojection form as

$$p_0(\vec{r}) = \int_{\Omega_0} b(\vec{r}_0, \bar{t} = |\vec{r} - \vec{r}_0|) \frac{d\Omega_0}{\Omega_0}. \tag{12.77}$$

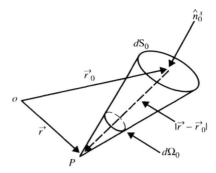

Figure 12.12. Diagram showing the solid angle $d\Omega_0$ subtended by detection element dS_0 with respect to point P at $\vec{r}$.

Here, $b(\vec{r}_0, \bar{t})$ is the backprojection term; $d\Omega_0$ is the solid angle subtended by detection element dS_0 with respect to reconstruction point P at $\vec{r}$ (Figure 12.12):

$$b(\vec{r}_0, \bar{t}) = 2p(\vec{r}_0, \bar{t}) - 2\bar{t}\frac{\partial p(\vec{r}_0, \bar{t})}{\partial \bar{t}}, \tag{12.78}$$

$$d\Omega_0 = \frac{dS_0}{|\vec{r} - \vec{r}_0|^2}\frac{\hat{n}_0^s \cdot (\vec{r} - \vec{r}_0)}{|\vec{r} - \vec{r}_0|}. \tag{12.79}$$

Factor $d\Omega_0/\Omega_0$ weighs the contribution from detection element dS_0 to the reconstruction. The reconstruction simply projects $b(\vec{r}_0, \bar{t})$ backward via a spherical surface centered at position $\vec{r}_0$. The first derivative with respect to time t actually represents a pure ramp filter k, which suppresses low-frequency signals.

The theory described above is based on an ideal infinite bandwidth in both time and space. In practice, if $k|\vec{r} - \vec{r}_0| \gg 1$ within the detection bandwidth, we have $\bar{t}\partial p(\vec{r}_0, \bar{t})/\partial \bar{t} \gg p(\vec{r}_0, \bar{t})$, which means that Eq. (12.78) can be simplified to

$$b(\vec{r}_0, \bar{t}) \approx -2\bar{t}\frac{\partial p(\vec{r}_0, \bar{t})}{\partial \bar{t}}. \tag{12.80}$$

A circular scanning configuration of PAT was implemented to image small-animal brains (Figure 12.13a). A Q-switched Nd:YAG laser provides light pulses (532-nm wavelength, 6.5-ns pulse duration, and 10-Hz pulse repetition frequency). The laser beam is expanded and homogenized to provide relatively uniform incident fluence, which is less than 10 mJ/cm^2 on the skin surface. Photoacoustic waves are coupled through water to an ultrasonic transducer with a center frequency of 3.5 MHz.

Since a circle—rather than a full spherical surface—is scanned, the reconstruction algorithm presented above is only approximately applicable. Nevertheless, good images are still attainable. Blood vessels in the cortical surface of small animals can be imaged transcranially with the scalp and the skull intact,

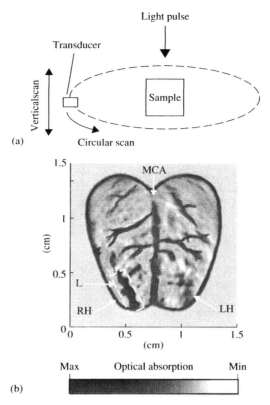

Figure 12.13. (a) Diagram of a circular-scanning PAT system for small-animal imaging; (b) a cross-sectional PAT image of a rat brain (RH represents right cerebral hemisphere; LH, left cerebral hemisphere; L, lesion; MCA, middle cerebral artery).

although the hair must be removed (Figure 12.13b). At this optical wavelength, the contrast of hemoglobin is high; the imaging depth is limited to about 1 cm, which is greater than the dimension of the entire brain of a small animal such as a mouse.

APPENDIX 12A. DERIVATION OF ACOUSTIC WAVE EQUATION

A longitudinal small-amplitude acoustic plane wave propagating in a homogeneous and nondissipative medium in the x direction is considered here. We examine the motion of a differential volume element $dV = dx\,dy\,dz$ at position x:

1. We derive the material equation. The excess pressure p is a function of the mass density ρ:

$$p = p(\rho). \qquad (12.81)$$

This equation can be expanded to the first order of the Taylor series around the equilibrium mass density ρ_0 as

$$p - p_0 = \left(\frac{\partial p}{\partial \rho}\right)(\rho - \rho_0), \tag{12.82}$$

where p_0 denotes the equilibrium pressure. The condensation parameter s is defined as

$$s = \frac{\rho - \rho_0}{\rho_0}, \tag{12.83}$$

which can be rewritten as

$$\rho = \rho_0(1 + s). \tag{12.84}$$

For a small-amplitude acoustic wave, we have $s \ll 1$. Substituting Eq. (12.84) into Eq. (12.82), we obtain

$$p - p_0 = \rho_0 \left(\frac{\partial p}{\partial \rho}\right) s. \tag{12.85}$$

2. We then derive the force equation. The force F due to pressure p experienced by the differential volume element is given by

$$F = -\left(\frac{\partial p}{\partial x}\right) dx\, dy\, dz. \tag{12.86}$$

From Newton's second law, we have

$$-\frac{\partial p}{\partial x} = \rho \frac{\partial u}{\partial t}, \tag{12.87}$$

where u denotes the medium velocity and t denotes time. Since $s \ll 1$, we replace ρ by ρ_0 and yield

$$-\frac{\partial p}{\partial x} = \rho_0 \frac{\partial u}{\partial t}. \tag{12.88}$$

This equation, termed the *linear inviscid force equation*, can be generalized to 3D space as $-\nabla p = \rho_0(\partial \vec{u}/\partial t)$. Substituting Eq. (12.85) into (12.88) yields

$$-\frac{\partial s}{\partial x} = \frac{1}{\partial p/\partial \rho} \frac{\partial u}{\partial t}. \tag{12.89}$$

3. We derive the continuity equation, based on the conservation of mass:

$$-\frac{\partial \rho}{\partial t} = \frac{\partial (\rho u)}{\partial x}, \tag{12.90}$$

This equation can be generalized to $-(\partial \rho/\partial t) = \nabla \cdot (\rho \vec{u})$. The right-hand side of Eq. (12.90) is expanded to

$$\frac{\partial (\rho u)}{\partial x} = \rho \frac{\partial u}{\partial x} + \frac{\partial \rho}{\partial x} u. \tag{12.91}$$

Since $s \ll 1$, we replace ρ by ρ_0 in the first term on the right-hand side to yield

$$\frac{\partial (\rho u)}{\partial x} = \rho_0 \frac{\partial u}{\partial x} + \frac{\partial \rho}{\partial x} u. \tag{12.92}$$

We will show later that the second term on the right-hand side is negligible. In this case, Eq. (12.90) becomes

$$-\frac{\partial \rho}{\partial t} = \rho_0 \frac{\partial u}{\partial x}, \tag{12.93}$$

which is the linearized continuity equation. Substituting Eq. (12.84) into Eq. (12.93), we obtain

$$-\frac{\partial s}{\partial t} = \frac{\partial u}{\partial x}. \tag{12.94}$$

4. Last, differentiating Eq. (12.89) with respect to x and differentiating Eq. (12.94) with respect to t, and then taking the difference between them, we obtain

$$\frac{\partial^2 p}{\partial x^2} = \frac{1}{\partial p/\partial \rho} \frac{\partial^2 p}{\partial t^2}. \tag{12.95}$$

This is a wave equation with

$$\frac{\partial p}{\partial \rho} = v_s^2, \tag{12.96}$$

where v_s is the speed of sound in the medium. Thus, we have

$$\frac{\partial^2 p}{\partial x^2} = \frac{1}{v_s^2} \frac{\partial^2 p}{\partial t^2}, \tag{12.97}$$

which can be generalized to the 3D case as

$$\nabla^2 p = \frac{1}{v_s^2} \frac{\partial^2 p}{\partial t^2}. \tag{12.98}$$

This is the basic acoustic wave equation that describes the propagation of an acoustic wave in a homogeneous nondissipative medium.

As promised, we now show that the second term on the right-hand side of Eq. (12.92) is negligible. The ratio of the second term to the first term on the right-hand side of Eq. (12.92) is

$$\frac{u}{\rho_0}\frac{\partial \rho}{\partial u} = \frac{u}{\rho_0}\left(\frac{\partial p/\partial u}{\partial p/\partial \rho}\right). \qquad (12.99)$$

From Eq. (12.97), we obtain

$$\left(\frac{\partial p}{\partial x} + \frac{1}{v_s}\frac{\partial p}{\partial t}\right)\left(\frac{\partial p}{\partial x} - \frac{1}{v_s}\frac{\partial p}{\partial t}\right) = 0. \qquad (12.100)$$

Therefore, we have $|(\partial p/\partial t)/(\partial p/\partial x)| = v_s$, which is substituted into Eq. (12.88) to yield

$$\left|\frac{\partial p}{\partial u}\right| = \rho_0 v_s. \qquad (12.101)$$

Substituting Eqs. (12.96) and (12.101) into Eq. (12.99) yields

$$\left|\frac{u}{\rho_0}\frac{\partial \rho}{\partial u}\right| = \frac{|u|}{v_s}. \qquad (12.102)$$

If $|u| \ll v_s$ (subsonic medium velocity of a small-amplitude wave), then

$$\left|\frac{\partial \rho}{\partial x}u\right| \ll \left|\rho_0 \frac{\partial u}{\partial x}\right|,$$

which means that the second term on the right-hand side of Eq. (12.92) is negligible.

APPENDIX 12B. GREEN FUNCTION APPROACH

The general Green function approach is summarized here. The acoustic wave equation with a source term $q(\vec{r}', t')$ is

$$\nabla'^2 p(\vec{r}', t') - \frac{1}{v_s^2}\frac{\partial^2 p(\vec{r}', t')}{\partial t'^2} = -q(\vec{r}', t'). \qquad (12.103)$$

On the basis of the reciprocity relation, Green's function $G(\vec{r}, t; \vec{r}', t')$ satisfies the following equation:

$$\nabla'^2 G(\vec{r}, t; \vec{r}', t') - \frac{1}{v_s^2}\frac{\partial^2 G(\vec{r}, t; \vec{r}', t')}{\partial t'^2} = -\delta(\vec{r} - \vec{r}')\delta(t - t'). \qquad (12.104)$$

Multiplying Eq. (12.103) by G and Eq. (12.104) by p, subtracting them, and then integrating over $\vec{r}'$ in the volume of interest V' and over t' from 0 to t^+,

we obtain

$$\int_0^{t^+} dt' \int_{V'} d\vec{r}' \left[G(\vec{r},t;\vec{r}',t')\nabla'^2 p(\vec{r}',t') - p\nabla'^2 G - \frac{1}{v_s^2}\left(G\frac{\partial^2 p}{\partial t'^2} - p\frac{\partial^2 G}{\partial t'^2}\right)\right]$$
$$= -\int_0^{t^+} dt' \int_{V'} d\vec{r}' G(\vec{r},t;\vec{r}',t')q(\vec{r}',t') + p(\vec{r},t). \tag{12.105}$$

Applying Green's theorem to Eq. (12.105) yields

$$p(\vec{r},t) = \int_0^{t^+} dt' \int_{V'} d\vec{r}' G(\vec{r},t;\vec{r}',t')q(\vec{r}',t')$$
$$+ \int_0^{t^+} dt' \int_{S'} d\vec{S}' \cdot [G\nabla' p(\vec{r}',t') - p\nabla' G] \tag{12.106}$$
$$+ \frac{1}{v_s^2} \int_{V'} d\vec{r}' \left(p\frac{\partial G}{\partial t'} - G\frac{\partial p}{\partial t'}\right)\bigg|_0^{t^+}.$$

Here, S' encloses V'. Choosing $t^+ = t + 0^+$, we have $p(\vec{r}',t^+) = 0$ and

$$\frac{\partial p(\vec{r}',t')}{\partial t'}\bigg|_{t'=t^+} = 0,$$

due to causality. Thus, Eq. (12.106) becomes

$$p(\vec{r},t) = \int_0^{t^+} dt' \int_{V'} d\vec{r}' G(\vec{r},t;\vec{r}',t')q(\vec{r}',t')$$
$$+ \int_0^{t^+} dt' \int_{S'} d\vec{S}' \cdot [G\nabla' p(\vec{r}',t') - p\nabla' G] \tag{12.107}$$
$$+ \frac{1}{v_s^2} \int_{V'} d\vec{r}' \left(p|_{t'=0}\frac{\partial G}{\partial t'}\bigg|_{t'=0} - G|_{t'=0}\frac{\partial p}{\partial t'}\bigg|_{t'=0}\right).$$

The first integral on the right-hand side depends on the source $q(\vec{r}',t')$; the second depends on the boundary condition; the third depends on the initial condition.

PROBLEMS

12.1 Show that complex conjugation of the temporal spectrum is equivalent to time reversal of the temporal function.

12.2 Derive the total acoustic energy transported through an enclosing spherical shell that is concentric to a delta heated sphere. Estimate how much optical energy is converted into mechanical energy.

12.3 Derive and plot the pressure as a function of time observed outside a sphere excited by **(a)** a delta pulse and **(b)** a Gaussian pulse.

12.4 Derive and plot the pressure as a function of time observed outside a thin spherical shell excited by **(a)** a delta pulse and **(b)** a Gaussian pulse.

12.5 Derive and plot the pressure as a function of time observed outside a line object first excited by **(a)** a delta pulse and **(b)** a Gaussian pulse.

12.6 Derive and plot the pressure as a function of time observed outside a cylindrical object excited by **(a)** a delta pulse and **(b)** a Gaussian pulse.

12.7 Derive and plot the pressure as a function of time observed outside an optically absorbing slab object excited by **(a)** a delta pulse and **(b)** a Gaussian pulse.

12.8 Derive and plot the pressure as a function of time observed outside a thin disk object excited by **(a)** a delta pulse and **(b)** a Gaussian pulse. Set the observation point first in the plane of the disk and then on the axis of the disk.

12.9 Derive and plot the pressure as a function of time observed outside a thin ring object excited by **(a)** a delta pulse and **(b)** a Gaussian pulse. Set the observation point first in the plane of the ring and then on the axis of the ring.

12.10 Using velocity potential, derive and plot the pressure as a function of time observed outside a slab object excited by **(a)** a delta pulse and **(b)** a Gaussian pulse.

12.11 Make a movie in MATLAB showing the pressure propagation from a sphere in response to a delta-pulse excitation.

12.12 Make a movie in MATLAB showing the pressure propagation from a spherical shell in response to delta-pulse excitation.

12.13 Make a movie in MATLAB showing the pressure propagation from two spheres in response to delta-pulse excitation.

12.14 Fourier-transform the pressure versus time observed outside a slab in response to delta-pulse excitation.

12.15 Fourier-transform the pressure versus time observed outside a sphere in response to delta-pulse excitation.

12.16 Sketch approximately the ideal wavefronts that are produced by a pulsed laser beam incident on a semiinfinite medium when the optical beam diameter is **(a)** much greater than, **(b)** much smaller than, and **(c)** comparable to the optical penetration depth.

12.17 Derive the pressure as a function of time produced from a periodic array of slabs by delta-pulse excitation. Then take its Fourier transformation.

12.18 Prove that the focal length of a planoconcave acoustic lens can be approximately expressed as $f = \delta/(1 - 1/n)$, where δ is the radius of curvature, and $n = v_{s1}/v_{s2}$, where v_{s1} is the acoustic velocity in the lens and v_{s2} is the acoustic velocity in the surrounding medium. Assume that the diameter of the lens is small compared with the radius of curvature.

12.19 Derive and plot Eq. (12.61).

12.20 Derive from Eq. (12.74) to Eq. (12.76) and then to Eq. (12.77).

12.21 Implement the delay-and-sum reconstruction algorithm and test it with dummy data generated from the forward solution.

12.22 Implement the general reconstruction algorithm and test it with dummy data generated from the forward solution.

READING

Cho ZH, Jones JP, and Singh M (1993): *Foundations of Medical Imaging*, Wiley, New York. (See Appendix 12A, above.)

Diebold GJ, Sun T, and Khan MI (1991): Photoacoustic monopole radiation in 1-dimension, 2-dimension, and 3-dimension, *Phys. Rev. Lett.* **67**(24): 3384–3387. (See Sections 12.6 and 12.7, above.)

Feng DZ, Xu Y, Ku G, and Wang LHV (2001): Microwave-induced thermoacoustic tomography: Reconstruction by synthetic aperture, *Med. Phys.* **28**(12): 2427–2431. (See Section 12.11, above.)

Hoelen CGA and de Mul FFM (1999): A new theoretical approach to photoacoustic signal generation, *J. Acoust. Soc. Am.* **106**(2): 695–706. (See Sections 12.6–12.9, above.)

Maslov K, Stoica G, and Wang LHV (2005): In vivo dark-field reflection-mode photoacoustic microscopy, *Opt. Lett.* **30**(6): 625–627. (See Section 12.10, above.)

Morse PM and Feshbach H (1999): *Methods of Theoretical Physics*, McGraw-Hill, New York. (See Appendix 12B, above.)

Wang LHV (2003): Ultrasound-mediated biophotonic imaging: A review of acousto-optical tomography and photo-acoustic tomography, *Disease Markers* **19**(2–3): 123–138. (See Section 12.2, above.)

Wang XD, Pang YJ, Ku G, Xie XY, Stoica G, and Wang LHV (2003): Noninvasive laser-induced photoacoustic tomography for structural and functional in vivo imaging of the brain, *Nature Biotechnol.* **21**(7): 803–806. (See Section 12.12, above.)

Xu MH and Wang LHV (2005): Universal back-projection algorithm for photoacoustic computed tomography, *Phys. Rev. E* **71**(1): 016706. (See Sections 12.3–12.5 and 12.12, above.)

FURTHER READING

Anastasio MA, Zhang J, Pan XC, Zou Y, Ku G, and Wang LHV (2005): Half-time image reconstruction in thermoacoustic tomography, *IEEE Trans. Med. Imaging* **24**(2): 199–210.

Andreev VG, Karabutov AA, and Oraevsky AA (2003): Detection of ultrawide-band ultrasound pulses in optoacoustic tomography, *IEEE Trans. Ultrasonics Ferroelectrics Freq. Control* **50**(10): 1383–1390.

Arfken GB and Weber HJ (1995): *Mathematical Methods for Physicists*, Academic Press, San Diego.

Beard PC, Perennes F, Draguioti E, and Mills TN (1998): Optical fiber photoacoustic-photothermal probe, *Opt. Lett.* **23**(15): 1235–1237.

Bell AG (1880): On the production and reproduction of sound by light, *Am. J. Sci.* **20**: 305–324.

Born M and Wolf E (1999): *Principles of Optics: Electromagnetic Theory of Propagation, Interference and Diffraction of Light*, Cambridge Univ. Press, Cambridge, UK/New York.

Cox BT and Beard PC (2005): Fast calculation of pulsed photoacoustic fields in fluids using k-space methods, *J. Acoust. Soc. Am.* **117**(6): 3616–3627.

Esenaliev RO, Karabutov AA, and Oraevsky AA (1999): Sensitivity of laser opto-acoustic imaging in detection of small deeply embedded tumors, *IEEE J. Select. Topics Quantum Electron.* **5**(4): 981–988.

Finch D, Patch SK, and Rakesh (2003): Determining a function from its mean values over a family of spheres, *SIAM J. Math. Anal.* **35**(5): 1213–1240.

Gusev VE and Karabutov AA (1993): *Laser Optoacoustics*, American Institute of Physics, New York.

Haltmeier M, Scherzer O, Burgholzer P, and Paltauf G (2004): Thermoacoustic computed tomography with large planar receivers, *Inverse Problems* **20**(5): 1663–1673.

Hoelen CGA, de Mul FFM, Pongers R, and Dekker A (1998): Three-dimensional photoacoustic imaging of blood vessels in tissue, *Opt. Lett.* **23**(8): 648–650.

Karabutov AA, Podymova NB, and Letokhov VS (1996): Time-resolved laser optoacoustic tomography of inhomogeneous media, *Appl. Phys. B—Lasers Opt.* **63**(6): 545–563.

Karabutov AA, Savateeva EV, and Oraevsky AA (2003): Optoacoustic tomography: New modality of laser diagnostic systems, *Laser Phys.* **13**(5): 711–723.

Karabutov AA, Savateeva EV, Podymova NB, and Oraevsky AA (2000): Backward mode detection of laser-induced wide-band ultrasonic transients with optoacoustic transducer, *J. Appl. Phys.* **87**(4): 2003–2014.

Kostli KP, Frenz M, Weber HP, Paltauf G, and Schmidt-Kloiber H (2000): Optoacoustic infrared spectroscopy of soft tissue, *J. Appl. Phys.* **88**(3): 1632–1637.

Kostli KP, Frauchiger D, Niederhauser JJ, Paltauf G, Weber HP, and Frenz M (2001): Optoacoustic imaging using a three-dimensional reconstruction algorithm, *IEEE J. Select. Topics Quantum Electron.* **7**(6): 918–923.

Kostli KP, Frenz M, Weber HP, Paltauf G, and Schmidt-Kloiber H (2001): Optoacoustic tomography: Time-gated measurement of pressure distributions and image reconstruction, *Appl. Opt.* **40**(22): 3800–3809.

Kostli KP and Beard PC (2003): Two-dimensional photoacoustic imaging by use of Fourier-transform image reconstruction and a detector with an anisotropic response, *Appl. Opt.* **42**(10): 1899–1908.

Kruger RA, Reinecke DR, and Kruger GA (1999): Thermoacoustic computed tomography-technical considerations, *Med. Phys.* **26**(9): 1832–1837.

Ku G and Wang LHV (2000): Scanning thermoacoustic tomography in biological tissue, *Med. Phys.* **27**(5): 1195–1202.

Ku G and Wang LHV (2001): Scanning microwave-induced thermoacoustic tomography: Signal, resolution, and contrast, *Med. Phys.* **28**(1): 4–10.

Ku G, Wang XD, Stoica G, and Wang LHV (2004): Multiple-bandwidth photoacoustic tomography, *Phys. Med. Biol.* **49**(7): 1329–1338.

Ku G and Wang LHV (2005): Deeply penetrating photoacoustic tomography in biological tissues enhanced with an optical contrast agent, *Opt. Lett.* **30**(5): 507–509.

Ku G, Wang XD, Xie XY, Stoica G, and Wang LHV (2005): Imaging of tumor angiogenesis in rat brains in vivo by photoacoustic tomography, *Appl. Opt.* **44**(5): 770–775.

Oraevsky AA, Jacques SL, and Tittel FK (1997): Measurement of tissue optical properties by time-resolved detection of laser-induced transient stress, *Appl. Opt.* **36**(1): 402–415.

Paltauf G and Schmidt-Kloiber H (2000): Pulsed optoacoustic characterization of layered media, *J. Appl. Phys.* **88**(3): 1624–1631.

Paltauf G, Viator JA, Prahl SA, and Jacques SL (2002): Iterative reconstruction algorithm for optoacoustic imaging, *J. Acoust. Soc. Am.* **112**(4): 1536–1544.

Viator JA, Au G, Paltauf G, Jacques SL, Prahl SA, Ren HW, Chen ZP, and Nelson JS (2002): Clinical testing of a photoacoustic probe for port wine stain depth determination, *Lasers Surg. Med.* **30**(2): 141–148.

Wang LHV, Zhao XM, Sun HT, and Ku G (1999): Microwave-induced acoustic imaging of biological tissues, *Rev. Sci. Instrum.* **70**(9): 3744–3748.

Wang XD, Pang YJ, Ku G, Stoica G, and Wang LHV (2003): Three-dimensional laser-induced photoacoustic tomography of mouse brain with the skin and skull intact, *Opt. Lett.* **28**(19): 1739–1741.

Xu MH, Ku G, and Wang LHV (2001): Microwave-induced thermoacoustic tomography using multi-sector scanning, *Med. Phys.* **28**(9): 1958–1963.

Xu MH and Wang LHV (2002): Time-domain reconstruction for thermoacoustic tomography in a spherical geometry, *IEEE Trans. Med. Imaging* **21**(7): 814–822.

Xu MH and Wang LHV (2003): Analytic explanation of spatial resolution related to bandwidth and detector aperture size in thermoacoustic or photoacoustic reconstruction, *Phys. Rev. E* **67**(5): 056605.

Xu Y and Wang LHV (2001): Signal processing in scanning thermoacoustic tomography in biological tissues, *Med. Phys.* **28**(7): 1519–1524.

Xu Y, Feng DZ, and Wang LHV (2002): Exact frequency-domain reconstruction for thermoacoustic tomography—I: Planar geometry, *IEEE Trans. Med. Imaging* **21**(7): 823–828.

Xu Y, Xu MH, and Wang LHV (2002): Exact frequency-domain reconstruction for thermoacoustic tomography—II: Cylindrical geometry, *IEEE Trans. Med. Imaging* **21**(7): 829–833.

Xu Y and Wang LHV (2004): Time reversal and its application to tomography with diffracting sources, *Phys. Rev. Lett.* **92**(3): 033902.

Zhang HF, Maslov K, Stoica G, and Wang LHV (2006): Functional photoacoustic microscopy for high-resolution and noninvasive *in vivo* imaging, *Nature Biotechnol.* **24**(7): 848–851.

Zhang J, Anastasio MA, Pan XC, and Wang LHV (2005): Weighted expectation maximization reconstruction algorithms for thermoacoustic tomography, *IEEE Trans. Med. Imaging* **24**(6): 817–820.

CHAPTER 13
Ultrasound-Modulated Optical Tomography

13.1. INTRODUCTION

Ultrasound-modulated optical tomography (UOT), first demonstrated in the 1990s, is another hybrid method that combines optical contrast and ultrasonic resolution as does photoacoustic tomography. UOT is based on the ultrasonic modulation of coherent laser light in a scattering medium. The medium is irradiated by both a laser beam and a focused ultrasonic wave. The ultrasound-modulated component of the reemitted light, which carries information about the local optical and acoustic properties, is used to provide tomographic imaging. Consequently, the image contrast is related to the optical and acoustic properties, whereas the spatial resolution is determined primarily by the ultrasonic wave. Because all the ultrasound-modulated light—regardless of the number of scattering events experienced—contributes to the imaging, UOT is capable of imaging deeply into the optical quasidiffusive or diffusive regime.

13.2. MECHANISMS OF ULTRASONIC MODULATION OF COHERENT LIGHT

Three mechanisms have been identified to account for ultrasonic modulation of light in a scattering medium:

1. *Incoherent Modulation of Light Due to Ultrasound-Induced Variations in Optical Properties of Medium.* As an ultrasonic wave propagates in a scattering medium, the medium is compressed or rarefied depending on the location and time, which causes the mass density to vary. The variations in the mass density further modulate the optical properties—including the absorption coefficient, scattering coefficient, and index of refraction—of the medium. Consequently, the reemitted light intensity varies with the

---

Biomedical Optics: Principles and Imaging, by Lihong V. Wang and Hsin-i Wu
Copyright © 2007 John Wiley & Sons, Inc.

ultrasonic wave. Although this mechanism does not require light coherence, ultrasonic modulation of low-coherence light is much weaker than that of high-coherence light.

2. *Variations in Optical Phase in Response to Ultrasound-Induced Displacements of Scatterers.* The displacements of scatterers modulate the free-path lengths (hence the phases) of light traversing the ultrasonic field. The modulated optical phases in the free-path lengths are accumulated along each complete path. Consequently, the reemitted light, which forms a speckle pattern, fluctuates with the ultrasonic wave.

3. *Variations in Optical Phase in Response to Ultrasonic Modulation of Index of Refraction of Background Medium.* The modulated index of refraction modulates the free-path phases of light traversing the ultrasonic field. The modulated optical phases contribute to speckle fluctuations as in the second mechanism.

Mechanisms 2 and 3 require light coherence. Both analytical and Monte Carlo models for the two coherent mechanisms have been developed. Only the former, however, is introduced here. In theory, the models for the coherent mechanisms can be unified by using the dielectric constant and further extended to include the incoherent mechanism.

In the analytical model, a plane ultrasonic wave irradiates a homogenous isotropic scattering medium. We assume that (1) the optical wavelength is much shorter than the mean free path (weak-scattering approximation) and (2) the ultrasound-induced change in the optical path length is much less than the optical wavelength (weak-modulation approximation). Owing to the weak-scattering approximation, ensemble-averaged correlations between electric fields from different paths are negligible compared with those from the same paths.

The autocorrelation function $G_1(\tau)$ of the scalar electric field of the scattered light can be expressed as

$$G_1(\tau) = \int_0^\infty p(s) \langle E_s(t) E_s^*(t+\tau) \rangle \, ds. \tag{13.1}$$

Here, $\langle \rangle$ denotes ensemble and time averaging, E_s denotes the unit-amplitude electric field of the scattered light of a path of length s, and $p(s)$ denotes the probability density function of s. The contributions to $G_1(\tau)$ from Brownian motion and from the ultrasonic field are independent and hence are treated separately. For brevity, only the latter is considered here.

In the following model, a coherent optical plane wave is incident normally on a slab of thickness d, and a point detector detects the transmitted light. The diffusion theory with a zero-boundary condition provides a solution to $p(s)$. From Eq. (13.1), we obtain

$$G_1(\tau) = \frac{(d/l_t') \sinh(\{\varepsilon[1-\cos(\omega_a \tau)]\}^{1/2})}{\sinh((d/l_t')\{\varepsilon[1-\cos(\omega_a \tau)]\}^{1/2})}, \tag{13.2}$$

where ω_a is the acoustic angular frequency, and the other parameters are

$$\varepsilon = 6(\delta_n + \delta_d)(n_0 k_0 A)^2, \tag{13.3}$$

$$\delta_n = (\alpha_{n1} + \alpha_{n2})\eta^2, \tag{13.4}$$

$$\alpha_{n1} = \frac{1}{2} k_a l'_t \arctan(k_a l'_t), \tag{13.5}$$

$$\alpha_{n2} = \frac{\alpha_{n1}}{k_a l'_t / \arctan(k_a l'_t) - 1}, \tag{13.6}$$

$$\delta_d = \frac{1}{6}. \tag{13.7}$$

Here, n_0 is the background index of refraction; k_0 is the magnitude of the optical wave vector in vacuo; A is the acoustic amplitude, which is proportional to the acoustic pressure; k_a is the magnitude of the acoustic wavevector; l'_t is the optical transport mean free path; η is the elastooptical coefficient, related to the adiabatic piezooptical coefficient of the material $\partial n/\partial p$ (derivative of refractive index n with respect to pressure p), the mass density ρ, and the speed of sound v_s: $\eta = (\partial n/\partial p)\rho v_s^2$; δ_n and δ_d are related to the average contributions per free path (or per scattering event) to the ultrasonic modulation of light via index of refraction and displacement, respectively.

Whereas δ_n increases with $k_a l'_t$, δ_d remains constant at $\frac{1}{6}$; thus, the ratio of δ_n to δ_d increases with $k_a l'_t$. The correlation between the two modulation mechanisms is neglected here for simplicity.

According to the Wiener–Khinchin theorem, the power spectral density $S(\omega)$ of the modulated speckle is related to $G_1(\tau)$ through the following Fourier transformation:

$$S(\omega) = \int_{-\infty}^{+\infty} G_1(\tau) \exp(i\omega\tau)\,d\tau. \tag{13.8}$$

Frequency ω is relative to the angular frequency of the unmodulated light (ω_0) because $\exp(-i\omega_0\tau)$, which is dropped for convenience, is implicit in $G_1(\tau)$. Therefore, $\omega = 0$ in $S(\omega)$ corresponds to absolute angular frequency ω_0.

Since $G_1(\tau)$ is an even periodic function of τ, the spectral intensity at frequency $n\omega_a$ can be calculated by

$$I_n = \frac{1}{T_a} \int_0^{T_a} \cos(n\omega_a \tau) G_1(\tau)\,d\tau, \tag{13.9}$$

where $n = 0, \pm 1, \pm 2 \ldots$, and T_a is the acoustic period. The frequency spectrum I_n is symmetric about ω_0. We define the one-sided modulation depth as

$$M_1 = \frac{I_1}{I_0}. \tag{13.10}$$

In the weak-modulation approximation, $(d/l'_t)\varepsilon^{1/2} \ll 1$; thus, Eq. (13.2) can be simplified to

$$G_1(\tau) = 1 - \frac{1}{6}\left(\frac{d}{l'_t}\right)^2 \varepsilon[1 - \cos(\omega_a \tau)]. \qquad (13.11)$$

Thus, we have

$$M_1 = \frac{1}{12}\left(\frac{d}{l'_t}\right)^2 \varepsilon \propto A^2, \qquad (13.12)$$

which indicates a quadratic relationship between M_1 and A. If the modulated light is measured experimentally using a Fabry–Perot interferometer, this quadratic dependence is observed. Otherwise, the detected apparent modulation depth (M')—defined as the ratio of the observed AC signal to the observed DC signal—can have a different dependence on A. In some cases, the AC signal originates from the beat between the electric field components at the fundamental frequency of the modulated light ($\omega_0 \pm \omega_a$) and the electric field component at the unmodulated optical angular frequency (ω_0). As a result, we have approximately $M' \propto (I_1/I_0)^{1/2} = M_1^{1/2} \propto A$, which indicates that M' is proportional to A.

13.3. TIME-RESOLVED FREQUENCY-SWEPT UOT

If a single-frequency ultrasonic wave is used in UOT, the axial resolution along the ultrasonic axis is much worse than the lateral resolution because of the elongated ultrasonic focal zone. Ultrasonic frequency sweeping (chirping), however, can improve the axial resolution. A time-resolved frequency-swept UOT system is shown in Figure 13.1. A frequency-swept signal is produced from a function generator and then amplified by a power amplifier followed by a transformer. The instantaneous frequency of the signal is

$$f_s(t) = a_s + bt, \qquad (13.13)$$

where a_s denotes the starting frequency, b denotes the sweep rate, and t denotes time. Here, the frequency sweeps from 7.0 to 10.0 MHz at a rate of 297 MHz/s. The amplified signal is applied to an ultrasonic transducer, which transmits a focused ultrasonic wave vertically into a scattering medium in a glass tank. An ultrasound absorber is placed at the bottom of the tank to minimize reflection from the water–glass interface.

After being broadened to 15 mm in diameter, a laser beam illuminates the scattering medium perpendicularly to the ultrasonic beam. The ultrasonic beam modulates the laser light with the following instantaneous frequency distribution along the ultrasonic axis:

$$f_s(t,z) = a_s + b\left(t - \frac{z-z_0}{v_s}\right) \quad \text{for} \quad t \geq \frac{z-z_0}{v_s}, \qquad (13.14)$$

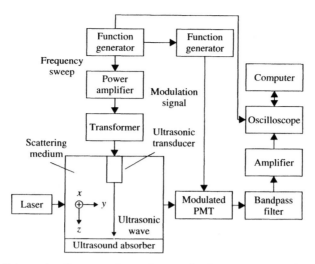

Figure 13.1. Schematic of an experimental setup for frequency-swept UOT. The z axis is along the acoustic axis; the y axis is along the optical axis; the x axis points into the paper.

Here, z denotes the ultrasonic axis and z_0 denotes a reference point along the ultrasonic axis at time zero. A PMT converts the received transmitted light into an electric signal. The gain of the PMT is modulated for heterodyne detection by a reference signal produced by another function generator. The reference modulation signal, also frequency-swept, has an instantaneous frequency given by

$$f_r(t) = a_r + bt, \quad (13.15)$$

where a_r denotes the starting frequency.

The heterodyned signal has the following frequency distribution:

$$f_h(z) = |f_s(t, z) - f_r(t)| = \left| a_s - a_r - \frac{b(z - z_0)}{v_s} \right|, \quad (13.16)$$

which is independent of time t. The heterodyned signal at the output of the PMT is bandpass-filtered and then amplified. The bandwidth of the filter Δf_h is determined by the desired range on the z axis to be imaged (region of interest) Δz as follows:

$$\Delta f_h = \frac{b}{v_s} \Delta z. \quad (13.17)$$

The signal from the amplifier is digitized by an oscilloscope and then transferred to a computer for postprocessing.

An object made of rubber is placed in the middle plane of the tank. The thickness of the tank along the laser beam is 17 cm. The scattering coefficient and the anisotropy of the scattering medium are 0.16 cm^{-1} and 0.73, respectively.

The object is translated in the tank along the x axis with a step size of 1 mm. A time-domain signal is recorded at each stop and then Fourier-transformed into a spectrum by a computer. Each spectrum is further converted into a 1D raw image of the scattering medium along the ultrasonic axis (z axis) on the basis of Eq. (13.16).

Two sample frequency spectra are depicted in Figure 13.2. Figure 13.2a shows a spectrum when the object is far from the ultrasonic axis. Figure 13.2b shows a spectrum when the object blocks part of the ultrasonic axis. As can be seen, the frequency components corresponding to the location of the object disappear.

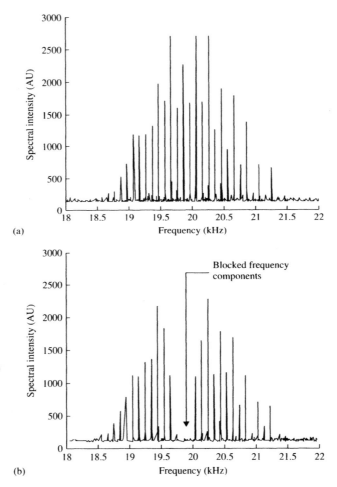

Figure 13.2. Frequency spectra of the heterodyned frequency-swept ultrasound-modulated optical signal when the object is (a) far from and (b) on the ultrasonic axis (AU = arbitrary units).

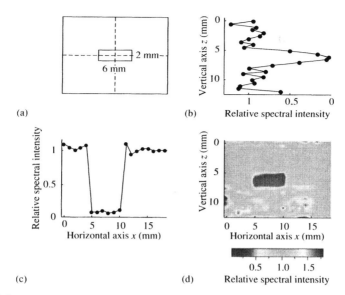

Figure 13.3. (a) Schematic cross-sectional view of the buried object in a scattering medium; (b) signals along the vertical dashed line in panel (a); (c) signals along the horizontal dashed line in panel (a); (d) 2D image of the scattering medium.

This figure demonstrates the one-to-one correspondence between the heterodyne frequency and the position along the ultrasonic axis. The image contrast reflects the spatial variation in the optical and acoustic properties.

Combining all the 1D spectra yields a 2D image (Figure 13.3). The first spectrum, which is taken when the object is far from the ultrasonic axis, is used as a reference. All spectra are divided by the reference spectrum point-by-point to yield relative spectra, which are 1D images. All 1D images are pieced together to form a 2D image. Signals along the dashed lines in Figure 13.3a are plotted in Figures 13.3b and 13.3c, respectively. As can be seen, the edge resolution in both directions is approximately 0.5 mm. The z-axis resolution is determined by the ultrasonic sweep parameters, whereas the x-axis resolution is determined by the ultrasonic focal diameter.

In summary, a frequency-swept (chirped) ultrasonic wave can encode laser light traversing the acoustic axis with various frequencies. Decoding the transmitted light provides resolution along the acoustic axis. This scheme is analogous to MRI.

13.4. FREQUENCY-SWEPT UOT WITH PARALLEL-SPECKLE DETECTION

The frequency-swept UOT based on a single-element photodetector in the previous section is demonstrated only in the quasiballistic regime. By improving the

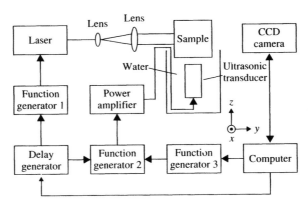

Figure 13.4. Schematic of a multiple-speckle UOT system. The z axis is along the acoustic axis; the y axis is along the optical axis; the x axis points out of the paper.

SNR with a CCD camera that detects multiple optical speckle grains in parallel, one can image in the diffusive regime. In this detection scheme, both the laser and the ultrasonic transducer are modulated with chirp functions. Imaging along the ultrasonic axis can be accomplished by electronically scanning the time delay between these two chirp functions.

An experimental setup is shown in Figure 13.4. A focused ultrasonic transducer with a 2.54-cm focal length in water, and a 1-MHz center frequency generates an ultrasonic wave with a peak pressure of $\sim 2 \times 10^5$ Pa at the focus. A laser emits a lightbeam modulated at 1 MHz with 690-nm wavelength, 12-mW average power, and 7-cm coherence length. The laser beam is expanded to 1.6×0.3 cm in cross section and then projected onto the sample, which is partially immersed in water for acoustic coupling. The light transmitted through the sample generates a speckle pattern, which is detected by a high-speed 12-bit CCD camera. The average speckle grain size is adjusted to match the CCD pixel size. Three function generators share the same timebase to ensure synchronization. Function generators 1 and 2 produce chirp functions to modulate the diode laser and to excite the ultrasonic transducer, respectively. A delay generator controls the time delay between the trigger signals to these two function generators.

If the chirp signal from function generator 2 were not amplitude-modulated, the frequency of the heterodyned signal along the ultrasonic axis (z axis) would be

$$f_h(z, \tau) = b \left(\tau - \frac{z}{v_s} \right), \qquad (13.18)$$

where b is the frequency sweep rate and τ is the time delay between the two chirps from function generators 2 and 1.

To implement the source-synchronized lock-in technique, the chirp signal from function generator 2 is amplitude-modulated by a reference sinusoidal wave of frequency $f_h(z, \tau)$ from generator 3. After lowpass filtering by the CCD, the signal from each CCD pixel can be expressed as

$$I_i(\phi_r) \propto I_b + I_m \cos(\phi_s + \phi_r). \tag{13.19}$$

Here, I_b denotes the background signal, I_m denotes the signal related to the ultrasound-modulated light component, ϕ_s denotes the initial phase of the speckle grain, and ϕ_r denotes the initial phase of the reference sinusoidal wave. The apparent modulation depth $M' = I_m/I_b$, which is related to the local optical and acoustic properties, is recovered for imaging. The initial phase ϕ_r is set sequentially to 0°, 90°, 180°, and 270°. The corresponding four frames of CCD images are acquired to calculate M' by

$$M' = \frac{1}{2I_b}\sqrt{[I_i(90°) - I_i(270°)]^2 + [I_i(0°) - I_i(180°)]^2}. \tag{13.20}$$

Each pixel of the CCD camera produces an M'; the average of the M' values from all 256 × 256 CCD pixels represents a single point (pixel) in the final image.

From Eq. (13.18), the spatial location being imaged is given by

$$z = v_s \left(\tau - \frac{f_h(z, \tau)}{b}\right). \tag{13.21}$$

At the same time, the ultrasound-modulated light from other spatial locations results in AC signals in the CCD pixels and hence is rejected by the CCD camera. One-dimensional images are obtained along the z axis by electronically varying τ. Further, 2D images are obtained by mechanically scanning the ultrasonic transducer along the x axis.

The spatial resolution z_R along the z axis is determined by

$$z_R \approx \frac{v_s}{\Delta f}, \tag{13.22}$$

where the speed of sound v_s is ~1500 m/s in most soft tissues and Δf denotes the frequency span of the chirp. Therefore, z_R is inversely proportional to Δf.

13.5. ULTRASONICALLY MODULATED VIRTUAL OPTICAL SOURCE

The original ultrasound-modulated optical signal can be considered a virtual light source. The virtual source is initially localized but is broadened with light propagation. If imaged near the acoustic axis, the virtual source can be seen clearly. Figure 13.5 shows a series of images at various z coordinates associated with different $f_h(z, \tau)$ values [Eq. (13.18)]. Ultrasonic modulation of light locally improves the spatial resolution of imaging because scanning a virtual small light source inside a highly scattering medium can produce a better image of the scanned cross section than scanning an actual small light source outside the medium.

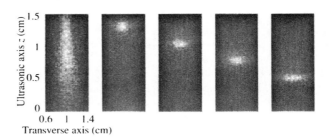

Figure 13.5. Demonstration of ultrasound-modulated light as a virtual source. The left frame represents the entire virtual source acquired without chirping. The following frames represent the virtual sources at various z values acquired with chirping Courtesy of Atlan et al. (2003).

13.6. RECONSTRUCTION-BASED UOT

Axial resolution along the ultrasonic axis can be achieved by reconstruction as in X-ray CT. In X-ray CT, a cross-sectional image of a sample is reconstructed from the transmitted X-ray intensities, which are acquired from multiple linear and angular scans around the object. In UOT, ultrasound-modulated optical signals are acquired while the ultrasonic beam is scanned linearly and angularly around the object. Subsequently, a filtered backprojection algorithm reconstructs an image of the cross section formed by the scanned ultrasonic axis.

A reconstruction-based UOT system that can operate in either reflection or transmission configuration is shown in Figure 13.6a. The CCD camera and the incident laser beam are on the same side of the sample for the reflection configuration and on opposite sides for the transmission configuration. After being expanded to 20 mm in diameter, light from a diode laser (690-nm wavelength, 11-mW power) illuminates the sample. The power density used is much lower than the $\sim$200-mW/cm^2 safety limit. An ultrasonic wave from a focused ultrasonic transducer is coupled into the sample through water in which the sample is partially immersed. Since the ultrasonic transducer has a 38-mm focal length in water and a 1-MHz center frequency, the focal zone is close to 2.8 mm in diameter and $\sim$20 mm in length; the peak pressure at the focus is $\sim 10^5$ Pa. The speckle pattern generated by the reemitted light is detected by a 12-bit CCD camera with 256 $\times$ 256 pixels. Ultrasound-modulated optical signals are extracted using parallel-speckle detection without chirping.

Linear and angular scans are required for the data acquisition. For experimental convenience, the buried object is translated horizontally and rotated about the optical axis while the imaging system is held stationary. A coordinate system is affixed to the buried object: the y axis is along the optical axis, and the z axis is initially along the ultrasonic axis. As shown in Figure 13.6b, the xz coordinates rotate with the buried object, whereas the measurement coordinates (x', y', z') are stationary with the z' axis parallel to the ultrasonic beam.

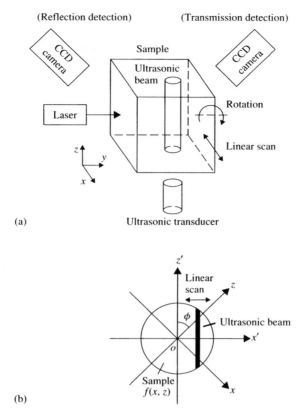

Figure 13.6. (a) Schematic of a reconstruction-based UOT system; (b) coordinate systems.

The detected ultrasound-modulated optical signal can be expressed as an integration of the signal originating from the z' axis:

$$s(\phi, x') = \int s_{\phi,x'}(z')\,dz', \qquad (13.23)$$

which is a Radon transform (see Chapter 8). The integrand can be expressed as

$$s_{\phi,x'}(z') = C_1 Q_{\phi,x'}(z') M_{\phi,x'}(z') G_{\phi,x'}(z'). \qquad (13.24)$$

Here, C_1 denotes a constant; $Q_{\phi,x'}(z')$ denotes the optical fluence rate; $M_{\phi,x'}(z')$ denotes the ultrasonic modulation depth, which is related to the optical and the ultrasonic properties; and $G_{\phi,x'}(z')$ denotes the Green function that describes the transport of the original ultrasound-modulated light to the detector. In the diffusive regime, $Q_{\phi,x'}(z')$ and $G_{\phi,x'}(z')$ have a weak dependence on z'.

From projection data $s(\phi, x')$, a filtered backprojection algorithm is used to reconstruct an image of the sample as follows:

$$\hat{f}(x, z) = \int_0^\pi \int_{-\infty}^{+\infty} S(\phi, k)|k| \exp(ikx')dk\, d\phi. \tag{13.25}$$

Here, k is the spatial frequency in the x' direction; $S(\phi, k)$ is the spatial Fourier transform of $s(\phi, x')$; $|k|$ is referred to as the *Ram–Lak filter*.

A tissue sample is imaged using the transmission configuration. Here, the step size of the linear scan along the x' axis is 1.2 mm, and the step size of the angular scan is 5°. By placing the CCD camera at an angle of 25° with respect to the optical axis, contributions from ballistic photons can be ruled out and the capability of nonballistic imaging can be demonstrated. Figure 13.7a shows a photograph of the middle cross section (xz plane) of a 14-mm-thick chicken breast tissue sample containing a chicken blood vessel, which measures $\sim 8 \times 3$ mm^2 on the xz plane and ~ 2 mm along the y axis. Figure 13.7b shows the reconstructed image, which clearly reveals the buried object.

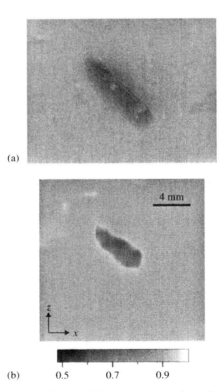

Figure 13.7. (a) Photograph of the middle xz cross section of a 14-mm-thick chicken breast tissue sample in which a blood vessel is buried; (b) reconstructed 2D image using the transmission configuration.

13.7. UOT WITH FABRY–PEROT INTERFEROMETRY

UOT can also be implemented with a long-cavity scanning confocal Fabry–Perot interferometer (CFPI), which provides a large etendue (the product of the detection area and the acceptance solid angle) and a short response time (Figure 13.8a). As shown in Figure 13.8b, the sample is gently pressed to a semicylindrical shape through a slit along the x axis; the orthogonal ultrasonic and optical beams are confocal below the sample surface. Reemitted light is collected on the opposite side of the ultrasound beam from the incident lightbeam. This configuration minimizes the effect of unmodulated light from the shallow region and enhances the ultrasonic modulation of some of the quasiballistic light that still exists at small imaging depths.

A focused ultrasonic transducer (15-MHz center frequency, 15-MHz bandwidth, 4.7-mm lens diameter, and 4.7-mm focal length) is driven by a pulser. The peak ultrasonic pressure at the focal spot measures 3.9 MPa, which is within the ultrasound safety limit at this frequency for biological tissue with no well-defined gas bodies. The laser light (532-nm wavelength, 100-mW power) has a 0.1-mm focal diameter in a clear medium. The sample is mounted on a three-axis (x', y', z') translation stage. The ultrasonic transducer and the sample are immersed in

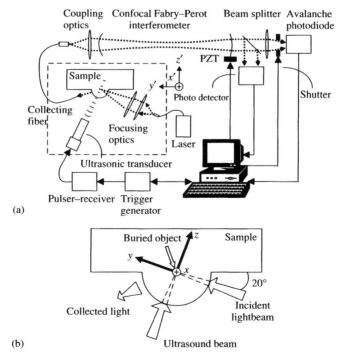

Figure 13.8. (a) Schematic of a CFPI-based UOT system (PZT represents a piezoelectric transducer made of lead zirconate titanate); (b) top view of the sample.

water for acoustic coupling. The light-focusing optics and the collection optical fiber (0.6-mm core diameter) are also immersed in the same water tank. The optical-fiber output is coupled to the CFPI (50-cm cavity length, 0.1-mm^2 sr etendue and >20 finesse), which operates in transmission mode.

A beamsplitter splits off some light for cavity tuning. Initially, one of the CFPI mirrors is scanned by more than one free spectral range of the cavity to search for a cavity length that matches the center frequency of the unmodulated light. Then, the mirror is displaced by a calibrated amount so that the cavity length matches the frequency of the positive sideband of the ultrasound-modulated light (15 MHz greater than the center frequency of the unmodulated light).

Next, an avalanche photodiode (APD) detects the light filtered through the CFPI; the output signal is sampled at 100 MHz by a data acquisition board. The entire system is controlled by computer. The APD signal is acquired during ultrasound propagation through the sample as a function of time. Converting the propagation time into z via the acoustic speed $v_s \approx 1500$ m/s yields the distribution of the ultrasound-modulated optical intensity along the ultrasonic axis $I_1(z)$, which provides a 1D image.

In each operational cycle, the resonant frequency of the CFPI is tuned first; then, both the ultrasonic transducer and the data-acquisition board are triggered by a trigger generator, and 4000 APD signals are acquired in one second. Averaging over 10 cycles produces a 1D image of satisfactory SNR. Two-dimensional images are obtained by further scanning of the sample along the x direction.

A typical profile of $I_1(z)$—which peaks at the intersection between the optical and ultrasonic axes because ultrasonic modulation is related to both optical fluence and ultrasound intensity—is shown in Figure 13.9. Chicken breast tissue is pressed through a 4-mm-wide slit to form a cylindrical tissue bump with a

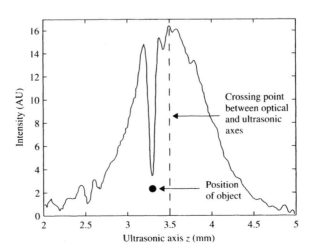

Figure 13.9. Ultrasound-modulated light intensity along the ultrasonic axis (AU = arbitrary units).

2-mm radius. A long black latex rod with a 60-μm diameter—which is transparent for ultrasound but absorptive for light—is placed along the x axis below the sample surface. When the ultrasound pulse passes through the object, the optical contrast produces a dip in $I_1(z)$.

The axial and lateral resolutions are investigated by imaging two chicken breast tissue samples (Figure 13.10). The samples are in semi-cylindrical shapes with 3.2- and 3-mm radii of curvature, respectively; each contains a 0.1-mm-thick black latex object (Figures 13.10b and 13.10d) at the axis of the semicylinder.

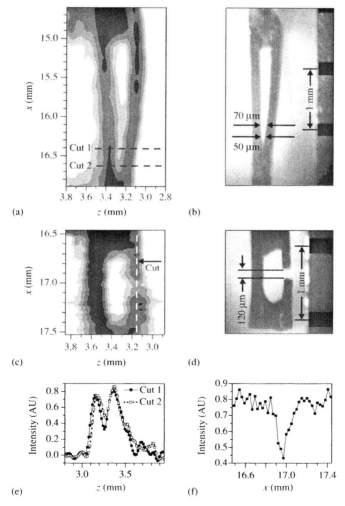

Figure 13.10. (a) Image and (b) photograph of an object; (c) image and (d) photograph of another object; (e) 1D axial profiles of intensity from the data in part (a); (f) 1D lateral profile of intensity from the data in part (c) (AU represents arbitrary units).

A typical 1D image of the background tissue is used as a reference $I_1^{\text{ref}}(z)$. Each 1D image of the object $I_1^{\text{obj}}(z)$ is converted into a relative profile $I_1^{\text{rel}}(z)$ by

$$I_1^{\text{rel}}(z) = \frac{I_1^{\text{obj}}(z) - I_1^{\text{ref}}(z)}{I_1^{\text{ref}}(z)}. \tag{13.26}$$

These relative profiles form 2D images are shown in gray scale with five equally spaced gray levels in Figures 13.10a and 13.10c. Figure 13.10e presents two $I_1^{\text{rel}}(z)$ profiles taken from Figure 13.10a along two cut lines. Along cut line 1, the actual gap is 70 μm and is resolved with a 55% contrast; along cut line 2, the actual gap reduces to 50 μm, and the contrast decreases to 40%. Similarly, Figure 13.10f represents the 1D lateral profile of intensity versus x taken from Figure 13.10c along the cut line. The actual gap along the cut line is 120 μm and is resolved with a 50% contrast. If a gap resolvable with a 50% contrast is defined as the resolution, the estimated axial and lateral resolutions are 70 and 120 μm, respectively.

PROBLEMS

13.1 Show that given $(d/l'_t)\varepsilon^{1/2} \ll 1$, Eq. (13.2) can be simplified to Eq. (13.11) and the modulation depth can be approximated by Eq. (13.12).

13.2 Plot the ratio of δ_n to δ_d as a function of $k_a l'_t$ using the following parameters: $\partial n/\partial p = 1.466 \times 10^{-10}$ m²/N, $\rho = 1000$ kg/m³, and $v_s = 1488$ m/s.

13.3 Plot the modulation depth as a function of l'_t using the following parameters: $\partial n/\partial p = 1.466 \times 10^{-10}$ m²/N, $\rho = 1000$ kg/m³, $v_s = 1488$ m/s, ultrasonic frequency $f_a = 1$ MHz, $n_0 = 1.33$, $\lambda_0 = 500$ nm, $d = 5$ cm, and $A = 0.01$ nm. Convert A into pressure.

13.4 Plot I_2/I_1 versus A using the following parameters: $\partial n/\partial p = 1.466 \times 10^{-10}$ m²/N, $\rho = 1000$ kg/m³, $v_s = 1488$ m/s, ultrasonic frequency $f_a = 5$ MHz, $n_0 = 1.33$, $\lambda_0 = 500$ nm, $d = 5$ cm, and $k_a l'_t = 1$.

13.5 In frequency-swept ultrasound-modulated optical tomography, the source that is applied to the ultrasonic transducer sweeps from 7 to 10 MHz in 10 ms. The reference signal that is applied to the PMT sweeps from 7.01 to 10.01 MHz in 10 ms. Derive the formula that converts the frequency of the heterodyned signal to the position along the ultrasonic axis.

13.6 Derive the phase of the chirped sinusoidal function that provides frequency sweep $f_s(t) = a_s + bt$.

13.7 (a) Derive Eq. (13.20). (b) Extend it to a three-phase method, where the three phases are 90° apart. (c) Extend it to a two-phase method, where the two phases are 180° apart.

13.8 A mechanical index—which is used to estimate the likelihood of mechanical bioeffects—is defined as $MI = P/(C_{MI}\sqrt{f})$. Here, P is the local peak-rarefactional pressure in MPa, f is the acoustic frequency in MHz, and C_{MI} is a coefficient equal to $1 \text{ MPa}/\sqrt{\text{MHz}}$. Calculate the MI for the ultrasonic parameters used in this chapter and compare with the current safety standards.

13.9 Given $\mu_a = 0.03 \text{ cm}^{-1}$, $\mu_s = 100 \text{ cm}^{-1}$, and $g = 0.9$, compute μ_{eff} in both cm^{-1} and dB/cm. Compare μ_{eff} with the ultrasonic attenuation coefficient at frequency $f = 3$ MHz in biological tissue, which is approximately $0.5f$ dB/(cm MHz).

13.10 Implement the Radon transformation in C/C++.

13.11 Implement the inverse Radon transformation in C/C++.

READING

Atlan M, Forget BC, Ramaz F, and Boccara AC (2003): Private communication. (See Section 13.5, above.)

Li J and Wang LHV (2004): Ultrasound-modulated optical computed tomography of biological tissues, *Appl. Phys. Lett.* **84**(9): 1597–1599. (See Section 13.6, above.)

Sakadzic S and Wang LHV (2004): High-resolution ultrasound-modulated optical tomography in biological tissues, *Opt. Lett.* **29**(23): 2770–2772. (See Section 13.7, above.)

Wang LHV, Jacques SL, and Zhao XM (1995): Continuous-wave ultrasonic modulation of scattered laser-light to image objects in turbid media, *Opt. Lett.* **20**(6): 629–631. (See Section 13.1, above.)

Wang LHV (2001): Mechanisms of ultrasonic modulation of multiply scattered coherent light: An analytic model, *Phys. Rev. Lett.* **8704**(4). (See Section 13.2, above.)

Wang LHV (2003): Ultrasound-mediated biophotonic imaging: A review of acousto-optical tomography and photo-acoustic tomography, *Disease Markers* **19**(2–3): 123–138. (See Section 13.2, above.)

Wang LHV and Ku G (1998): Frequency-swept ultrasound-modulated optical tomography of scattering media, *Opt. Lett.* **23**(12): 975–977. (See Section 13.3, above.)

Yao G, Jiao S, and Wang LHV (2000): Frequency-swept ultrasound-modulated optical tomography in biological tissue by use of parallel detection, *Opt. Lett.* **25**(10): 734–736. (See Section 13.4, above.)

FURTHER READING

Abbott JG (1999): Rationale and derivation of MI and TI—a review, *Ultrasound Med. Biol.* **25**(3): 431–441.

Atlan M, Forget BC, Ramaz F, Boccara AC, and Gross M (2005): Pulsed acousto-optic imaging in dynamic scattering media with heterodyne parallel speckle detection, *Opt. Lett.* **30**(11): 1360–1362.

Blonigen FJ, Nieva A, DiMarzio CA, Manneville S, Sui L, Maguluri G, Murray TW, and Roy RA (2005): Computations of the acoustically induced phase shifts of optical paths

in acoustophotonic imaging with photorefractive-based detection, *Appl. Opt.* **44**(18): 3735–3746.
Bossy E, Sui L, Murray TW, and Roy RA (2005): Fusion of conventional ultrasound imaging and acousto-optic sensing by use of a standard pulsed-ultrasound scanner, *Opt. Lett.* **30**(7): 744–746.
Granot E, Lev A, Kotler Z, Sfez BG, and Taitelbaum H (2001): Detection of inhomogeneities with ultrasound tagging of light, *J. Opt. Soc. Am. A* **18**(8): 1962–1967.
Gross M, Goy P, and Al-Koussa M (2003): Shot-noise detection of ultrasound-tagged photons in ultrasound-modulated optical imaging, *Opt. Lett.* **28**(24): 2482–2484.
Gross M, Ramaz F, Forget BC, Atlan M, Boccara C, Delaye P, and Roosen G (2005): Theoretical description of the photorefractive detection of the ultrasound modulated photons in scattering media, *Opt. Express* **13**(18): 7097–7112.
Hisaka M, Sugiura T, and Kawata S (2001): Optical cross-sectional imaging with pulse ultrasound wave assistance, *J. Opt. Soc. Am. A* **18**(7): 1531–1534.
Hisaka M (2005): Ultrasound-modulated optical speckle measurement for scattering medium in a coaxial transmission system, *Appl. Phys. Lett.* **87**(6): 063504.
Kempe M, Larionov M, Zaslavsky D, and Genack AZ (1997): Acousto-optic tomography with multiply scattered light, *J. Opt. Soc. Am. A* **14**(5): 1151–1158.
Leutz W and Maret G (1995): Ultrasonic modulation of multiply scattered-light, *Physica B* **204**(1–4): 14–19.
Lev A, Kotler Z, and Sfez BG (2000): Ultrasound tagged light imaging in turbid media in a reflectance geometry, *Opt. Lett.* **25**(6): 378–380.
Lev A and Sfez BG (2002): Direct, noninvasive detection of photon density in turbid media, *Opt. Lett.* **27**(7): 473–475.
Lev A and Sfez BG (2003): Pulsed ultrasound-modulated light tomography, *Opt. Lett.* **28**(17): 1549–1551.
Lev A, Rubanov E, Sfez B, Shany S, and Foldes AJ (2005): Ultrasound-modulated light tomography assessment of osteoporosis, *Opt. Lett.* **30**(13): 1692–1694.
Leveque S, Boccara AC, Lebec M, and Saint-Jalmes H (1999): Ultrasonic tagging of photon paths in scattering media: Parallel speckle modulation processing, *Opt. Lett.* **24**(3): 181–183.
Leveque-Fort S (2001): Three-dimensional acousto-optic imaging in biological tissues with parallel signal processing, *Appl. Opt.* **40**(7): 1029–1036.
Li H and Wang LHV (2002): Autocorrelation of scattered laser light for ultrasound-modulated optical tomography in dense turbid media, *Appl. Opt.* **41**(22): 4739–4742.
Li J and Wang LHV (2002): Methods for parallel-detection-based ultrasound-modulated optical tomography, *Appl. Opt.* **41**(10): 2079–2084.
Li J, Ku G, and Wang LHV (2002): Ultrasound-modulated optical tomography of biological tissue by use of contrast of laser speckles, *Appl. Opt.* **41**(28): 6030–6035.
Li J, Sakadzic S, Ku G, and Wang LHV (2003): Transmission- and side-detection configurations in ultrasound-modulated optical tomography of thick biological tissues, *Appl. Opt.* **42**(19): 4088–4094.
Li J and Wang LHV (2004): Ultrasound-modulated optical computed tomography of biological tissues, *Appl. Phys. Lett.* **84**(9): 1597–1599.
Mahan GD, Engler WE, Tiemann JJ, and Uzgiris E (1998): Ultrasonic tagging of light: Theory, *Proce. Natl. Acad. Sci. USA* **95**(24): 14015–14019.
Marks FA, Tomlinson HW, and Brooksby GW (1993): A comprehensive approach to breast cancer detection using light: Photon localization by ultrasound modulation and

tissue characterization by spectral discrimination. *Proc. Soc. Photo-opt. Instrum. Eng.* **1888**: 500–510.
Murray TW, Sui L, Maguluri G, Roy RA, Nieva A, Blonigen F, and DiMarzio CA (2004): Detection of ultrasound-modulated photons in diffuse media using the photorefractive effect, *Opt. Lett.* **29**(21): 2509–2511.
Sakadzic S and Wang LHV (2002): Ultrasonic modulation of multiply scattered coherent light: An analytical model for anisotropically scattering media, *Phys. Rev. E* **66**(2): 026603.
Sakadzic S and Wang LHV (2004): High-resolution ultrasound-modulated optical tomography in biological tissues, *Opt. Lett.* **29**(23): 2770–2772.
Schenk JO and Brezinski ME (2002): Ultrasound induced improvement in optical coherence tomography (OCT) resolution, *Proc. Natl. Acad. USA* **99**(15): 9761–9764.
Selb J, Pottier L, and Boccara AC (2002): Nonlinear effects in acousto-optic imaging, *Opt. Lett.* **27**(11): 918–920.
Solov'ev AP, Sinichkin YP, and Zyuryukina OV (2002): Acoustooptic visualization of scattering media, *Opt. Spectrosc.* **92**(2): 214–220.
Sui L, Roy RA, DiMarzio CA, and Murray TW (2005): Imaging in diffuse media with pulsed-ultrasound-modulated light and the photorefractive effect, *Appl. Opt.* **44**(19): 4041–4048.
Wang LHV and Zhao XM (1997): Ultrasound-modulated optical tomography of absorbing objects buried in dense tissue-simulating turbid media, *Appl. Opt.* **36**(28): 7277–7282.
Wang LHV (1998): Ultrasonic modulation of scattered light in turbid media and a potential novel tomography in biomedicine, *Photochem. Photobiol.* **67**(1): 41–49.
Wang LHV and Ku G (1998): Frequency-swept ultrasound-modulated optical tomography of scattering media, *Opt. Lett.* **23**(12): 975–977.
Wang LHV (2001): Mechanisms of ultrasonic modulation of multiply scattered coherent light: A Monte Carlo model, *Opt. Lett.* **26**(15): 1191–1193.
Wang LHV (2003): Ultrasound-mediated biophotonic imaging: A review of acousto-optical tomography and photo-acoustic tomography, *Disease Markers* **19**(2–3): 123–138.
Yao G and Wang LHV (2000): Theoretical and experimental studies of ultrasound-modulated optical tomography in biological tissue, *Appl. Opt.* **39**(4): 659–664.
Yao G, Jiao SL, and Wang LHV (2000): Frequency-swept ultrasound-modulated optical tomography in biological tissue by use of parallel detection, *Opt. Lett.* **25**(10): 734–736.
Yao G and Wang LHV (2004): Signal dependence and noise source in ultrasound-modulated optical tomography, *Appl. Opt.* **43**(6): 1320–1326.
Yao Y, Xing D, and He YH (2001): AM ultrasound-modulated optical tomography with real-time FFT, *Chinese Sci. Bulle.* **46**(22): 1869–1872.
Yao Y, Xing D, He YH, and Ueda K (2001): Acousto-optic tomography using amplitude-modulated focused ultrasound and a near-IR laser, *Quantum Electron.* **31**(11): 1023–1026.
Zhu Q, Durduran T, Ntziachristos V, Holboke M, and Yodh AG (1999): Imager that combines near-infrared diffusive light and ultrasound, *Opt. Lett.* **24**(15): 1050–1052.

APPENDIX A

Definitions of Optical Properties

TABLE A.1. Basic Optical Properties

Parameter	Definition	Symbol	Unit	Typical Value
Absorption coefficient	Probability of photon absorption by a medium per unit (infinitesimal) path length	μ_a	cm^{-1}	0.1
Scattering coefficient	Probability of photon scattering by a medium per unit (infinitesimal) path length	μ_s	cm^{-1}	100
Anisotropy	Average of cosine of scattering polar angle by single scattering	g	—	0.9
Index of refraction	Ratio of speed of light in vacuum to phase velocity in medium; only the real part is considered here	n	—	1.38

TABLE A.2. Derived Optical Properties

Parameter	Definition	Symbol	Unit	Typical Value
Albedo	$a = \mu_s/(\mu_a + \mu_s)$	a	—	0.999
Diffusion coefficient	$D = 1/[3(\mu_a + \mu_s')]$; the coefficient linking the current to the gradient of fluence in Fick's law	D	cm	0.033

Biomedical Optics: Principles and Imaging, by Lihong V. Wang and Hsin-i Wu
Copyright © 2007 John Wiley & Sons, Inc.

TABLE A.2. (*Continued*)

Parameter	Definition	Symbol	Unit	Typical Value
Effective attenuation coefficient	$\mu_{\text{eff}} = \sqrt{\mu_a/D}$; exponential decay rate of fluence far from the source	μ_{eff}	cm^{-1}	1.74
Extinction coefficient	$\mu_t = \mu_a + \mu_s$; probability of photon interaction with a medium per unit (infinitesimal) path length, where interaction includes both absorption and scattering (also referred to as *total interaction coefficient*)	μ_t	cm^{-1}	100.1
Mean free path	$l_t = 1/\mu_t$; mean free path length between interactions	l_t	cm	0.01
Penetration depth	$\delta = 1/\mu_{\text{eff}}$; exponential decay constant of fluence far from the source	δ	cm	0.575
Reduced scattering coefficient	$\mu'_s = \mu_s(1-g)$; probability of equivalent isotropic photon scattering by a medium per unit (infinitesimal) path length in diffusive regime (also referred to as *transport scattering coefficient*)	μ'_s	cm^{-1}	10
Transport albedo	$a' = \mu'_s/(\mu_a + \mu'_s)$	a'	—	0.990
Transport interaction coefficient	$\mu'_t = \mu_a + \mu'_s$	μ'_t	cm^{-1}	10.1
Transport mean free path	$l'_t = 1/\mu'_t$	l'_t	cm	0.099

APPENDIX B

List of Acronyms

1D	one dimension, or one-dimensional
2D	two dimensions, or two-dimensional
3D	three dimensions, or three-dimensional
AC	alternating current
APD	avalanche photodiode
ART	algebraic reconstruction technique
CCD	charge-coupled device
CDF	cumulative distribution function
CFPI	confocal Fabry–Perot interferometer
CNR	contrast-to-noise ratio
CT	computed tomography
CW	continuous wave
dB	decibel
DC	direct current
DOP	degree of polarization
DOT	diffuse optical tomography
DRR	depth-to-resolution ratio
EEM	excitation-emission matrix
ESF	edge spread function
FFT	Fast Fourier transform
FOV	field of view
FWHM	full width at half maximum
GVD	group velocity dispersion
IDM	inverse distribution method
IFT	inverse Fourier transform
IR	infrared
LSF	line spread function
MCML	Monte Carlo modeling of light transport in multilayered scattering media

Biomedical Optics: Principles and Imaging, by Lihong V. Wang and Hsin-i Wu
Copyright © 2007 John Wiley & Sons, Inc.

MCP	microchannel plate
MRI	magnetic resonance imaging
MTF	modulation transfer function
NA	numerical aperture
NIR	near-infrared
OCT	optical coherence tomography
OD	optical density
PAM	photoacoustic microscopy
PAT	photoacoustic tomography
PDF	probability density function
PDI	polarization-difference imaging
PMT	photomultiplier tube
PSF	point spread function
PZT	lead (Pb) zirconate titanate (piezoelectric transducer)
rms	root-mean-squared
RTE	radiative transfer equation
SIRT	simultaneous iterative reconstruction technique
SLD	superluminescent diode
SNR	signal-to-noise ratio
STF	system transfer function
SVD	singular-value decomposition
TPM	two-photon microscopy
UOT	ultrasound-modulated optical tomography
US	ultrasonography
UV	ultraviolet

INDEX

Absorbance, 139, 164
Absorbers, primary, 6, 7
Absorption
  coefficient. *See* Absorption coefficient
  cross section, 5
  cross section (two photon), 171
  efficiency, 5
  illustration, 4
  origins, 5
  spectrum, 1
  spectra of primary absorbers (plot), 7
Absorption coefficient
  conversion from fluence to specific absorption, 85
  conversion from specific absorption to fluence, 54
  conversion to pressure, 286
  definition, 5, 343
  hemoglobin, 6
  map, 267
  sensing, 140, 145, 146
  spectra of primary absorbers (plot), 7
Acceptance angle
  antenna theorem, 161, 162
  collimated transmission method, 136, 137
  OCT, 212
  solid, 161, 335
Acoustic focusing, 307
Acoustic lens, 304, 307, 319
Acoustic pressure, 285
Acoustic transit time, 303
Acoustic wave equation, 313, 316
Acoustooptic modulator, 161
Activatable retarder, 155
Adjoint method, 274
Agar gel, 135, 305
Airy disc, 176
Albedo, 88, 123
  definition, 343

Algebraic reconstruction technique (ART), 265, 275
  illustration, 276
A-line. *See also* A-scan
  definition, 186
  Fourier-domain OCT, 198, 199, 201, 202
  time-domain OCT, 187
Amino acid, 9
Amplitude reflectivity, 167, 187, 199
  density, 199
Analyzer, polarization. *See* Polarization analyzer
Angiogenesis, 1, 7
Angle-biased sampling, 211, 212
Angular wavenumber. *See* Propagation constant
Anisotropy
  definition, 47, 343
  formula, 87
  Mie theory, 20, 33
  plot, 24
  similarity relation, 111
  structural, 38
ANSI Standard C, 40, 67
Antenna theorem, 161, 162
  illustration, 161
Anti-Stokes transition, 5
APD. *See* Avalanche photodiode
Arm-length difference, 183, 186, 206
Arm-length mismatch. *See* Arm-length difference
ART. *See* Algebraic reconstruction technique
A-scan. *See also* A-line
  definition, 186
  PAM, 305, 306
  time-domain OCT, 186
Attenuation coefficient, ultrasonic, 339
Autocorrelation function, 184, 201, 324
Auxiliary angle, 221
Avalanche photodiode (APD), 252, 336

Biomedical Optics: Principles and Imaging, by Lihong V. Wang and Hsin-i Wu
Copyright © 2007 John Wiley & Sons, Inc.

Axial resolution
  OCT, 181, 186, 189, 193, 207
  PAM, 305, 307
  UOT, 326, 329, 332, 337
Azimuthal angle (illustration), 18

Backprojection, 311–334
Ballistic imaging, 2, 153
Ballistic light. *See* Ballistic photon
Ballistic photon, 153, 161, 334
  arrival time, 154, 158
  polarization, 157
  spatial frequency, 156, 160
Ballistic regime, 115
  definition, 114
  OCT, 186
Ballistic transmittance, 8, 135. *See also*
  Unscattered transmittance
Bandwidth
  axial resolution, 193
  chirp, 327
  coherence length, 184
  coherence-gated holographic
    imaging, 160
  ideal, 312
  interference signal, 206
  SLD, 209, 237, 238
  spatial frequency, 161
  ultrasound, 284, 304, 335
Bar chart, 305
Basis set, 88
Beam splitter
  confocal microscope, 165
  DOT, 252
  Michelson interferometer, 181
  nonpolarizing, 237–239
  OCT, 182
  optical heterodyne imaging, 161
  polarizing, 238, 239
  UOT, 336
Beamforming, 309
Beat frequency, 160, 161, 206
Beer's law
  absorption, 5
  ballistic imaging, 153
  depth, 99
  generalized, 164
  primary beam, 147
  probability, 45
  scattering, 8
  thickness, 135, 136
  time, 98, 115
Bessel equation, transformed, 100

Bessel function, 21, 166
  modified, 70, 74
  spherical, 21, 28
Binary tree, 74
Biochemical information, 1, 146
Bioluminescence, 2
Biomarker, 2
Bipolar pressure profile, 299
Birefringence, 154, 158, 242. *See also*
  Polarization
  circular, 9
  dextrorotatory, 9
  levorotatory, 9
  linear, 9
  negative, 9
  positive, 9
Blood flow, 2, 206
Blue sky, 18
Body temperature, 285, 286
Boltzmann equation. *See* Radiative transfer
  equation
Born approximation, 147, 262, 271
Boundary
  refractive-index-matched. *See*
    Refractive-index-matched boundary
  refractive-index-mismatched. *See*
    Refractive-index-mismatched boundary
Boundary condition
  Cauchy, 102
  Dirichlet, 103, 265, 266
  DOT, 273
  extrapolated (illustration), 102
  Green's function approach, 317
  homogeneous, 293
  Mie theory, 26, 32, 33
  RTE, 97, 101
  semiinfinite medium, 106, 107
  slab, 120
Brain, 249, 312, 313
Breast, 249
Broad beam, 67, 156, 199, 326
Brownian motion, 324
B-scan, 186, 305, 306

Carbon fiber, 305
Carrier, 189
Carrier frequency, 206, 239
Cauchy boundary condition, 102
Cauchy's contour integration, 290
Causality, 98, 117, 291, 293, 317
CCD
  holography, 158
  reflectometry, 140, 145
  speckle imaging, 330–332, 334

INDEX 349

CDF. *See* Cumulative distribution function
Cell nuclei, 1, 8
Center frequency
   OCT, 187, 197
   PAM, 304
   PAT, 284, 312
   UOT, 330, 332, 335, 336
Central limit theorem, 55
Chirping, 208, 209, 326, 329, 332
Circular polarization, 9, 222, 224, 229
   Mueller matrix, 227
Circular polarizer, 231, 232, 234
CNR. *See* Contrast-to-noise ratio
Coefficient
   absorption. *See* Absorption coefficient
   extinction. *See* Extinction coefficient
   molar extinction. *See* Molar extinction coefficient
   scattering. *See* Scattering coefficient
   total interaction. *See* Extinction coefficient
Coherence gating, 153, 158, 160, 186
Coherence length, 184–186, 190, 330
Coherence time, 184
Coherence-gated holographic imaging, 153, 158
   illustration, 159
Collagen, 1, 9, 10, 38, 242
Collimated transmission method, 135, 139
   illustration, 136
Comparison of imaging modalities, 2, 284
Compensator. *See* Retarder
Complex conjugation, 89, 311
Complex expression. *See* Phasor representation
Compressibility, 285
Computed tomography (CT), 1, 163, 332
Concentration of hemoglobin. *See* Hemoglobin concentration
Condensation parameter, 314
Condenser lens, 164, 304
Confocal microscopy
   fluorescence (illustration), 166
   optical, 154, 164
   photoacoustic. *See* Photoacoustic microscopy
   system (illustration), 165
Conical lens, 304
Conjugate-gradient method, 275
Conservation of energy, 85, 88, 223
Conservation of mass, 314
Constant fraction discriminator, 252
Continuity equation, 314, 315
Contrast
   comparison, 2
   definition, 12
   DOT, 269

functional imaging, 1
illustration, 13
molecular imaging, 2
   OCT, 181, 219
   PAT, 313
   UOT, 329, 338
Contrast-to-noise ratio (CNR), 13
CONV program, 67, 77
Converging spherical wave, 299, 300, 311
Conversion
   between optical wavelength and photon energy, 14
   from pencil beam to isotropic source, 106
   from temperature to pressure, 286
Convolution
   arbitrary source, 98
   broad beam, 67
   coherent, 168
   CONV program, 77
   DOT, 251, 263
   fluorescence source, 147
   Fourier-domain OCT, 201
   Gaussian beam, 69
   incoherent, 167–170
   infinitely wide beam, 54
   LSF, 14
   numerical example, 77
   numerical solution, 72
   object function, 11
   PAT, 302
   top-hat beam, 71
   truncation error, 76
Correlation, 324, 325
Critical angle, 49, 104
Critical depth, 122, 128, 129
Cross section
   absorption, 5
   absorption (two photon), 171
   scattering. *See* Scattering cross section
Cross-correlation theorem, 184
Cross-interference term, 200
C-scan, 190
CT. *See* Computed tomography
Cumulative distribution function (CDF), 41, 42, 45, 60
Current density
   conservation of energy, 117
   definition, 84
   direction, 85, 97
   effect on radiance (illustration), 91
   fractional change, 96, 97, 117
   projection, 108
Cytoplasm, 8

Dark-field confocal photoacoustic microscopy (PAM), 303
Dark-field illumination, 307
Data-acquisition board, 239
dB. *See* Decibel
Decibel (dB), 139, 209
Deconvolution, 202, 205
Defocus distance, 166
Degeneracy, 97
Degree of circular polarization (DOCP), 223
Degree of linear polarization (DOLP), 223
Degree of polarization (DOP), 219, 223, 224, 236, 237
Delay-and-sum detection, 309
Delta heating, 289, 290, 295, 299
　slab, 293
　sphere, 297
Demodulation, 191, 193, 196
Depolarizing medium, 223
Depth of focus. *See* Focal zone
Depth-priority scanning, 190
Depth-to-resolution ratio (DRR), 13, 181
Dermis, 305
Determinant, 241
Diattenuator, 225
Dichroic mirror, 165
Dichroism, 225
Differential path length, 277
Diffraction limit, 156, 165, 170
Diffraction theory, 165
Diffuse optical tomography (DOT), 2, 249, 283
　DC, 250, 252
　DC (illustration), 253
　frequency domain, 250, 253
　frequency domain (illustration), 253, 255
　reconstructed image (plot), 268
　time domain, 250, 251
　time domain (illustration), 251
Diffuse reemittance
　relative, 39
Diffuse reflectance
　angularly resolved, 56
　angularly resolved (plot), 57
　approximations (illustration), 107
　diffusion step, 123, 124
　diffusion theory, 106
　diffusion theory (plot), 109
　experimental data (plot), 142
　far, 140, 143, 144
　hybrid, 124
　hybrid (plot), 130, 131
　image source (plot), 112
　isotropic source in slab (plot), 126, 128
　Monte Carlo data (plot), 142

　Monte Carlo step, 123, 124
　oblique incidence, 143
　optical fibers, 144
　pencil beam and isotropic source (plot), 111
　pencil beam on slab (plot), 127
　projection of current density, 108
　relative, 39, 106
　representation, 51
　similarity relation (plot), 110, 114
　slab, 121, 125
　source depth (plot), 113
　time-resolved, 145
　total, 52, 55, 145
　weight recording, 49
Diffuse transmittance
　angularly resolved, 56
　angularly resolved (plot), 57
　isotropic source in slab (plot), 126
　Monte Carlo step, 123
　pencil beam on slab (plot), 127
　relative, 39
　representation, 51
　slab, 121, 125
　total, 52, 55
　weight recording, 49
Diffusion approximation
　boundary condition, 102
　directional and temporal broadening, 97
　expansion of radiance, 88, 105
　high albedo, 88
　P_1 approximation, 89
　similarity relation, 97
Diffusion coefficient, 120, 145, 256
　definition, 97, 343
　oblique incidence, 143
Diffusion equation
　approximated RTE, 83
　background, 262, 269
　derivation, 96
　DOT, 249, 267
　excitation, 147
　expression, 97
　fluorescence, 147
　impulse response, 98
　linearity, 257
　numerical methods, 273
　photon density, 256
Diffusion expansion of radiance, 88, 105
Diffusion theory
　accuracy, 122
　accuracy and speed, 106
　breakdown, 99
　derivation, 88
　diffuse reflectance (plot), 109

fluence, 56
fluorescence, 147
 hybrid, 119
 oblique incidence, 143
 validation, 110
 zero-boundary condition, 145, 324
Diffusive regime
 boundary condition, 106
 definition, 114, 115
 DOT, 249
 effective reflection coefficient, 105
 OCT, 186
 PAT, 283
 penetration depth, 56
 UOT, 323, 330, 333
Diffusivity
 optical, 256
 thermal, 284
Digital holography, 158, 175
Dimensionless step size, 40, 43, 45, 46, 48
Dipole moment, 23
Dipole radiation, 24, 311
Dirac delta function, 2, 11, 38, 201
Direct method, 274
Direction cosines, 39, 43, 47, 49
Dirichlet boundary condition, 103, 265, 266
Dispersion compensation, 209
Divergence, 86, 154, 288
Diverging spherical wave, 29, 288, 299, 310
DOCP. *See* Degree of circular polarization
DOLP. *See* Degree of linear polarization
DOP. *See* Degree of polarization
Doppler
 effect, 2, 206
 frequency. *See* Doppler shift
 OCT, 206
 shift, 186, 206, 239
DOT. *See* Diffuse optical tomography
Dot product, 89
Double refraction, 9
Dynamic focusing, 190
Dynamic range, 251, 252

Early-photon imaging, 154
Edge spread function (ESF), 11
 illustration, 11
EEM. *See* Excitation-emission matrix
Effective attenuation coefficient, 98, 120, 143
 definition, 344
Effective path length, 258
Effective reflection coefficient, 104, 105, 120
Eigenequation, 231
Eigenpolarization, 225, 226, 231, 242
Eigenvalue, 231, 232

Eigenvector, 231–233
Elastic scattering, 5, 85, 165
Elastooptical coefficient, 325
Elliptical polarization, 220, 222
 illustration, 221
Ellipticity angle, 220
Empirical formula
 center shift of diffuse reflectance, 143
 effective reflection coefficient, 120
 Grueneisen parameter, 285
Encoding ambiguity, 200
Energy density, 85, 286
Energy flow (illustration), 84
Ensemble averaging, 37, 210, 211, 236, 324
Envelope of interference fringes, 186
 axial resolution, 189
 demodulation, 191
 expression, 189
 GVD, 208, 209
 number of periods, 193, 194
Equation
 acoustic wave. *See* Acoustic wave equation
 Bessel (transformed). *See* Bessel equation, transformed
 Boltzmann. *See* Radiative transfer equation
 continuity. *See* Continuity equation
 diffusion. *See* Diffusion equation
 eigen. *See* Eigenequation
 force. *See* Force equation
 Helmholtz. *See* Helmholtz equation
 inviscid force. *See* Inviscid force equation
 Maxwell. *See* Maxwell equations
 motion. *See* Equation of motion
 photoacoustic. *See* Photoacoustic equation
 radiative transfer. *See* Radiative transfer equation
 telegraphy. *See* Telegraphy equation
 thermal. *See* Thermal equation
 thermal expansion. *See* Thermal expansion equation
Equation of motion, 288
ESF. *See* Edge spread function
Etendue, 335, 336
Excitation
 definition, 3
 illustration, 4
 nonlinear optical, 169
 one-photon, 170
 one-photon (illustration), 170
 time, 4
 two-photon. *See* Two-photon excitation
 two-photon (illustration), 170
Excitation-emission matrix (EEM), 146

Expansion of radiance, diffusion. *See* Diffusion expansion of radiance
Extended trapezoidal rule, 72
　integrand evaluation (illustration), 73
Extinction coefficient
　ballistic imaging, 153
　collimated transmission method, 135
　definition, 6, 344
　formula, 8, 44
　OCT, 212
Extracellular fluid, 8
Extraordinary ray, 9
Extrapolated boundary
　DOT, 265
　illustration, 102
　image point, 106
　oblique incidence, 143
　refractive-index-matched, 102
　refractive-index-mismatched, 105
　slab, 120
Extrapolation (illustration), 73

Fabry–Perot interferometer, 326, 335
Far field, 24, 259, 307
Fast axis, 155, 226, 231, 238
Fast Fourier transformation (FFT), 202
FFT. *See* Fast Fourier transformation
Fick's law, 102, 106, 343
　diffuse reflectance, 108
　formula, 97
Field of view (FOV), 13, 304
Filter wheel, 256
Finite-difference method, 273
Finite-element method, 273
First photon-tissue interaction, 76
First-order diffraction term, 160
Flowchart for tracking photons, 40
Fluence, 53
　conversion to pressure, 286
　definition, 84
　depth resolved, 56
　distribution (plot), 58
　plot, 78, 79
　relative, 39
Fluence rate
　boundary value, 145
　conversion to pressure, 287
　definition, 84
　diffusion theory, 119
　primary beam, 147
Fluorescence
　characteristics, 9
　confocal imaging (illustration), 166
　definition, 3

emission spectrum, 146
excitation spectrum, 146
illustration, 4
incoherent, 9
lifetime, 3, 4, 10, 146, 147
modeling, 147
origins, 9
quantum yield. *See* Quantum yield
red shift, 9
spectroscopy, 146
spectrum, 1, 9
Stokes shift, 9
time scales, 4
Fluorophore, endogenous, 10
Flux, 56
　energy, 85
　photon, 171
Focal plane, 156, 157, 306
Focal zone, 190, 326, 332
Focused ultrasonic transducer, 304, 330, 332, 335
Force equation, 288, 314
Forcing function, 263, 266
Forward problem, 249, 272–274
　perturbation, 262–264
Fourier optics, 156
Fourier space-gated imaging, 156
Fourier transformation
　autocorrelation function, 325
　differential equation, 99, 100, 290
　Doppler OCT, 207
　Fourier-domain OCT, 201
　PAT, 310
　spatial, 12, 156, 160
　temporal, 184, 191, 250
　UOT, 328
Fourier transformer
　inverse, 195
　spatial, 157, 195
　temporal, 195
Fourier-domain OCT, 198
　signal processing (illustration), 204, 205
　system (illustration), 198
Fourier-domain optical delay line
　illustration, 197
FOV. *See* Field of view
Frame rate, 13, 199
Frequency sweeping. *See* Chirping
Frequency-division multiplexing, 252
Frequency-swept gating, 160, 202, 326, 329
Fresnel reflection, 43, 103, 105, 122
Full-field image, 164
Functional imaging, 1, 7, 249
Fused-silica fiber, 209

Gabor holography, 173, 174
 illustration, 173
Gaussian beam, 69, 190
 fluence (plot), 79
Gaussian envelope, 189, 208
Gaussian line shape, 184, 189
Gaussian quadratures, 125
Gene expression, 2
Glucose, 9
Grating, 140, 145, 195, 307
Grating lobe, 309
Grating-lens pair, 195
Green's function, 316. *See also* Impulse response
 DOT, 263, 265, 269, 270, 274
 PAT, 288, 289, 291, 293, 310, 311
 pencil beam, 38, 67, 68
 point source, 98, 99
 PSF, 10
 UOT, 333
Green's function approach
 DOT, 263, 265, 269, 270
 illustration, 263
 PAT, 288, 289
 summary, 316
Green's second identity, 270
Green's theorem
 diffusion theory, 98
 DOT, 263
 PAT, 292, 310, 317
Group delay, 188, 195, 197, 198
Group velocity, 188
Group velocity dispersion, 207
Group-path-length mismatch, 198
Grueneisen parameter, 285

Hankel function, 21, 29, 33
Hankel transform, 166
Heated slab
 illustration, 293
 snapshots of pressure (plot), 298
Heated sphere
 illustration, 298
 pressure versus time (plot), 299
 snapshots of pressure (plot), 302
Heating function, 287, 289
Heaviside function, 295, 299
Helmholtz equation, 257, 270
 scalar, 27, 29
 vector, 30

Hemoglobin
 concentration, 1, 6, 7, 145
 deoxygenated, 6, 283
 oxygen saturation, 1, 6, 7
 oxygenated, 6, 283
 PAT, 313
 primary absorber, 6
 spectrum of molar extinction coefficient (plot), 6
 two forms, 6
Henyey–Greenstein phase function
 hybrid model, 119, 123
 Monte Carlo method, 46, 47
 OCT, 211
 sensing, 137
Heterodyne detection, 160, 161, 186, 254, 327
Heterodyne frequency, 161, 329, 330
Heterodyne imaging, optical. *See* Optical heterodyne imaging
High resolution, 2, 13, 164, 181, 284
Highpass filter, 191
High-speed shutter, 154
Histogram, 252
Hologram, 158, 159, 171–173
 reconstruction (illustration), 172, 173, 175
 recording (illustration), 172, 173
Holographic imaging, 153, 158
Holography, 158, 171
Homogeneous boundary condition, 293
Hooke's law, 287
Hybrid model, 119, 122
 illustration, 123
Hypermetabolism, 1, 7

IDM. *See* Inverse distribution method
Image function, 11
Image plane, 164
Image reconstruction
 CT, 163
 DOT, 249, 252, 261, 263, 267
 PAT, 309
 UOT, 332
Image source, 106, 108, 120, 143
 illustration, 107
 slab (illustration), 121
Imaging modalities, comparison of, 2, 284
Implicit function, 269
Implicit photon capture, 42
Impulse function. *See* Dirac delta function
Impulse heating. *See* Delta heating
Impulse response, 68, 76, 251, 256 *See also* Green's function
 pencil beam, 38, 67

Impulse response, (*Continued*)
  point source, 98
  PSF, 10
Index of refraction. *See* Refractive index
  definition, 343
Inelastic scattering, 4
Infinitely narrow photon beam. *See* Pencil beam
Initial condition, 317
Initial photoacoustic pressure, 284, 309
  slab, 293
  sphere, 297
Inner product, 89
Instantaneous frequency, 326
Intensity, 84
Interference fringes
  antenna theorem, 161
  Doppler OCT, 206, 207
  Michelson interferometry, 183
  time-domain OCT, 186, 194
Interferogram, 171, 183. *See also* Interference fringes
  spectral, 198–200, 202, 204, 205
Interferometer, Fabry–Perot. *See* Fabry–Perot interferometer
Interferometry, Michelson. *See* Michelson interferometry
Internal conversion, 4
Interpolation (illustration), 73
Intralipid® solution, 135, 305
Inverse distribution method (IDM)
  definition, 41
  illustration, 41
  proof, 42
  scattering angles, 46–48
  step size, 44, 45
Inverse Fourier transformation
  differential equation, 100, 290
  Fourier-domain OCT, 198, 200, 201, 204, 205
  PAT, 311
  spatial, 160
  temporal, 186, 189, 195, 250
Inverse method, 274
Inverse problem, 249, 262, 263, 272
Inverse Radon transformation, 163
Inviscid force equation, 288, 314
IQ detection, 254
Isosbestic point, 6
Isothermal compressibility, 285
Isotropic scattering
  definition, 47
  primary beam, 147
  similarity relation, 123
  UOT, 324
Isotropic source, 92
  array, 121
  normal and oblique incidence (illustration), 141
  plane, 99
  point, 108
  slab (plot), 126, 128
  volume, 122
Iterative method, 265, 272

Jablonski energy diagram, 4
  illustration, 4, 170
Jacobian matrix, 264, 274
Jones calculus, 219, 235
Jones matrix, 219, 240
  conversion to Mueller matrix, 235
  definition, 230
  OCT, 237
  requisite independent measurements, 240
  rotator, polarizer, and retarder, 230
Jones reversibility theorem, 240, 241
Jones vector, 219, 229, 238, 239
  conversion to Stokes vector, 236
Jones–Mueller transformation, 235

k clock, 202
Keratin, 9
Kerr effect, 154
Kerr gate (illustration), 155
Krönecker delta function, 89
Krönecker tensor product, 235, 236

Lateral resolution. *See* Transverse resolution
Law
  Beer's. *See* Beer's law
  Fick's. *See* Fick's law
  Hooke's. *See* Hooke's law
  Newton's second. *See* Newton's second law
  Snell's. *See* Snell's law
Left circular polarization, 222, 224, 227, 229
Legendre polynomials, 89
  associated, 29, 89
Leith–Upatnieks holography (illustration), 174, 175
Lifetime, fluorescence. *See* Fluorescence lifetime
Line spread function (LSF), 11
  illustration, 11
Linear inverse algorithm, 261, 266
Linear polarization
  decomposition, 49
  definition, 221

Jones vector, 229
Mueller matrix, 227
Poincaré sphere, 224
principal axes, 9
Linear polarizer
  eigenpolarization, 225
  Jones matrix, 230, 231
  part of a circular polarizer, 233
  polarization-difference imaging, 157
  time gate, 155
Linearity, 67, 251
Local oscillator, 161, 254
Lock-in, 254, 330
  illustration, 254
LSF. *See* Line spread function

Magnetic resonance imaging (MRI), 1, 2, 329
Markov chain, 50
MATLAB
  conventional and confocal microscopes, 169
  heated slab, 295, 296
  heated sphere, 300
  Mie theory, 22
  null plane, 259
  OCT, Fourier-domain, 202
  OCT, time-domain, 194
  Rayleigh theory, 20
Maximum imaging depth
  ballistic imaging, 153
  comparison, 2
  definition, 13
  DOT, 249
  OCT, 181
  PAM (plot), 305
  PAT, 284, 313
Maximum-amplitude projection
  definition, 306
  plot, 306
Maxwell equations, 17, 26
MCML, 40, 58, 67
MCP. *See* Microchannel plate
Mean absorption length, 2, 5
Mean free path, 8, 115, 154, 324
  definition, 344
  typical value, 2
Mean path length of flight, 277
Mean time of flight, 276
Mechanisms of ultrasonic modulation, 323
Medical imaging modalities, 1
Medium displacement, 288
Medium velocity, 314, 316
Melanin, 6, 8
  primary absorber, 6

Michelson interferometry, 181, 185
  illustration, 182
Microchannel plate (MCP), 155, 252
Mie theory, 7, 17
  derivation, 26
  particle sizing, 145
  phase function, 47
Mitochondria, 8, 10
Mixer, 161, 254, 255
Modal matrix, 232
Modes of DOT, 249
Modulation depth
  DOT, 250, 257, 258
  UOT, 325, 326, 331, 333
Modulation transfer function (MTF), 12, 305
Molar concentration, 6, 7
Molar extinction coefficient, 6, 10
  spectrum (plot), 6
Molecular conformation, 1
Molecular imaging, 2
Monochromator, 140
Monte Carlo method, 37
  accuracy and speed, 106
  benchmark, 110, 138
  broad beam, 67
  equivalence to RTE, 83
  flowchart, 40
  hybrid, 119
  oblique incidence reflectometry, 141
  OCT, 210
  UOT, 324
MRI. *See* Magnetic resonance imaging
MTF. *See* Modulation transfer function
Mueller calculus, 219, 235
Mueller matrix, 219
  conversion from Jones matrix, 235
  definition, 224
  image, 242
  measurement, 227
  OCT, 237
  requisite independent measurements, 240
  rotator, polarizer, and retarder, 225
Mueller OCT, 219
  images of tendon (plot), 242
  system (illustration), 239
Multimode fiber, 186
Muscle fiber, 1, 9, 38
Myelin, 9

NA. *See* Numerical aperture
NAD(P)H, 10
NADH, 10
Nd:YAG, 304, 312
Near field, 259

Nepers, 139
Neutral-density filter, 139, 251, 256
Newton's second law, 288, 314
Nonballistic light. See Nonballistic photon
Nonballistic photon, 115, 153
  arrival time, 154, 158
  polarization, 157
  spatial frequency, 156, 160
Nondepolarizing medium, 223, 230
Nonionizing radiation, 2
Nonlinear inverse algorithm, 267, 272
Nonlinear optical excitation, 169
Nonlinear problem, 272
Nonlinear reconstruction, 272
Nonradiative relaxation, 3
Null line (plot), 261
Null plane, 259, 261
  plot, 261
Numerical aperture (NA)
  definition, 166
  OCT, 190
  PAM, 303, 304
  two-photon microscopy, 171
Nyquist criterion, 194, 200, 254

Object function, 11
Objective lens
  confocal microscopy, 165
  conventional microscopy, 164
  OCT, 190, 237–239
  two-photon microscopy, 171
Oblique-incidence reflectometry, 140
  illustration, 141
  white-light (illustration), 144
OCT. See Optical coherence tomography
Offset-reference holography. See Leith–Upatnieks holography
Operator, 274
  autocorrelation, 201
  linear, 159, 266
  real-part, 159
Optical absorption, 1, 2, 6, 283, 307
Optical attenuator, variable, 252
Optical coherence tomography (OCT), 181, 283
  axial resolution versus bandwidth (plot), 193
  class I signal, 210
  class II signal, 211
  classes of signals (illustration), 210
  degree of polarization, 236
  demodulation (illustration), 192, 196
  Doppler. See Doppler OCT
  Fourier-domain. See Fourier-domain OCT
  Monte Carlo modeling, 210
  Mueller. See Mueller OCT
  polarization-sensitive, 219
  scattering versus depth (plot), 213
  signal versus depth (plot), 212
  system (illustration), 185, 191
  time-domain, 185
Optical coordinates, 166
Optical delay line, 195
Optical density, 139, 256
Optical fiber
  fused-silica, 209
  multimode, 186
  single-mode, 185, 237, 239
Optical heterodyne imaging, 154, 160
  illustration, 162
Optical imaging, motivation for, 1
Optical microscopy, conventional, 164
  axial PSF, 167
  dark-field, 303
  illustration, 165
  lateral PSF, 167
  PSF, 166
  PSF (plot), 168
Optical properties, 2, 85, 259
Optical sectioning, 164, 169
Optically thick medium, 154
Optically thin medium, 154
Optimal coordinates, 51, 53
Ordinary ray, 9
Orientation angle of major axis, 220
Orthogonality, 34, 89, 265. See also Orthonormality
Orthonormality, 89, 230, 232. See also Orthogonality
Outgoing spherical wave. See Diverging spherical wave
Oximetry, 7
Oxygen saturation of hemoglobin. See Hemoglobin oxygen saturation

PAM. See Photoacoustic microscopy
Parallel Mueller OCT, 237
Parallel-speckle detection, 329
Paraxial approximation, 166
PAT. See Photoacoustic tomography
Path-length difference, 183, 186
Path-length distribution, 186
Path-length mismatch. See Path-length difference
PDF. See Probability density function
Pencil beam, 2
  conversion to isotropic source, 106, 107, 112, 122
  conversion to isotropic source (plot), 129
  conversion to photon cloud, 115

cylindrical symmetry, 39
definition, 38
diffuse reflectance, 106
diffuse reflectance (plot), 109, 111
fluence (plot), 78, 79
Green's function for broad beam, 67, 68
hybrid model, 119
normal and oblique incidence, 140
normal and oblique incidence (illustration), 141
oblique-incidence reflectometry, 143
OCT, 211
slab (plot), 127, 130, 131
Penetration depth, 56, 249
definition, 99, 344
scale for grid, 51
water, 6
Perturbation, 262–264, 269, 272
Phase
delay in GVD, 207
delay in OCT, 188, 197
delay in photon-density wave, 258
delay in retarder, 226
lag, 220
lead, 220
shift, 197
shifter. See Retarder
velocity, 188, 343
Phase function
Henyey–Greenstein. See Henyey–Greenstein phase function
Mie theory, 47
Phasor representation, 159
holography, 158, 172
Jones vector, 229
Michelson interferometry, 182
Mie theory, 29
photon density, 256, 257
Phosphor screen, 155
Phosphorescence, 3, 4
illustration, 4
Photoacoustic effect, 283
Photoacoustic equation, 287–289, 309
Photoacoustic microscopy (PAM), 303
image (plot), 306
imaging depth (plot), 305
spatial resolution (plot), 305
system (illustration), 304
Photoacoustic tomography (PAT), 2, 283
brain image (plot), 313
reconstruction (illustration), 310, 312
Photoacoustic wave, 283, 284, 287, 304, 312
Photobleaching, 170

Photocathode, 155, 252
Photocurrent, 163, 182, 183, 209
Photoelectric effect, 155
Photoelectron, 155
Photomultiplier tube (PMT)
frequency-domain DOT, 254, 256
frequency-swept UOT, 327
optical heterodyne imaging, 161
time-domain DOT, 252
Photon cloud, 115
Photon current, 97
Photon density, 256, 273
definition, 85
perturbation, 262
snapshot (plot), 3
wave, 257–259, 267
Photon energy, 4, 10, 171, 183, 256
relation with wavelength, 14
Photon packet
absorption, 46
boundary crossing, 48
launching, 43
moving, 46
representation, 42
scattering, 46
step size, 44
termination, 50
Photon propagation regimes, 114
ballistic. See Ballistic regime
diffusive. See Diffusive regime
quasiballistic. See Quasiballistic regime
quasidiffusive. See Quasidiffusive regime
Physical depth distribution, 186
Physical quantities, 50, 72, 83
interpolation and extrapolation (illustration), 73
Picosecond time analyzer, 252
Piezooptical coefficient, 325
Pinhole
coherence-gated holographic imaging, 158
confocal microscopy, 165
spatiofrequency filtered imaging, 156, 157
two-photon microscopy, 169, 170
Planck constant, 14, 85
PMT. See Photomultiplier tube
Poincaré sphere, 224
Point spread function (PSF)
confocal and conventional microscopy (plot), 168
confocal and two-photon microscopy, 170
confocal microscopy, 165, 167
conversion to LSF, 11
convolution, 11, 168
definition, 10

Point spread function (PSF) (*Continued*)
  lens, 166
  OCT, 189, 207, 208
Poisson distribution, 136
Polar angle
  illustration, 18
  sampling, 46
Polarimetry, 219
Polarizability, 17, 18, 23
Polarization, 157, 219, 236
Polarization analyzer
  measurement of Mueller matrix, 227
  measurement of Stokes vector, 222, 223, 228
  polarization-difference imaging, 157, 158
  polarization-difference imaging (illustration), 157
Polarization ellipse, 220, 222, 224
  illustration, 221
Polarization homogeneous medium, 231
Polarization inhomogeneous medium, 231
Polarization origins, 9
Polarization state, 219
Polarization-difference imaging, 154, 157
  illustration, 157
Polarization-sensitive OCT, 219
Polarizer, 225, 230
Polarizing element, 225, 226, 231
Polystyrene sphere, 135
Positive lens, 304
Power spectral density, 184, 200, 325
Primary absorbers, 6, 7
Primary beam, 147
Primary scatterers, 8
Probability density function (PDF)
  definition, 60
  free path length, 44, 45
  path length, 324
  phase function, 87, 137
  sampling, 41
Projection data, 163, 334
Projection image, 154, 156, 163. *See also* Shadowgram
Propagation constant
  GVD, 207
  Michelson interferometry, 182
  Mie theory, 27
  OCT, 187, 189
  photon density wave, 257, 258
  Rayleigh theory, 17
Pseudorandom number, 41
  azimuthal angle, 47
  polar angle, 47
  Russian roulette, 50

specular reflection, 49
step size, 122
PSF. *See* Point spread function
Pulse oximetry, 7
Pupil, 166

Quantum efficiency of detector, 182
Quantum yield, 146, 147, 171
  definition, 3
  fluorophores, 10
Quasiballistic photon, 153
Quasiballistic regime
  definition, 114, 115
  OCT, 186
  UOT, 329
Quasidiffusive regime, 115
  definition, 114, 115
  OCT, 186
  PAM, 307
  PAT, 283
  UOT, 323

Radiance, 83, 88
  diffusion expansion (illustration), 91
  illustration, 84
Radiative transfer equation (RTE), 83, 88
  derivation (illustration), 86
Radon transformation, 163, 333
  illustration, 164
Raman scattering, 4, 5
  illustration, 4
  spectrum, 1
Ram-Lak filter, 334
Ramp filter, 312
Random walk, 14, 37
Raster scanning, 154, 305
Rayleigh criterion, 259
Rayleigh range, 190
Rayleigh theory, 7, 17, 23
  scattering efficiency (plot), 24
Real image, 160, 173–175
Reciprocity
  acoustic wave, 288, 291, 307, 316
  confocal microscopy, 167
  grating-lens pair, 197
  optical fluence rate, 98
  polarization, 227
Reconstruction of a hologram, 158, 173, 174
Red shift, 9
Reduced interaction coefficient. *See* Transport interaction coefficient
Reduced scattering coefficient
  definition, 94, 344
  fluorescence excitation, 147
  map, 267

Mie theory, 21
oblique-incidence reflectometry, 140
PAM, 305
particle sizing, 145
similarity relation, 123
wavelength dependence, 170
Reemission, 39, 52, 106
imaging configuration, 249
modeling, 50
photon termination, 50
Reemittance, 49, 50
Reference-intensity term, 200
Reflectance, 51
Reflection mode, 158, 165
Reflectometry, 140
Refractive index
glass, 44
tissue, 8, 44
water, 8, 44
Refractive index, relative, 18, 104
Refractive-index-matched boundary, 106
boundary condition, 101
definition, 55
effect on fluence, 56
effect on penetration depth, 57
fluence distribution (plot), 58
photon entry, 122
Refractive-index-mismatched boundary, 55, 119
boundary condition, 103
computation time, 132
effect on fluence, 56
effect on penetration depth, 57
fluence distribution (plot), 58
Regularization, 275
Relative refractive index, 18, 104
Relaxation (illustration), 4
Resonant frequency, 336
Retarder, 226, 231
activatable, 155
half-wave, 154, 155, 226, 237
part of a circular polarizer, 233
quarter-wave, 226, 228, 237–239
variable, 237
Retina, 2, 9, 181, 209
Riccati-Bessel function, 21
Right circular polarization, 222, 224, 227
illustration, 221
Rotator, 225, 230
RTE. *See* Radiative transfer equation
Russian roulette, 43, 50, 123, 132
Rytov approximation, 278

Sample-intensity term, 200
Sampling of a random variable, 41

Scalar wave, 259
Scatterers, primary, 8
Scattering
biological structures (illustration), 8
coefficient, 8, 21, 343
cross section, 8, 18, 25, 87
efficiency, 8, 19, 20, 33
efficiency (plot), 24
mean free path, 8
media, 2
medium, layers (illustration), 38
medium, slab (illustration), 121
optical depth, 136
origins, 7
Secondary electron, 155
Self-interference term, 200
Sensitivity, 199, 209
Sensitivity matrix, 264
Serial Mueller OCT, 237
system (illustration), 238
Shadowgram, 154, 156. *See also* Projection image
Shift invariance. *See* Translation invariance
Shutter speed, 155
Side lobe, 308
Signal-to-noise ratio (SNR), 13, 250
definition, 12
effect on spatial resolution, 259
UOT, 330, 336
Similarity relation, 97, 106, 114, 123, 147
diffuse reflectance (plot), 110
Simultaneous iterative reconstruction technique (SIRT), 265, 275
sinc function, 163
Single-mode fiber, 185, 237, 239
Single-photon counting, 145, 155, 251, 252
Singly backscattered light, 153, 210, 212, 213, 239
Singularity, 29, 290
Singular-value decomposition (SVD), 265
Sinogram. *See* Projection data
SIRT. *See* Simultaneous iterative reconstruction technique
Size parameter, 18, 21
Slab, heated, 293, 302
SLD. *See* Superluminescent diode
Slow axis, 226
Snell's law, 49, 104, 137, 140
photon-density wave, 259
SNR. *See* Signal-to-noise ratio
Soft tissue contrast, 2
Sound velocity. *See* Speed of sound
Source density function, relative, 123
plot, 129

360 INDEX

Spatial filter, 157, 158, 238
Spatial frequency, 12
   ballistic imaging, 156, 161
   grating-lens pair, 196
   holography, 160, 175
   PAM, 305
   spectrum, 156
   UOT, 334
Spatial modulation frequency (plot), 305
Spatial resolution
   ballistic imaging, 153
   comparison, 2
   confocal microscopy, 165
   definition, 11
   DOT, 249, 259
   OCT, 181, 237
   PAM, 307
   PAM (plot), 305
   PAT, 284
   streak camera, 155
   time gate, 156
   tradeoff with FOV, 13
   two-photon microscopy, 169
   UOT, 323, 331
Spatiofrequency filtered imaging, 154, 156
   illustration, 156
Specific absorption, 53, 72, 285
   definition, 85
   relative, 39
Specific absorption rate, 287
   definition, 85
Specific energy deposition. *See* Specific absorption
Specific heat capacity, 285
Specific power deposition. *See* Specific absorption rate
Speckle
   coherence-gated holographic imaging, 158
   grain, 236
   OCT, 210, 236, 284
   UOT, 324, 325, 329, 332
Spectral intensity, 325
Spectral interferometry, 198
Spectral radiance, 83
Spectral resolution, streak camera, 155
Spectrograph, 145
Spectrometer, 155, 198, 202
Spectrophotometer, 139, 140
Spectrophotometry, 139
Spectroscopy, 2, 135
   fluorescence, 146
   white-light, 144

Specular reflectance
   example, 44, 56
   formula, 43
   incident beam, 143, 147
   photon packet, 48
   unscattered reflectance, 52
Specular reflection. *See* Fresnel reflection
Speed of sound
   comparison with speed of light, 181
   formula, 315
   soft tissue, 305, 331
   UOT, 325, 336
   water, 284
Sphere, heated, 297, 303
Spherical harmonics, 88
Spherical polar coordinates (illustration), 18
Standard error, 55
Step heating, 289
Step size, 45, 122
STF. *See* System transfer function
Stokes parameters, 222, 223
Stokes shift, 9
Stokes transition, 5
Stokes vector, 219, 222
   construction of Mueller matrix, 237
   conversion from Jones vector, 236
   polarization-difference imaging, 158
Streak camera, 145, 155, 251
   illustration, 155
Stress confinement, 285
Stress relaxation time, 284
Subsonic medium velocity, 316
Superluminescent diode (SLD), 186, 209, 237, 238
Surface normal, 108, 265, 271
SVD. *See* Singular-value decomposition
Synthetic aperture, 307, 309
System transfer function (STF), 12

Targeted contrast agent, 2
Taylor expansion, 137
   fluence rate at boundary, 102
   matrix form, 273
   pressure, 314
   propagation constant, 187, 207
Telegraphy equation, 117
Temporal PSF, 251
Temporal resolution, 155
Temporal spread function, 252
Tendon, 242
Thermal coefficient of volume expansion, 285
Thermal confinement, 285, 287, 289

Thermal diffusion, 284
Thermal diffusivity, 284
Thermal equation, 287
Thermal expansion equation, 287
Thermal relaxation time, 284
Thermoelastic expansion, 283
Time gate, 154
Time of flight, 250, 252, 258
Time reversal, 227, 291, 311
Time-domain OCT, 185
Time-gated imaging, 153, 154
   illustration, 154
Time-resolved measurement, 145
Tomography, 1
Top-hat beam, 71
   fluence (plot), 78
Top-hat function, 303
Total interaction coefficient. *See* Extinction coefficient
Total transmittance, 55
Translation invariance, 67
   spatial, 11, 98, 263
   temporal, 68, 302
Translation stage, 256, 304, 335
Transmission mode, 165, 336
Transmittance, 5, 51, 158
Transport albedo, 106, 108, 123, 143
   definition, 344
Transport interaction coefficient, 94
   definition, 344
Transport mean free path
   conversion to isotropic source, 123
   definition, 344
   diffusion scale, 56, 97, 115, 132
   far diffuse reflectance, 140
   formula, 94
   PAM, 305
   scale for grid, 51
   UOT, 325
Transport mean free time, 96, 97
Transport scattering coefficient. *See* Reduced scattering coefficient
Transposition symmetry, 240, 242
Transverse resolution, 181, 309
   OCT, 190
   PAM, 305, 307
   UOT, 326, 329, 331, 337
Transverse scanning, 156, 161
Transverse-priority scanning, 190
Turbid media, 2
Two-photon excitation, 169, 170
Two-photon microscopy, 154, 169
   illustration, 169

Ultrafast laser, 154
Ultrasonic array. *See* Ultrasonic transducer array
Ultrasonic attenuation coefficient, 339
Ultrasonic coupling, 304, 330, 336
Ultrasonic pressure, 335. *See also* Acoustic pressure
Ultrasonic scattering, 283
Ultrasonic transducer, 283
   array, 307
   array (illustration), 308
   PAM, 304, 306, 307
   PAT, 312
   UOT, 326, 330, 332, 335
Ultrasonography, 1
   analog of OCT, 181
   comparison, 2, 284
   Doppler flow, 206
   lack of polarization, 219
Ultrasound absorber, 326
Ultrasound imaging. *See* Ultrasonography
Ultrasound-coupling gel, 304
Ultrasound-modulated optical tomography (UOT), 323
Unscattered absorption, 39, 46, 53
Unscattered reflectance, 51, 52
Unscattered transmittance, 51, 52, 55
Unscattered transmitted photons, 136
UOT. *See* Ultrasound-modulated optical tomography
UOT, Fabry-Perot interferometry
   image (plot), 336, 337
   system (illustration), 335
UOT, frequency-swept
   image (plot), 329
   spectrum (plot), 328
   system (illustration), 327
   virtual source (plot), 332
UOT, multiple-speckle
   system (illustration), 330
UOT, reconstruction-based
   image (plot), 334
   system (illustration), 333
USAF-1951 target, 305
User times, 130, 132

Variable optical attenuator, 252
Variance reduction technique, 42, 211
Vasculature, 305, 306
Vector wave, 259
Velocity potential, 287
Velocity resolution, 207
Vibrational relaxation, 4, 9, 170

Video rate, 13
Virtual image, 160, 173–175
Virtual optical source, 331

Water
    absorption, 6
    compressibility, 285
    dispersion compensation, 209
    Grueneisen parameter, 285
    primary absorber, 6
    refractive index, 8
    refractive-index-matching, 101, 135
    speed of sound, 284
    ultrasonic coupling, 304, 312, 330, 332, 336
Wavelength of photon-density wave, 258

Waveplate. *See* Retarder
Weak-modulation approximation, 324, 326
Weak-scattering approximation, 324
Weight of a photon packet, 43
White-light spectroscopy, 144
Wideband light source, 160
Wiener-Khinchin theorem, 184, 189, 201, 325

X-ray
    CT, 163, 332
    projection imaging, 156
    radiography, 1

Zero-boundary condition, 145, 324